The Nazi Dictatorship

THE
NAZI DICTATORSHIP

Problems and Perspectives of Interpretation

THIRD EDITION

Ian Kershaw

Edward Arnold
A member of the Hodder Headline Group
LONDON NEW YORK MELBOURNE AUCKLAND

Edward Arnold is a division of Hodder Headline PLC
338 Euston Road, London NW1 3BH

© 1985, 1989, 1993 Ian Kershaw

First published in the United Kingdom 1985
Third edition 1993

10 9 8 7 6 5 4
99 98 97 96 95 94

Distributed in the USA by Routledge, Chapman and Hall, Inc.
29 West 35th Street, New York, NY 10001

British Library Cataloguing in Publication Data
Kershaw, Ian
Nazi Dictatorship: problems and perspectives of interpretation.—3Rev. ed
I. Title
943.086

ISBN 0-340-55047-3

Library of Congress Cataloging-in-Publication Data
Kershaw, Ian.
The Nazi Dictatorship : Problems and Perspectives of Interpretation /
Ian Kershaw.—3rd ed.
   p.   cm.
Includes bibliographical references and index.
ISBN 0-340-55047-3 $16.95
1. National socialism—History.
2. Germany—Politics and Government—1933-1945. I. Title.
DD256.5.K47   1993
321.9'4'0943—dc20   92-47083
                                               CIP

Typeset in 10/11 Times by Colset Private Ltd, Singapore
Printed and bound in the United Kingdom by J W Arrowsmith Ltd, Bristol

Contents

Preface to the Third Edition   vii

Abbreviations   xi

1 Historians and the Problem of Explaining Nazism   1

2 The Essence of Nazism: Form of Fascism, Brand of Totalitarianism, or Unique Phenomenon?   17

3 Politics and Economics in the Nazi State   40

4 Hitler: 'Master in the Third Reich' or 'Weak Dictator'?   59

5 Hitler and the Holocaust   80

6 Nazi Foreign Policy: Hitler's 'Programme' or 'Expansion without Object'?   108

7 The Third Reich: 'Social Reaction' or 'Social Revolution'?   131

8 'Resistance without the People'?   150

9 'Normality' and Genocide: The Problem of 'Historicization'   180

10 Shifting Perspectives: Historiographical Trends in the Aftermath of Unification   197

Suggestions for Further Reading   218

Index   227

Preface to the Third Edition

Scholarly research on Nazism has produced a literature so immense that even experts cannot hope to master all of it. It has also raised in acute fashion a number of complex theoretical problems of interpretation. Not surprisingly, therefore, students specializing in modern German history frequently have difficulty in orientating themselves and finding their way through the labyrinth of interpretation and counter-interpretation. It was this which, some 10 years ago, prompted me to think that a book which tried to single out the key interpretational problems related to the Nazi Dictatorship – providing a concise guide to why there is a problem and how historians of different persuasion have tackled it, and then, on the basis of the most recent research, attempting a clear evaluation of the positions – might have some uses. I decided to confine the selected 'problem areas' to the period of the Dictatorship itself. However desirable it might have been to include them, this meant omitting themes relating to the history of Nazism before 1933. Contentious and important issues touching on the origins and rise of Nazism, such as whether there was a German 'special path' (*Sonderweg*) of development leading to the Third Reich, relations with 'big business' and Nazism before 1933, or the social composition of the Nazi movement, demand extensive analysis of their own, and I only alluded to them as background to the problems under consideration. This remains the case in the present edition.

I have been most gratified at the positive response to the book since the appearance of the first edition in 1985. But historiography, like history itself, does not stand still. It is in the nature of the historical scholarship of any era that over a period of years new research themes become dominant, innovative approaches offer fresh ways forward, important publications remould interpretations. Such changes in historiography are, if anything, more rapid and more pronounced in the case of the Third Reich, even almost half a century after Hitler's death, than they are with regard to more remote historical epochs. Some 10 years on from the first writing of the book, I am, then, often reminded of the classic piece of Irishman's advice to the traveller asking for directions: 'if I were you, I wouldn't start from here'. But to have started from elsewhere would, in effect, have meant completely rewriting the book. So I had to restrict myself to making many minor, and a few more substantial, changes to the original text. Wherever possible, I have tried to incorporate references to the most important recent publications. Sometimes, these publications were of such significance that, had I been writing the text for the first time, they would have warranted more extended consideration and prompted different treatment in places. Peter Hayes' study of IG-Farben, which puts the debate on the 'primacy of ideology' on to a new plane,[1] and Dieter Rebentisch's analysis of the Nazi state during the war, demanding a reconsideration of Hitler's role in decision-making,[2] are two such works which immediately come to mind. In one of the original chapters, that on 'Hitler and the Holocaust', I did find it necessary to reshape some passages completely in the light of Philippe Burrin's reassessment of the timing and decision-making processes in the genesis of Nazi genocide against the Jews.[3] In addition, I have contributed a wholly new chapter on German resistance. I had initially thought of including a chapter on this topic in the first edition, but then changed my mind. I think in retrospect that this was a mistake. The problems of interpretation – not least the definition of what constituted resistance – remain considerable, and still warrant, in my view, detailed examination and evaluation.

One theme running throughout the book is the influence of what, in the first chapter, I call the 'historical-philosophical', the 'political-ideological', and, not least, the 'moral dimension' on the historiography of the Third Reich. Historical writing on Nazi Germany offers demonstration in practice of Croce's dictum that 'all history is contemporary history' – that each generation writes history anew as historiography is inevitably affected and constantly refashioned by changing patterns of thought related to changing external circumstances. The historiography of the Third Reich continues in this respect to be subjected to rapid transformation and to more than usual turbulence. To mention one obvious example: even when the Second Edition of this book appeared in 1989, the German Democratic Republic was still 'a going concern', and its marxist-leninist historiography had to be reckoned with in exploring the different themes of interpretational conflict. Yet, within months the GDR became itself no more than a 'historical episode', its historiographical tradition effectively terminated – something to be treated only in the past tense.

Within the Federal Republic, the fortieth anniversary in 1985 of the end of the war, and of the Third Reich, prompted the major debates which flared up soon afterwards about the place of Nazism in German history. In the Second Edition, I included two new chapters describing and analysing these debates. I have discarded one of these chapters, on the '*Historikerstreit*' ('historians' dispute'), for this Third Edition. An extensive literature on the '*Historiker-streit*' has become available in the meantime. In any case, the dispute shed more heat than light before dying down almost as quickly as it had arisen. However, one related strand of the '*Historikerstreit*', the question of the 'historicization' of National Socialism – whether the Third Reich could be treated as a 'normal' part of Germany's past – was, I thought at the time, intellectually more significant and likely to be of more lasting importance than the polemics of the '*Historikerstreit*' itself. This has proved to be the case and, therefore, only slightly amended, I have retained the treatment of this issue in chapter 9, ' "Normality" and Genocide'. I have extended the discussion, moreover, in a new final chapter which, somewhat speculatively, tries to take some account of the way the aftermath of German Unification is affecting perspectives on the place of the Third Reich in German history.

[1] Peter Hayes, *Industry and Ideology. IG Farben in the Nazi Era* (Cambridge, 1987).
[2] Dieter Rebentisch, *Führerstaat und Verwaltung im Zweiten Weltkrieg* (Stuttgart, 1989).
[3] Philippe Burrin, *Hitler et les Juifs. Genèse d'un Génocide* (Paris, 1989; Eng, trans. *Hitler and the Jews. The Genesis of the Holocaust*, forthcoming, 1993).

I would like to thank once again all those friends and colleagues who contributed directly and indirectly to the initial making of this book, and to its subsequent revisions. To the late Martin Broszat and to Hans Mommsen, my main mentors and intellectual inspiration, I owe a quite special debt of gratitude. As regards the preparation of the Third Edition, I would particularly like to thank Saul Friedländer, who considerably influenced my thinking on the final chapter, and above all my Sheffield colleague Stephen Salter, whose constructive criticism of drafts of the new chapters was invaluable. I am most grateful now, as before, for the excellent help and guidance I have received from my editor at Edward Arnold. I remain indebted to the Leverhulme Foundation and the British Academy for the funding of the original research for the book, and to the Alexander von Humboldt-Stiftung for continuing support over the years.

Ian Kershaw
Sheffield, September 1992

Abbreviations

AfS	*Archiv für Sozialgeschichte*
AHR	*American Historical Review*
APZ	*Aus Politik und Zeitgeschichte (Beilage zur Wochenzeitung 'das Parlament')*
BAK	Bundesarchiv, Koblenz
CEH	*Central European History*
EcHR	*Economic History Review*
GG	*Geschichte und Gesellschaft*
GWU	*Geschichte in Wissenschaft und Unterricht*
HWJ	*History Workshop Journal*
HZ	*Historische Zeitschrift*
IMT	*International Military Tribunal (Trial of the Major War Criminals [Nuremberg, 1949], 42 vols.)*
JCH	*Journal of Contemporary History*
JMH	*Journal of Modern History*
MGM	*Militärgeschichtliche Mitteilungen*
NPL	*Neue Politische Literatur*
PVS	*Politische Vierteljahresschrift*
VfZ	*Vierteljahrshefte für Zeitgeschichte*

1

Historians and the Problem of Explaining Nazism

More than four decades after the destruction of the Third Reich, leading historians are far from agreement on some of the most fundamental problems of interpreting and explaining Nazism. Of course, great progress has been made since the historical writing of the immediate post-war era, when historians were attempting to write 'contemporary history' even before the dust had begun to settle on the wreckage of Hitler's Europe in a climate determined by the horrific disclosures of the Nuremberg Trials and the full realization of the bestiality of the regime. In such a climate, it is hardly surprising that recrimination from the Allied side and a proneness to apologetics from the German side featured strongly in writing on the immediate past. The lengthier time perspective and a vast outpouring of high-class scholarly research by a new generation of historians – especially from the 1960s on, following the opening up of the captured German records which in the meantime had been returned to Germany – brought major advances in knowledge of many vital aspects of Nazi rule. But as soon as the detailed scholarly monographs are placed in the context of overarching interpretative questions about Nazism, the limits of consensus are rapidly reached. A synthesis of polarized interpretations, often advocated and pleaded for, is nowhere in sight. The debate continues unabated, conducted with great vigour and frequently, too, with a rancour going beyond the bounds of conventional historical controversy. This was most vividly illustrated by the explosion of feeling which accompanied the '*Historikerstreit*' (or 'historians' dispute') – a major public controversy about the place of the Third Reich in German history involving Germany's leading historians which flared up in 1986.

Of course, debate and controversy are the very essence of historical study, the prerequisite for progress in historical research. However, Nazism raises questions of historical interpretation which either have a flavour of their own or highlight in marked fashion wider issues of historical explanation. The special features of historians' fundamental disagreements about interpreting Nazism are framed, in my view, by the inevitable merging of three dimensions – a historical–philosophical, a political–ideological, and a moral dimension – which are inseparable both from the historian's subject matter and from the historian's understanding of his present-day role and task in studying and writing about Nazism. These special features are, I would further argue, conditioned by and a reflection of a central element in the political consciousness of both post-war German states: mastering the Nazi past – *Vergangenheitsbewältigung*, coming to grips with and learning from Germany's recent history.

The radically different approaches to the Nazi past in East and West

1

Germany have naturally lent a peculiar colouring to historical writing about Nazism. But since the problem of facing the past has been tackled in a less unilinear fashion in the Federal Republic than it was in the German Democratic Republic, the controversies about interpreting Nazism have above all been West German controversies. To say this is, of course, in no sense to underestimate the major, often path-breaking, contribution made to German history by non-German historians. Often it has been, in fact, the very detachment (with correspondingly different perspective) of foreign historians both from the burden of 'mastering the past' and from the intellectual currents of West German society which has provided the springboard for fresh impulses and new methods. The important mark of international scholarship will be clearly apparent in the following chapters. Nevertheless, it is a basic contention of this book that the contours of the debates have generally been established by German historians, especially of the Federal Republic, and have been shaped in great measure by West German historians' perception of their task in helping to shape 'political consciousness' and thereby overcome the past.

It has been said of the Federal Republic that it is even more than Israel or South Vietnam 'a State born of contemporary history, a product of catastrophe erected to overcome the catastrophe'.[1] In such a society, the historian of the recent past clearly has a much more overtly *political* role than, for example, in Britain. It is not going too far to say that through his interpretation of the recent past the historian is seen, and sees himself, in certain ways as the guardian or the critic of the present. The inseparability of historical research on Nazism from 'political education' contributes in part to the latent feeling of some historians that, above all in grasping the essence of the Nazi system, there *ought* to be clarity. This feeling was expressed by the (then) Chancellor of the Federal Republic, Helmut Schmidt, when he addressed the German Historians' Annual Conference in 1978 and complained that a surfeit of theory had produced for many present-day Germans a picture of Nazism still lacking 'a clear contour'.[2] The same argument marks the tone – a mixture of anger and sorrow – of some historians, whose interpretation dominated the 1950s and 1960s, in reacting to a 'revisionist' challenge to established orthodoxy which goes so far as to subject to radical questioning 'basic scholarly findings which had been taken to be certain, indeed uncontested'.[3]

The connection between the changing perspective of historical research and the shaping of current political consciousness is recognized as an explicit one – by 'traditionalists' and 'revisionists' alike.[4] As the '*Historikerstreit*' has again clearly demonstrated, conflicting interpretations of Nazism are part of a continuing reappraisal of West Germany's political identity and political future. The contemporary historian and his work are public property. This forms the basic framework and colours the nature of the historical controversies we shall be appraising.

---

[1] Ernst Nolte, *Marxismus, Faschismus, Kalter Krieg* (Stuttgart, 1977), p. 217.
[2] Cited in Walther Hofer, '50 Jahre danach. Über den wissenschaftlichen Umgang mit dem Dritten Reich', *GWU* 34 (1983), p. 2.
[3] Hofer, p. 2.
[4] 'Revisionism' is not only a 'dirty' word, but also a shifting and confusing one. Thus some of those who in the 1970s criticized the 'revisionists' were, in the '*Historikerstreit*', themselves decried as 'revisionists'.

The extent of the literature on Nazism is so vast that even experts have difficulty in coping. And it is clear to see that students specializing in modern German history are frequently unable to assimilate the complex historiography of Nazism and to follow interpretational controversies carried out for the most part in the pages of German scholarly journals or in scholarly monographs. My book was written with this in mind. It offers no description of the development of historiography, no history of the history of Nazism, so to say.[5] Rather, it is an attempt to examine the nature of a number of central problems of interpretation, relating specifically to the period of the Dictatorship itself, which confront present-day historians of Nazi Germany.[6]

The structure of the book is largely pre-shaped by the interlocking and inter-related themes which form the basis of the controversies. The following chapter seeks to analyse the wide-ranging and sharply opposed interpretations of the nature of Nazism: whether it can be most satisfactorily viewed as a form of fascism, as a brand of totalitarianism, or as a unique product of recent German history – a political phenomenon 'of its own kind'. Directly related to the fascism debate is the heated controversy about Nazism and capitalism, in particular over the role of German industry, which forms the subject of the subsequent chapter. The key issue which has emerged in recent years has been how to interpret the position, role, and significance of Hitler himself in the Nazi system of rule, a complex problem explored later in three separate chapters on the power structure of the Third Reich and the framing of anti-Jewish and foreign policy. The focus is then moved from the government of the Third Reich to society under Nazi rule, seeking to examine the extent to which Nazism altered, even revolutionized, German society, and attempting to evaluate the complex issue of German resistance to Hitler. This is followed by an analysis of the important recent debate which has developed about the 'historicization' of the Third Reich – whether the Nazi era can be dealt with at all like other periods of the past, as 'history'. Finally, I try to consider some of the ways in which historiographical trends are shifting in the wake of German unification. Within each chapter I try to summarize adequately the differing interpretations and the current state of research, and then offer an evaluation. I have not seen it as my task to attempt to sit on the fence and adopt a neutral stance in reviewing the controversies – in any case an impossibility. I hope to represent the views I am summarizing as fairly as I can, but also to participate in the debate, not 'referee' it, advancing my own position in each case.

The varied approaches to the history of the Third Reich encountered in this book share a common aim: to offer an adequate *explanation* of Nazism. To explain the past is the task of all historians, but the daunting nature and complexity of this task in the case of Nazism will become apparent in the pages which follow. Arguably, indeed, an *adequate* explanation of Nazism is an intellectual impossibility. In Nazism, we have a phenomenon which seems

---

[5] For a good, if often rather agnostic, historiographical survey, see Pierre Ayçoberry, *The Nazi Question* (London, 1981).

[6] For discussions of the literature and state of research on Nazism, see Klaus Hildebrand, *Das Dritte Reich* (Munich/Vienna, 1979), Engl. trans., *The Third Reich* (London, 1984); Andreas Hillgruber, *Endlich genug über den Nationalsozialismus und Zweiten Weltkrieg? Forschungsstand und Literatur* (Düsseldorf, 1982); and John Hiden and John Farquharson, *Explaining Hitler's Germany. Historians and the Third Reich* (London, 1983). Gerhard Schreiber's *Hitler. Interpretationen* (Darmstadt, 1984) offers the best historiographical survey of works on Hitler.

scarcely capable of subjection to rational analysis. Under a leader who talked in apocalyptic tones of world power or destruction and a regime founded on an utterly repulsive ideology of race-hatred, one of the most culturally and economically advanced countries in Europe planned for war, launched a world conflagration which killed around 50 million people, and perpetrated atrocities – culminating in the mechanized mass murder of millions of Jews – of a nature and scale as to defy imagination. Faced with Auschwitz, the explanatory powers of the historian seem puny indeed. How can he hope to write adequately and 'objectively' about a system of government which produced horror of such monumentality? How is he to go about his task? He can hardly confine himself in neo-Rankean terms to recovering from the sources the story of 'what it was actually like'. And can he hope to 'understand' (in the historicist tradition) such a criminal regime and its inhumane leader? Or is his task to lay bare the evil of Nazism to provide witness for the present and warning for the future? If so, how is this to be done? Can or should the historian strive to attain the 'detachment' from his subject matter which is usually taken to be the very essence of 'objective' historical writing? Simply to pose such questions suggests some of the reasons why no explanation of Nazism can be intellectually wholly satisfying. Ultimately, nevertheless, the merit of any interpretational approach must rest in the extent to which it might be seen to *contribute* towards a potentially improved explanation of Nazism. The aim of this book will have been served if its evaluation of varying interpretations of the Nazi Dictatorship suggests which approaches have a better potential than others (or differently expressed: are less inadequate than others) to offer an explanation of the processes of dynamic radicalization in the Third Reich which led to war and genocide on an unparalleled scale.

Before considering the historical–philosophical, political–ideological, and moral dimensions underlying the controversies which we shall be examining, a final preliminary point must be made. It is an obvious enough point, but bears repeating nevertheless: the inadequacies of the source materials. For, despite the vastness of surviving archival relics of the Third Reich, the documentation is patchy in the extreme and serious problems of interpretation are in part linked to fundamental deficiencies in the nature of the sources. Much crucial documentation was, of course, deliberately destroyed by the Nazis towards the end of the war, or lost through bombing raids. But the problem extends beyond the mere physical loss of record material. It reaches to the huge gaps in the documentary sources at the most critical and sensitive points, which themselves are an inevitable product of the way the Nazi system of government functioned. Nowhere are the gaps more apparent or more frustrating than those surrounding Hitler himself and his role in the government of the Third Reich. Thus, the increasing breakdown of any formalized central government machinery in the Third Reich, together with Hitler's extraordinarily unbureaucratic style of rule where decisions were seldom formally registered, has left a huge void in the documentation of the sphere of central decision-making. The immense bureaucratic remnants of the Third Reich stop, therefore, short of Hitler. It is difficult to know what material from the government was even reaching Hitler, let alone whether he read it and how he reacted to it. As Dictator of Germany, Hitler is for the historian largely unreachable, cocooned in the silence of the sources. For this very reason, fundamental conflicts of interpretation about Hitler's place in the Nazi system of rule can neither be avoided nor conclusively resolved on the basis of the available evidence.

The deficiencies of the sources form a relatively minor part of the problem of interpreting Nazism. A more significant role in shaping the character of the controversies over the Nazi Dictatorship has been played by historians' divergent, often quite contradictory, conceptions and methods of historical writing, when applied to the study of Nazism.

The Historical–Philosophical Dimension

Two points can be made at the outset. The first is, that the differences in historical approach, method, and philosophy are by no means peculiar to the study of Nazism, though the problems involved in interpreting Nazism bring out these issues of historical philosophy in a particularly forceful fashion. The second point is, that the intensity and rigour of the debate on historical method stems from the specifically German tradition of historical writing and the challenge to this tradition, applied to the terrain of the Third Reich. Though non-German historians have often made significant contributions, the debate on historical method is largely and characteristically a West German affair. In what follows, therefore, we need to turn our attention to the course and nature of German historiography, and to the radically opposed views on the form and purpose of historical writing advanced by current leading West German historians.

The contours of post-war German historiography have been shaped by a number of specific factors distinguishing Germany from the historiographical development of other countries. Underpinning the whole process has been the need to come to terms with the Nazi past. This has been fundamental in shaping the particularly close connection in post-war German historical scholarship between the problems of interpreting the course and character of recent German history and far-reaching questions of historical method and philosophy. Broadly speaking, the development since the war of historical studies in West Germany – the GDR has to be excluded from this categorization – can be divided into three phases: a period of continued and partly refurbished historicism, lasting until the early 1960s; a transitional phase of transformation, extending into the mid 1970s; and a phase continuing to the present time, despite some stiff challenges and certain regressive tendencies, in which new forms of structurally based 'social history' aligned to the social sciences and closely interwoven with parallel developments in international scholarship, can be said to have established themselves.[7]

The historicist tradition had exerted a dominance over historical philosophy and writing in Germany since the time of Ranke incomparably greater than that of any philosophy of history in any other country.[8] It rested on an idealistic – in the philosophical sense – concept of history as cultural

[7] For this periodization, see Jörn Rüsen, 'Theory of History in the Development of West German Historical Studies: A Reconstruction and Outlook', *German Studies Review* 7 (1984), pp. 14–18. I am indebted to Prof. Rüsen for his comments and suggestions on this section and profited, too, from the excellent essay of Bernd Faulenbach, 'Deutsche Geschichtswissenschaft nach 1945', *Tijdschrift voor Geschiednis* 94 (1981), pp. 29–57. See also, Georg G. Iggers, *Deutsche Geschichtswissenschaft* (Munich, 1971), ch. 8 (revised and extended from his *The German Conception of History* (Middletown, Connect., 1968)), and Wolfgang J. Mommsen, 'Gegenwärtige Tendenzen in der Geschichtsschreibung der Bundesrepublik', *GG* 7 (1981), pp. 149–88.
[8] Iggers, p. 11.

development formed by men's 'ideas' as revealed through their actions, from which their intentions, motives, and 'self-reflection' could be deduced. Historical writing concentrated on the task of trying to explain actions by 'understanding' intuitively the intentions which lay behind them. In practice, this led to a heavy emphasis on the uniqueness of historical events and personages, on the overwhelming importance of will and intention in the historical process, and on the power of the State as an end in itself (and consequently the elevation of the Prussian–German national State).

For a historical profession which had concentrated heavily upon the nature and role of the State as a 'positive' factor in history, it was an extreme shock after 1945 to have to deal 'not only with the break-up of a State . . . but with the break-up of a State burdened with State crimes of inconceivable extent'.[9] Nevertheless, the collapse of the Third Reich brought no fundamental change in the historicist tradition and dominance in historical writing. As in 1918 and 1933, continuity was the essential hallmark. The two foremost historians of post-war Germany, Friedrich Meinecke and Gerhard Ritter, had both been reared and had written in the historicist tradition, and their ideas were deeply embedded in the German idealistic tradition of historical and political thought. Neither had been a Nazi. In fact, both had had their brushes with the Nazis: Meinecke had been removed from his post as editor of the *Historische Zeitschrift* in 1935; Ritter was, as an associate of Carl Goerdeler, imprisoned in 1944 following the attempt on Hitler's life. Meinecke's influential book, *Die deutsche Katastrophe*, which appeared in 1946, and Ritter's more strongly apologetic *Europa und die deutsche Frage*, published in 1948, formed in essence attempts to justify German idealism and the national political tradition. According to such a view, Nazism had emerged from a sort of parasitic sub-growth, traceable to the negative forces which had first come to the fore in the French Revolution, and existing alongside the generally healthy and positive development of the German State. Though there were menacing signs in the late nineteenth century, it was above all a disastrous series of events triggered by the First World War which brought in the whole of Europe and not just in Germany a collapse of moral and religious values, the dominance of materialism, the growth of barbarism, and the corruption of politics as machiavellianism and demagogy. Nazism was, therefore, according to such an interpretation, the terrible outcome of European, not specifically German, trends; it marked a decisive break with the 'healthy' German past rather than being a product of it. Meinecke spoke of 'the history of the degeneration of German mankind'.[10] Ritter found it 'almost unbearable' to think that 'the will of a single madman' had driven Germany into the Second World War.[11] Nazism was, therefore, more or less an accident in an otherwise commendable development. And the disaster which had befallen Germany could in no small measure be attributed to the 'demon' Hitler. (Such defensive attempts to interpret Nazism as part of a European disease were of course the direct counter to the crude interpretation of Anglo-American writers after the war, that Nazism could only be seen as the culmination of centuries of German cultural and political misdevelopment

---

9 The comment of Manfred Schlenke, cited in Iggers, pp. 356–7.

10 Friedrich Meineke, *Die deutsche Katastrophe* (Weisbaden, 1946), p. 28.

11 Gerhard Ritter, *Das deutsche Problem. Grundfragen deutschen Staatslebens gestern und heute* (Munich, 1962), p. 198. This was a new edition, with added introduction and conclusion, of Ritter's *Europa und die deutsche Frage. Betrachtungen über die geschichtliche Eigenart des deutschen Staatsdenkens* (Munich, 1948).

reaching back to Luther and beyond.[12])

The beginning of a rapid decline in the influence of historicism and a transformation in historical thinking was ushered in by the 'Fischer Controversy' of the early 1960s. By using wholly traditional methods of research, Fritz Fischer, in his *Griff nach der Weltmacht*, published in 1961, demonstrated the aggressive, expansionist war aims of Germany's élites in the First World War, and in so doing knocked the bottom out of the argument that a hitherto basically healthy development had somehow 'gone off the rails' *after* the war. Unwittingly, too, Fischer had opened up new areas of concern for historical research – especially the role of the 'traditional' élites and the continuities in social structures and domestic as well as foreign policy linking the Imperial with the Nazi era. The furore which Fischer's work provoked reflected plainly the extent of the culture shock for the older historical establishment.[13] The transformation process partly unleashed by the 'Fischer Controversy' was greatly furthered by the weakening of old rigidities through the expansion of the university system, challenges to the historical profession arising from the advances being made by the social sciences, and by the changes in the political and intellectual climate which accompanied the end of a long spell of conservative rule and the 'students' movement' of the late 1960s.[14]

Denuded of its historicist isolation, and in a political context where close cultural relations with other European countries and with the USA were actively and intensively promoted, German historical scholarship moved into the outside world. Structural history concepts, derived in particular from the French *Annales* school, and the influence of American political and social science began to transform historical approaches in West Germany.

New, more theoretical approaches to historical scholarship leaning heavily upon transatlantic developments in social and political science fought to establish themselves for the first time in German universities. The 'new social history' or 'historical–social science' approach, arguing for a theoretically based integrative discipline to build a structural analysis of the 'history of society', upturned the traditional emphasis in German historical scholarship by asserting that the concept of 'politics' needed to be subordinated to the concept of 'society', so that 'political history', while important in itself, could not alone provide a key to historical understanding and needed to be rooted in a wider (and theoretical) context.[15] The foundation of two new journals – *Geschichte*

[12] Classics of the genre are Rohan O'Butler, *The Roots of National Socialism* (London, 1941), and William Montgomery McGovern, *From Luther to Hitler. The History of Nazi-Fascist Philosophy* (London, 1946). Such anti-German distortions were massively popularized in William Shirer's bestseller, *The Rise and Fall of the Third Reich* (New York, 1960).

[13] Fritz Fischer, *Griff nach der Weltmacht* (Düsseldorf, 1961), Engl. trans., *Germany's Aims in the First World War* (London, 1966). For the 'Fischer Controversy' see the collections of essays in Hans W. Koch, ed., *The Origins of the First World War* (London, 1972), and, more recently, especially Volker Berghahn, 'Die Fischerkontroverse – 15 Jahre danach', *GG* 6 (1980), pp. 403–19.

[14] Rüsen, 'Theory of History', p. 16; see also, Hans-Ulrich Wehler, 'Geschichtswissenschaft heute', in Jürgen Habermas, ed., *Stichworte zur 'Geistigen Situation der Zeit'* (2 vols., Frankfurt am Main, 1979), vol. 2, pp. 739–42, Engl. trans., *Observations on 'The Spiritual Situation of the Age'* (Cambridge, Mass., 1984).

[15] See, for example, Hans-Ulrich Wehler, 'Anwendung von Theorien in der Geschichtswissenschaft', in Jürgen Kocka and Thomas Nipperdey, eds., *Theorie der Geschichte. Beiträge zur Historik* (Munich, 1979), vol. 3, pp. 17–39; Jürgen Kocka, 'Theorien in der Sozial- und Gesellschaftsgeschichte', *GG* 1 (1975), pp. 9–42; and the critical (untitled) review essay of K.G. Faber in *History and Theory* 16 (1977), pp. 51–66.

und Gesellschaft in 1975, and *Geschichtsdidaktik* in 1976 – embodying the methodology and publishing the research findings of these new approaches, could be said to reflect the fact that 'history as social science', innovative in the mid 1960s, had become established and institutionalized a decade later.

This progress was not, of course, unchallenged. The gauntlet laid down by representatives of the 'new social history' approach was taken up by leading historians who, though now divorced from classical historicism, still held fast to conventional historical method and spheres of interest. The debates about historical method between exponents of the two – seemingly irreconcilable – sides were at times fierce. And they have a direct relevance to the nature of the controversies about Nazism.

The leading protagonist of the 'history of society' approach, Hans-Ulrich Wehler, is not generally regarded as a specialist on Nazism, though his studies of Imperial Germany were expressly related to the question of continuity in the structures of German society between 1870 and 1945.[16] Among the foremost assailants of the 'new social history' and defenders of the merits of conventional political history – with heavy emphasis upon foreign and diplomatic history, the importance of the individual and his will and intention as against structural determinants, and the value of the traditional historical method of empirical research – were the late Andreas Hillgruber and Klaus Hildebrand, both renowned experts on the foreign policy of Nazi Germany.[17]

In a key-note article in 1973, Hillgruber made a plea for a return to a central emphasis on modern political history.[18] He fiercely attacked the 'exaggerated and modish claims of "social history" ', in which models replaced concrete evidence. The new social history approaches were in his view simply not suited to cast light on the international system and the still crucial determinant of the 'balance of power' in international affairs. He rejected the oversimplicity of theories of 'imperialism' or 'fascism', and ended with a broadside against the notion that there is no such thing as 'value-free scholarship', reasserting his view that the work of the scholar must remain independent of his political engagement. Hildebrand's line of attack was similar, though he was even more forthright in tone.[19] He hit out at the application of theory, since political action must be sought in the sources and in source criticism, in the evaluation of the particular situation, individual aspirations, decisions, accidental and surprising events. He denied that international relations could be regarded as a derivate of social developments, and argued that, compared with 'hegemony' and 'balance of power', the concepts of the 'new social history' were of limited value. The only legitimate procedure for the historian is to work from the particular to the general, not the other way round. The application of theory he

---

[16] Best known of all his works is Hans-Ulrich Wehler, *Das Kaiserreich, 1871–1918* (Göttingen, 1973), Engl. trans., *The German Empire, 1871–1918* (Leamington Spa, 1984). See also the penetrating critique by Thomas Nipperdey, 'Wehlers Kaiserreich', *GG* 1 (1975).
[17] Prominent among their many works: Andreas Hillgruber, *Hitlers Strategie, Politik und Kriegführung 1940–1941* (Frankfurt am Main, 1965), and his collected essays, *Deutsche Großmacht und Weltpolitik im 19. und 20. Jahrhundert* (Düsseldorf, 1977); Klaus Hildebrand, *Vom Reich zum Weltreich. Hitler, NSDAP und koloniale Frage 1919–1945* (Munich, 1969), and *The Foreign Policy of the Third Reich* (London, 1973).
[18] Andreas Hillgruber, 'Politische Geschichte in moderner Sicht', *HZ* 216 (1973), pp. 529–52.
[19] Klaus Hildebrand, 'Geschichte oder "Gesellschaftsgeschichte"? Die Notwendigkeiten einer politischen Geschichtsschreibung von den internationalen Beziehungen', *HZ* 223 (1976), pp. 328–57.

found methodologically dubious, potentially excluding many facets of reality, and he concluded by reasserting the view that the past is autonomous and not there to inform or instruct the present.

Wehler's replies argued that Hillgruber's approach, too, needed theoretical and conceptual underpinning, and that his reliance upon the aims of leadership groups, political ideas, and intentions led inexorably towards a political history of ideas which opened up no new vistas. Wehler emphasized the limitations of concentrating on archival sources alone for analysis of foreign-policy decision-making.[20] His response to Hildebrand was more sharply couched, accusing him of rhetorical exaggeration, straw-man attacks, and seeming in one place even to imply deliberate misquoting.[21] He saw Hildebrand's insistence on moving from the particular to the general as insufficient even for Hildebrand's own research on Nazism. In a later broadside, he attacked the approach to the history of Nazism as featured in Hildebrand's work as a 'weedy and mangled historicism'.[22] Hildebrand in return claimed that Wehler's comments demonstrated just how the relationship of society and Hitler, of structure and personality in the Third Reich, 'can be distorted and simplistically described through prejudice and lack of knowledge', claiming that Wehler's article lay outside the bounds of serious scholarship, simply accumulated statements of political opinion and personal insult, and had no use in the context of serious academic discussion.[23]

These uncompromising exchanges on theoretical approaches and methodological questions have a direct bearing on the nature of some of the key interpretative controversies about Nazism. They indicate the theoretical difficulties in reconciling a 'structural' approach to the history of Nazism with a personalistic one – a key problem in interpreting the role and place of Hitler in the Nazi system of government. Secondly, they point to some of the difficulties of the relationship of the historian to his sources – how he should approach and read his sources. Thirdly, they raise the complex question of the political stance of the historian, his relationship to the political circumstances in which he lives and works, and the relationship between theoretical–methodological and political–ideological positions.

On the first point, Wehler's theoretical, conceptual approach prompts an instinctive methodological preference and sympathy for the work of so-called 'revisionist' historians of Nazism such as Hans Mommsen, the late Martin Broszat, and Wolfgang Schieder, who, for the most part working without the conscious application of a great theoretical encumbrance, have approached complex problems such as the interrelationship of domestic and foreign policy

[20] Hans-Ulrich Wehler, 'Moderne Politik-geschichte oder "Große Politik der Kabinette"?', *GG* 1 (1975), pp. 344–69.

[21] Hans-Ulrich Wehler, 'Kritik und kritische Antikritik', *HZ* 225 (1977), pp. 347–84.

[22] Wehler, 'Geschichtswissenschaft heute', p. 745.

[23] Klaus Hildebrand, 'Monokratie oder Polykratie? Hitlers Herrschaft und das Dritte Reich', in Gerhard Hirschfeld and Lothar Kettenacker, eds., *Der 'Führerstaat': Mythos und Realität* (Stuttgart, 1981), p. 95 n. 74. The polemical debate is summarized in W.J. Mommsen, 'Gegenwärtige Tendenzen', pp. 165–8. For a different attack on Wehler's 'critical history' approach, ending in the suggestion that the 'search for national identity' might be a legitimate new theme which could help further a 'change of paradigm' in German historical scholarship, see Irmline Veit-Brause, 'Zur Kritik an der "Kritischen Geschichtswissenschaft": Tendenzwende oder Paradigmawechsel?', *GWU* 35 (1984), pp. 1–24.

in the Nazi State, the structure of the State machinery and decision-making processes, and not least the place and function of Hitler in the Nazis system, in
what can loosely be described as a 'structural–functionalist' way. Correspondingly, the limitations are strongly emphasized of explanations resting heavily
upon Hitler's conscious intentions and individual role in forming Nazi
policy.[24]

On the second point, the dispute about historical method has highlighted the
problem of how the historian builds his explanation from the sources. Quite
apart from the deficiencies in the source materials on Nazism which we noted
earlier, sources can often (as the late Tim Mason pointed out with express
reference to Hitler's intentions and aims) 'be read in very different ways, depen-
ding upon the different kinds of other historical knowledge which is brought
to bear upon these texts', and should not necessarily be read solely in what
appears to be the literal 'common-sense' way.[25] Hence, some of the controversies (particularly those surrounding Hitler) are between historians using
precisely the same documentary sources but starting from different premises
and conceptions – not only about what the Third Reich was like but also what
writing history is about – and reading them in radically different fashion.

The third point, the influence of political–ideological considerations on the
historiography of Nazism, raises a separate and important issue, to which I now
want to turn.

The Political–Ideological Dimension

Two separate, though related, areas need consideration: first, the ways in which
the division of Germany moulded the political–ideological premises of interpreting Nazism on both sides of the Wall; and second, the ways in which
political–ideological differences have shaped the changing patterns of writing
on Nazism within the Federal Republic itself.

In the German Democratic Republic, anchored in marxist–leninist principles,
anti-fascism was from the beginning an indispensable cornerstone of the state's
ideology and legitimacy. Historical work on 'Hitler-Fascism', therefore,
always had a direct political relevance. And since fascism was taken to be an
intrinsic product of capitalism, and the neighbouring West German state was
founded on the capitalist principles of the Western Allies, historical research
on fascism had the task not merely of educating East German citizens about the
horrors and evil of the past, but even more so of the dangers and evil of the
present and future – of the potential fascism seen as built into western capitalist
imperialism, especially in the Federal Republic.

The understanding of Nazism in the German Democratic Republic rested on
the long tradition in the Communist International of wrestling with the problem
of fascism in the 1920s and 1930s, culminating in Georgi Dimitroff's famous
formulation, definitively established at the Seventh Congress of the Comintern
in 1935, that fascism was 'the open terroristic dictatorship of the most reac-

---

[24] Wehler, 'Geschichtswissenschaft heute', pp. 731–2.
[25] Tim Mason, 'Intention and Explanation: A Current Controversy about the Interpretation of
National Socialism', in Hirschfeld and Kettenacker, pp. 23–42, here p. 31.

tionary, most chauvinist, and most imperialist elements of finance capital'.[26] The 'unmastered past' of the West German state – not least the survival in prominent places in the economy and in political life of persons with a more than dubious past in the Third Reich scarcely behind them – simply underlined for East German scholars the present-day relevance and political purpose of their historical scholarship. The introduction to a collection of essays summarizing the results of historical research in the German Democratic Republic on Nazism, stated categorically: 'The aim and concern of the book will have been satisfied if, as a first step on the way to comprehensive research on the historical and current political problems of fascism, it provides scholarly material for the present-day struggle against fascism and imperialism'.[27] And a contributor to the volume further emphasized: the attempt by capitalists to prop up their power with new methods – those of fascism – is a truth which 'has been taken to heart by marxist historians, who, with their research into the history of fascism, want to make a contribution to combating the reactionary forces which are ever reappearing in new guise and, on the basis of their historical experience, proceed from the standpoint that the anti-fascist struggle can only be carried to victory through the complete removal from power and overcoming of monopoly capital'.[28] One of the foremost GDR historians precisely summed up the point: 'For us, research on fascism means participation in the current class struggle'.[29]

The ideological framework within which historical research operated in West Germany was less openly stated, but was none the less apparent. The main aim in the formulation of the West German Constitution (the 'Basic Law') was to eliminate the potential for the creation of a 'totalitarian' system, not only such as had existed in the Third Reich, but as continued to exist in the Soviet Union and in the Soviet Zone of Germany. The constitution was intentionally both anti-fascist and anti-communist. As has been pointed out, 'the theory of totalitarianism which compares and even equates fascism and communism can therefore be seen as the dominant idea behind the basic constitutional law and even to some extent as the official ideology of the Federal Republic'.[30] The 'totalitarian' premise was thus implicitly widely accepted in Western Germany, even among Social Democrats, before the scholarly writings of German emigrants in the USA, notably Hannah Arendt and Carl Friedrich, established totalitarianism as the central concept in interpreting Nazism.[31] The 'totalitarian' approach dominated research on 'contemporary history' in the Federal Republic in the 1950s and early 1960s. The seminal works of Karl Dietrich Bracher on the end of the Weimar Republic and on the Nazi 'seizure

[26] Georgi Dimitroff, *Gegen Faschismus und Krieg. Ausgewählte Reden und Schriften* (Reclam edn., Leipzig, 1982), p. 50. The definition had already been formulated at the 13th Plenum of the Executive Committee of the Comintern in December 1933.

[27] Dietrich Eichholtz and Kurt Gossweiler, eds., *Faschismusforschung. Positionen, Probleme, Polemik* (Berlin (East), 1980), p. 18.

[28] Wolfgang Ruge, 'Monopolbourgeoisie, faschistische Massenbasis und NS-Programmatik', in Eichholtz and Gossweiler, pp. 125–55, here p. 155.

[29] Kurt Gossweiler, 'Stand und Probleme der Faschismusforschung in der DDR', *Bulletin des Arbeitskreises 'Zweiter Weltkrieg'* 1 (1976), p. 13.

[30] Wolfgang Wippermann, 'The Post-War German Left and Fascism', *JCH* 11 (1976), p. 192.

[31] Hannah Arendt, *The Origins of Totalitarianism* (New York, 1951); Carl Joachim Friedrich and Zbigniew Brzezinski, *Totalitarian Dictatorship and Autocracy* (Cambridge, Mass., 1956).

of power' are among the most prominent examples.[32] The central journal of 'contemporary history', the *Vierteljahrshefte für Zeitgeschichte*, first published in 1953, also saw its task residing not only in studying Nazism, but in undertaking research on totalitarian movements in general, including of course communism.[33]

The challenge to the dominant totalitarian theory and the revival of fascist theories in West Germany in the 1960s was carried out on two planes, those of academic scholarship and of ideological–political polemic. But, as always, there was an intrinsic connection between the two levels, which could never be completely separated. Slotting into the first major challenge to the dominant values of the conservative state run by the Christian Democrats in the mid 1960s and the growing crisis within German universities which broke in 1968, academic discussion of fascism and the scholarly rehabilitation of fascist theories of the inter-war years was quickly turned into political sloganizing by segments of the Left, while the shocked over-reaction of the Liberal and Conservative Right ensured the place of debate about fascism or totalitarianism as part of current political dialogue and conflict. We will go into the theories and criticisms of them in the next chapter. Here, it is a matter of illustrating the clear political overtones which the academic controversies carry. Moreover, not only the repercussions of the year of turmoil in 1968, but also the far more overt politicization of university faculties themselves in West Germany helped to delineate the contours of the debates. And whereas in the 1960s and early 1970s the expansion of universities on the whole promoted a sense of challenge to orthodoxy and establishment positions, the restrictions in growth in higher education and the *Berufsverbot* contributed towards a changed climate.[34] The dominance – supported by prolific, highly influential publications – of the conservative–liberal establishment in the historical profession was in no small measure reasserted. The tone of the conflict is well represented in the comments of two of the leading 'liberal–conservative' historians of Nazism, Karl Dietrich Bracher and Andreas Hillgruber.

In a short, widely-read textbook on post-war German history, published in the mid 1970s,[35] Andreas Hillgruber spoke of radical criticism in universities growing increasingly dependent upon 'the forces of doctrinaire marxism-leninism' orientated towards the model of the German Democratic Republic, and of a search among the 'New Left' for ideology and indoctrination (which, in labelling 'need for theory', he implicitly associated with the 'progressive' side of the theoretical–methodological debates within the historical discipline). He saw the 'primacy of domestic politics' hypothesis, which Wehler and others had derived from the work of Eckhart Kehr and deployed mainly as a heuristic

---

[32] Karl Dietrich Bracher, *Die Auflösung der Weimarer Republik* (Stuttgart, 1955); Karl Dietrich Bracher, Wolfgang Sauer, and Gerhard Schulz, *Die nationalsozialistische Machtergreifung. Studien zur Errichtung des totalitären Herrschaftssystems in Deutschland 1933-34* (Cologne/Opladen, 1960).

[33] See Iggers, p. 357.

[34] See Wehler, 'Geschichtswissenschaft heute', pp. 745 ff. Veit-Brause (pp. 1–3) argues that the shift in political climate towards conservatism was only a minor part of the revision of paradigms in West German historical writing, which she attributes in far greater extent to new intellectual insights putting the 'critical history' approach in question.

[35] Andreas Hillgruber, *Deutsche Geschichte 1945-1972* (Berlin, 1974), pp. 162-4. See Wehler's comments in 'Geschichtswissenschaft heute', pp. 747–8 and 'Moderne Politik-geschichte', p. 355.

device, as providing an 'apparent scholarly legitimation' of the alleged conviction of the 'New Left' that radical social change and even revolution was the only concern of the present.

The most eminent of all West German historians of the Third Reich, Karl Dietrich Bracher, also made his views absolutely clear on the changing nature of writing on 'contemporary history'.[36] The lively discussion of the 1960s, he wrote, had been stimulated, but also overshadowed and often distorted, by the politicization and institutional upheavals in German universities and higher education. Research tendencies towards interdisciplinary and comparative approaches had also made their contribution, especially the widening of historical method and the demand for a social science base to historical writing. A 'marxist renaissance' of the 'New Left' had increased the complexity and confusion of concepts, especially in the 'vehemently voiced demands for theory' and in the 'radical attack on previous patterns of interpretation which had essentially arisen from the effort at mastering the past after the catastrophes of 1933 and 1945'. As the approaches shaped by the experience of the Third Reich faded, they had been replaced by social-critical approaches and concepts which had placed the former interpretations under a cross-fire frequently carried out 'with crude weapons'. Previous research achievements were ignored or distorted, and there was a resort to political agitation in which 'the ideological struggle was carried out on the back and in the name of scholarship'. Under the demand for theory and revision, previous scholarly standards were also distorted. At its most obvious, the attack on liberal–democratic values had been articulated in the bitter assaults on the totalitarianism concept and in the boundless expansion of the general theory of fascism, which had rapidly degenerated from new scholarly approaches (such as those of Ernst Nolte) into marxist–communist agitatory formulations revamping those of the 1920s and 1930s, attacking the Western concept of democracy as 'late bourgeois' and 'late capitalist', and the liberal–democratic West German parliamentary state as simply 'restorative'. Ideological monocausal explanations had replaced the earlier openness of historical and political science. Non-marxist writers, too, had under the impetus of socio-economic methods and the 'sociologization of contemporary history' contributed to a changed language and style of interpretation. All in all, the tapping of new sources and the intensification of empirical research had widened the base for solid, specialized work. But this stood in an uneasy relationship to the 'tendency, through theorizing and ideologizing alienation from the history of persons and events, to show and put into effect as the dominant leading theme the contemporary criticism of capitalism and democracy'.

The controversies which we shall be exploring arose in this climate, overlain with political and ideological considerations. In a state which has not had a dominant marxist historiographical school, most of the debates we shall consider are controversies between historians of different kinds of liberal–democratic persuasion. The politicization of the debate is here more latent than overt. In so far as it comes into the open at all, it is darkly reflected in philosophical disputes about the relevance of present-day social and political values to the historian's writing, and whether these should be banished in the interests

[36] Karl Dietrich Bracher, 'Zeitgeschichte im Wandel der Interpretationen', *HZ* 225 (1977), pp. 635-55, here esp., pp. 635-8, 648-51, 654-5.

of a 'value-free' and 'objective' history.[37] There is general agreement on the historian's task of 'enlightenment' in the values of reason, freedom, and 'emancipation', but such a vague commitment to virtue not sin naturally leaves room for a multitude of often only semi-concealed ideological positions. And, as the above comments demonstrate, it also does not prevent the occurrence of slights and slurs as the accompaniment to scholarly controversy. One manifestation of this is the allegation that, in their attempted 'revision' of established interpretations of Nazism, historians are 'trivializing' the evil nature of the Nazi regime. This strikingly indicates the prominence, also, of the moral dimension, inescapable in writing on Nazism.

The Moral Dimension

The moral content of early post-war writing on Nazism was explicit. Historians of the victorious powers were only too anxious to find in Nazism a confirmation of all the worst traits in Germans present throughout the centuries, and from the evident mass support for Hitler in the 1930s deduced a peculiarly 'German disease' and an easy equation of Germans and Nazis. We have already noted the moral tone of the defence against this crude allegation in the works of Meinecke and Ritter, which reflected the not unnatural apologetic character of German historical writing in the post-war era. The emphasis on 'the other Germany' and the resistance plot of 1944 – as, for instance, in Gerhard Ritter's biography of Goerdeler – again indicates the dominance of the moral dimension in early German post-war writing on the Third Reich.[38]

Though more recent scholarship has totally departed from the indignation and resentment, condemnation and apology, which characterized the post-war era, a strong moral element remains as a latent presence. All serious scholars (Germans above all) demonstrate even by the very language they employ – as in the frequent use of terms such as 'criminality' or 'barbarity' in connection with the Nazi regime – their moral detestation for Nazism. This raises a point which numerous commentators have noted as a difficulty about interpreting Nazism. Whereas historians traditionally try to eschew moral judgement (with varying degrees of success) in attempting to reach a sympathetic 'understanding' (*Verstehen*) of their subject matter, this is clearly an impossibility in the case of Nazism and Hitler. Wolfgang Sauer put the dilemma in the following way: 'In Nazism, the historian faces a phenomenon that leaves him no way but

---

[37] See e.g., Thomas Nipperdey, 'Geschichte als Aufklärung', in M. Zöller, ed., *Aufklärung heute. Bedingungen unserer Freiheit* (Zurich, 1980), pp. 50–62; Jürgen Kocka, 'Legende, Aufklärung und Objektivität in der Geschichtswissenschaft', *GG* 6 (1980), pp. 449–55; and Jörn Rüsen, 'Geschichte als Aufklärung?', *GG* 7 (1981), pp. 189–218.

[38] See Gerhard Ritter, *Carl Goerdeler und die deutsche Widerstandsbewegung* (Munich, 1955) and Hans Rothfels, *The German Opposition to Hitler* (Chicago, 1948). See also Iggers, pp. 344–7. For surveys of more recent trends in the historiography of 'resistance', see Hildebrand, *Das Dritte Reich*, pp. 181–6, and Reinhard Mann, 'Widerstand gegen den Nationalsozialismus', *Neue politische Literatur* 22 (1977), pp. 425–42. For a comprehensive statement of the present state of research, see Jürgen Schmädeke and Peter Steinbach, eds., *Der Widerstand gegen den Nationalsozialismus. Die deutsche Gesellschaft und der Widerstand gegen Hitler* (Munich, 1985). Problems of interpreting 'resistance' to Hitler will always remain thorny ones, not least because of the normative political and moral connotations of the term 'resistance' (*Widerstand*), which is at the same time used as a concept of scholarly analysis. See chapter 8 for a full discussion of these problems.

rejection, whatever his individual position. There is literally no voice worth considering that disagrees on this matter. . . . Does not such fundamental rejection imply a fundamental lack of understanding? And if we do not understand, how can we write history? The term "understanding" has, certainly, an ambivalent meaning; we can reject and still "understand". And yet, our intellectual, and psychological, capacities reach, in the case of Nazism, a border undreamed of by Wilhelm Dilthey. We can work out explanatory theories, but, if we face the facts directly, all explanations appear weak'.[39] It may be that the problem is in practice less serious than Sauer imagined. After all, historians of many other political regimes and their leaders often have little enough chance to show 'sympathetic understanding' for the object of their studies.

Even so, the problem could not be highlighted more plainly than in the case of Hitler's Germany, although the universal moral condemnation of Nazism makes it all the more surprising that the question of its implicit moral trivialization in recent historical writing has been raised at all. Karl Dietrich Bracher appears to have started it, and his comments show that the allegation is not unconnected with the questions of historical method and political–ideological overtones which we discussed earlier. Bracher claimed that recent marxist and 'New Left' approaches – but also those of some well-established liberal 'bourgeois' (or, as he calls them, 'relativist') historians – amounted to a gross underestimation of the reality of Nazism. Accordingly, 'the ideological and totalitarian dimension of National Socialism shrinks to such an extent that the barbarism of 1933–45 disappears as a moral phenomenon'. As a result, 'it could well appear as if a new wave of trivialization or even apologetics was beginning'.[40] In similar vein, Klaus Hildebrand criticized those who 'theoretically fixed, are vainly concerned with functional explanations of the autonomous force in history and as a result frequently contribute towards its trivialization'.[41] The most forthright rejection of such allegations was voiced by Tim Mason, within the context of debates on Nazism: 'The debate has reached such a pitch of intensity that some historians are now accusing other historians of "trivializing" National Socialism in their work, of implicitly, unwittingly, furnishing an apologia for the Nazi regime. This is perhaps the most serious charge which can be made against serious historians of the subject', raising 'fundamental questions about the moral and political responsibility of the historian'.[42]

The interpretations which have given rise to these allegations of trivialization will concern us later in the book. It suffices for now to point out that the charge has been made in order to illustrate the inevitable moral undertones to any discussion about Nazism, particularly among German historians. In actual fact, though Bracher had some grounds for his charge in the more trite productions of the 'New Left' which saw no essential difference between fascism and other forms of 'bourgeois domination', it seems to me that it is a wholly unnecessary and unjustified slur when extended to serious historians of Nazism.

[39] Wolfgang Sauer, 'National Socialism: Totalitarianism or Fascism?', *AHR* 73 (1967-8), p. 408. See also Klaus Hildebrand, 'Der "Fall" Hitler', *NPL* 14 (1969), p. 379.
[40] Karl Dietrich Bracher, *Zeitgeschichtliche Kontroversen. Um Faschismus, Totalitarismus, Demokratie* (Munich, 1976), pp. 62-3.
[41] Hildebrand, 'Geschichte oder "Gesellschaftsgeschichte"?', p. 355.
[42] Mason, 'Intention and Explanation', p. 23.

However, the charge of 'trivialization' does raise pointedly the question of a moral purpose in writing about Nazism. Is the aim to learn the evil of Nazism by 'understanding' it? Is it a matter of condemning a uniquely evil phenomenon which by the nature of its uniqueness can never repeat itself and is gone for ever? Is it to draw lessons from this horror of the past about the fragility of modern democracy and the need to maintain a constant guard against the threat to liberal democracy from Right and Left? Is it to provide strategies for the recognition and prevention of the re-emergence of fascism? Is it to carry out simultaneously an act of remembrance and warning cast through hatred and anger? The latter seemed to be the position of the late Lucy Dawidowicz in a book solely about the morality of historical writing on the holocaust.[43] She spoke there of Nazism as 'the essence of evil, the daemon let loose in society, Cain in a corporate embodiment'. She held that 'nothing but the most lucid consciousness of the horror that happened can help avoid it for the future'. And she cited approvingly the words of Karl Jaspers: 'That which has happened is a warning. To forget it is guilt. It must be continually remembered. It was possible for this to happen, and it remains possible for it to happen again at any minute. Only in knowledge can it be prevented'.[44] At the same time, her distaste for the methods of marxist and structuralist historians (who are again accused of abdicating their professional responsibility) and her predilection for personalized history – for the 'attribution of human responsibility for the occurrence of historic events . . . to the movers and shakers who made events happen'[45] – raises once more in striking fashion the problem of how the historical method she favoured can produce the ends she desired.

We are back again to the interrelationship of the historian's method, the moral nature of his professional obligation, and the political–ideological framework in which this obligation is carried out.

[43] Lucy Dawidowicz, *The Holocaust and the Historians* (Cambridge, Mass., 1981). See the sharply critical reviews by Richard Bessel, *Times Higher Education Supplement*, 19 March 1982, p. 14, and Geoff Eley, 'Holocaust History', *London Review of Books*, 3–17 March 1982, p. 6.
[44] Dawidowicz, *Holocaust*, pp. 20–1.
[45] Dawidowicz, *Holocaust*, p. 146.

2

The Essence of Nazism: Form of Fascism, Brand of Totalitarianism, or Unique Phenomenon?

There has been debate since the 1920s about the nature and character of the Nazi phenomenon – how it ought to be located in the context of the strikingly new political movements which, since the Bolshevik Revolution of 1917 and five years later Mussolini's 'March on Rome', had been recasting the shape of Europe. While Comintern theorists in the 1920s were already categorizing Nazism as a form of fascism engendered by capitalism in crisis, bourgeois writers a little later were only beginning to associate Right and Left as the combined totalitarian enemies of democracy. The debates were, of course, considerably broadened during the years of Nazi rule: on the one hand through the finalizing of the Comintern definition of fascism in 1935 and through analyses of fascism by left-wing theorists exiled in the West; and on the other hand through a growing readiness in the western democracies and in the USA to view Nazism and Soviet Communism as two sides of the same totalitarian coin – a view seemingly confirmed by the Nazi–Soviet Non-Aggression Pact of 1939. Though this line was naturally played down from 1941 onwards, it re-emerged all the more strongly with the onset of the Cold War in the later 1940s. During the Cold War era, left-wing interpretations of Nazism as a form of fascism lost their influence, while totalitarianism theories enjoyed their hey-day and came gradually under fire – crumbling beneath the weight of accumulating detailed research – only in the period of growing detente, increasing introspection and criticism of western society and governments, and then the upheavals in universities and intellectual currents in the later 1960s. Revival of interest in fascism as a generic problem was reflected in a burgeoning output of studies not only from the Left but also from liberal writers, setting the 'totalitarianism' theorists on the defensive, though there was some retrenchment in the 1970s as some weaknesses of the comparative fascism approach became increasingly visible. The debate about whether Nazism can best be regarded as a type of fascism, or whether it should be seen as a prominent form of totalitarian rule has still not subsided, least of all in West German historiography.

The debate about fascism and totalitarianism has been kept alive, too, by its relationship to a third strand of interpretation which, in recent years, has proved highly influential: that Nazism can only be explained as a product of the peculiarities of Prussian–German development over the previous century or so. Such an interpretation has, however, itself been advanced in two quite distinct and opposed forms.

Social historians, concentrating on the *causes* of Nazism, have emphasized a specific path of modernization in Germany in which, far more so than was the case in western societies, pre-industrial, pre-capitalist, and pre-bourgeois authoritarian and feudal traditions survived in a society which was never truly bourgeois, existing in a relationship of tension with a modern, dynamic capitalist economy and finally exploding into violent protest when that economy collapsed in crisis. Less the nature of German capitalism than the strength of pre-modern forces in German society determined the road to Nazi victory in 1933. Though stressing the peculiarities of the German development, exponents of such an interpretation point to obvious parallels in other societies, for instance in Italy, and regard Nazism, for all its singular characteristics, as a form of fascism in terms of its socio-economic origins and formation, seeing also no necessary incompatibility with elements of the totalitarianism theory in terms of certain components of rule.[1]

This emphasis upon a 'failed bourgeois revolution' and the dominance of pre-industrial, neo-feudal structures in explaining a German 'special path' of development has, however, been subjected to a frontal attack.[2] The alternative position stresses in contrast the *bourgeois* character of late-nineteenth-century German society and politics and – implicitly rather than explicitly – the need to explain Nazism not through 'German peculiarities' but through the particular instabilities of the form of capitalism and capitalist state which existed in Germany. It might be thought that this line of argument – whatever its merits – only brought one back to a slightly different set of questions about 'peculiarities' in order to answer the obvious problem about why Germany alone of all the highly advanced industrial capitalist economies – Italy, though making great advances in industrialization before the War, could not rank with the major industrial economies – produced a fully-blown 'fascist' dictatorship. The recent heated (if somewhat artificial) debate on the 'special path' of Germany's development is concerned more with interpreting the Imperial period than the Third Reich. Despite its obvious connotations for understanding the origins of Nazism, it need occupy us no longer here – not least because historians on both sides of the debate fully accept that, for all its singular characteristics, Nazism belongs to a wider category of political movements which we call 'fascist'. The German 'peculiarities' under question in this con-

[1] Representative of this line of argument is Jürgen Kocka, 'Ursachen des Nationalsozialismus', *APZ* (21 June 1980), pp. 3–15.
[2] See David Blackbourn and Geoff Eley, *Mythen deutscher Geschichtsschreibung* (Frankfurt am Main/Berlin/Vienna, 1980), Engl. trans., *The Peculiarities of German History* (Oxford, 1984). For the sharp and polemical debate unleashed by this book, see e.g. the reviews by Hans-Ulrich Wehler, ' "Deutscher Sonderweg" oder allgemeine Probleme des westlichen Kapitalismus?', *Merkur* 5 (1981), pp. 478–87; Hans-Jürgen Puhle, 'Deutscher Sonderweg. Kontroverse um eine vermeintliche Legende', *Journal für Geschichte*, Heft 4 (1981), pp. 44–5; Wolfgang J. Mommsen, in *Bulletin of the German Historical Institute*, London 4 (1980), pp. 19–26; and the discussion forum *Deutscher Sonderweg – Mythos oder Realität* (Kolloquien des Instituts für Zeitgeschichte, Munich/Vienna, 1982). Directly relating this to the causes of fascism, and partly in reply to Kocka's article (see note 1), see also Geoff Eley, 'What produces Fascism: Preindustrial Traditions or a Crisis of the Capitalist State?', *Politics and Society* 12 (1983), pp. 53–82. Jürgen Kocka, 'German History before Hitler'. The Debate about the German *Sonderweg*', *JCH* 23 (1988), pp. 3–16, provides an excellent critique of the pros and cons of the *Sonderweg* argument. He concludes that while the term '*Sonderweg*' is itself misleading and dispensable, the notion of a divergence from the pattern of development of other 'advanced' western countries retains its value in explaining why Germany offered so few barriers to the fascist challenge.

troversy are those which set Germany apart from western parliamentary democracies, not from Italian or other manifestations of fascism.

A different and more exclusive emphasis upon the singularity of Nazism as the product of recent Prussian–German history is an important focus of the interpretation of some of the leading West German political historians in their analyses of the character and nature of Nazi rule. According to such an interpretation, Nazism was *sui generis* – altogether a unique phenomenon, emerging from the peculiar legacy of the Prusso-German authoritarian state and German ideological development, but owing its uniqueness above all to the person of Hitler, a factor of overriding importance in the history of Nazism and one which is incapable of being ignored, played-down, or substituted. So singular was Hitler's ideological and political contribution to the shaping and direction of the Nazi movement and then the Nazi State that any attempt to label National Socialism as 'fascism', thus placing it in comparison with other 'similar' movements, is meaningless and implies, moreover, the 'trivialization' of Hitler and Nazism. Rather, so completely interwoven was National Socialism with the rise, fall, political aims, and destructive ideology of this unique personality, that it is legitimate to speak of Nazism as 'Hitlerism'. Though excluding vehemently any possibility of regarding 'Hitlerism' as a type of fascism, exponents of this interpretation nevertheless attach one important strand of comparison, arguing that the form and nature of Nazi rule make it essential to regard Nazism as a brand of totalitarianism alongside Soviet Communism (in particular Stalinism).[3]

In this chapter I shall first summarize briefly the stages of development and the main variants of interpretation within the 'totalitarianism' and 'fascism' approaches. There is by now a wide literature examining and describing these approaches in detail, so that I shall offer as brief an outline as possible for purposes of orientation. Secondly, I shall attempt to evaluate the strengths and weaknesses of the concepts in their application to Nazism. Finally, in the light of discussion of totalitarianism and fascism I shall return to consider the argument for the singularity of Nazism in the context of the 'peculiarity' of German development.

Totalitarianism

It is mistaken to regard the totalitarianism concept as simply a product of the Cold War, though that was indeed the period of its full flourishing. Its usage is in fact almost as old as that of fascism, dating back to the 1920s. And though slightly later on the scene than fascist theorems, the totalitarianism approach came earlier to gain general acceptance as an 'established' and 'establishment' theory before being subjected to damaging challenge in the 1960s. I shall deal, therefore, with totalitarianism first.

The term was coined in Italy as early as May 1923 and used initially as an anti-fascist term of abuse. In order to turn the tables on his opponents, Mussolini usurped the term in June 1925, speaking of the 'fierce totalitarian

[3] See the essays by Karl Dietrich Bracher in his *Zeitg. Kontrov.*, Part I, and his 'The Role of Hitler: Perspectives of Interpretation', in Walter Laqueur, ed., *Fascism. A Reader's Guide* (Harmondsworth, 1979), pp. 193–212; Hildebrand, *Das Dritte Reich*, pp. 132 ff., 187 ff.; and Hillgruber, *Endlich genug?*, pp. 38–42.

will' of his Movement. Thereafter, it was used in positive self-depiction by Mussolini and other Italian fascists, then later by German legalists and by the Nazis. Gentile, the chief ideologue of Italian fascism, also employed the term on numerous occasions, though in a more étatist sense, implying an all-embracing state which would overcome the state–society divide of weak pluralist democracies. The two notions, this étatist one and Mussolini's implication of the dynamic revolutionary will of the Movement, existed side by side. The German usage was somewhat different, but related and with the same dual approach. Ernst Jünger was one of a number of writers already coining the notion of 'total war' and 'total mobilization' in the 1920s – a term with dynamic, revolutionary implications. Around the same time Carl Schmitt, Germany's foremost legal theorist, was developing the concept of power politics based on a friend–foe relationship, into which he fitted, as the historical antithesis to the liberal pluralization of the state, the 'total state of identity of state and society'. Both forms, therefore, the 'actionist' and the 'étatist', existed before the Nazis came to power and were incorporated into Nazi usage (though the word 'totalitarian' was, in fact, seldom used by the Nazi leadership).[4]

First usage of the word 'totalitarianism' to bracket together fascist and communist states seems to have been in England in 1929, although several years earlier Nitti, the former prime minister of Italy was among those making structural comparisons between Italian fascism and bolshevism. In the 1930s and 1940s the concept was also applied by notable left-wing analysts of fascism such as Borkenau, Löwenthal, Hilferding, and Franz Neumann as a tool for characterizing what they saw as the new and specific in fascism (or Nazism) alone, without the comparative element of extension to Soviet Communism. Franz Neumann, for example, built his application of the term in his masterly *Behemoth* on the contemporary fascist self-stylization and the notion of the collapse into chaos of the Schmitt 'total state' under the 'totalitarian' drive of the Nazi movement.[5] At the same time the dominant usage of the adjective 'totalitarian' to link fascism and Nazism with communism was already gaining ground in Anglo-Saxon countries in the 1930s, boosted by German exile writing, the Stalinist terror, and the Nazi–Soviet Pact. The way was being paved for the emergence of the fully-fledged totalitarian model of the early post-war era, popularized in different ways above all by Hannah Arendt and Carl Friedrich.

Hannah Arendt's *Origins of Totalitarianism* is a passionate and moving denunciation of inhumanity and terror – depersonalized and rationalized as

4 For the developing usage of the 'totalitarianism' concept, see Walter Schlangen, *Die Totalitarismus-Theorie. Entwicklung und Probleme* (Stuttgart/Berlin/Cologne/Mainz, 1976), chs. 1–3. For guidance on the early Italian usage, I am indebted to Professor Meir Michaelis (Hebrew Univ. of Jerusalem). See his informative paper: 'Anmerkungen zum italienischen Totalitarismusbegriff. Zur Kritik der Thesen Hannah Arendts und Renzo de Felices', *Quellen und Forschungen aus italienischen Archiven und Bibliotheken* (published by the German Historical Institute in Rome) 62 (1982) pp. 270–302, esp. pp. 292–7.
5 Franz Neumann, *Behemoth. The Structure and Practice of National Socialism* (London, 1942). The German edition (Cologne/Frankfurt am Main, 1977, based on the extended 1944 English edition) has a valuable 'Nachwort' by the editor, Gert Schäfer. See also Richard Saage, 'Das sozio-politische Herrschaftssystem des Nationalsozialismus. Reflexionen zu Franz Neumanns "Behemoth" ', *Jahrbuch des Instituts für Deutsche Geschichte, Tel Aviv* 10 (1981), pp. 342–62.

the execution of objective laws of history. Her emphasis on the radicalizing, dynamic, and structure-destroying inbuilt characteristics of Nazism has been amply borne out by later research. However, the book is less satisfactory on Stalinism than on Nazi Germany. Moreover, it offers no clear theory or satisfactory concept of totalitarian systems. And its basic argument explaining the growth of totalitarianism – the replacement of classes by masses and the emergence of a 'mass society' – is clearly flawed.[6]

Carl Friedrich's publications, written from a standpoint of constitutional theory, were even more influential than Hannah Arendt's. Every subsequent writer on totalitarianism has had to confront Friedrich's work, and especially his famous 'six-point syndrome' highlighting what he saw as the central characteristics of totalitarian systems (an official ideology, a single mass party, terroristic police control, monopoly control over the media, a monopoly of arms, and central control of the economy). The main weaknesses of the Friedrich model have frequently been pointed out. It is above all a static model, allowing little room for change and development in the inner dynamics of a system, and it rests on the exaggerated assumption of the essentially monolithic nature of 'totalitarian regimes'. His model has, therefore, come largely to be rejected even by those scholars still operating with a totalitarianism approach.[7]

Following the stabilization of the USSR in the post-Stalin era, totalitarianism theorists tended to concentrate attention far more on current eastern-bloc regimes rather than on the dead Nazi system, and divided into those who broadened the totalitarianism concept to include all manifestations of communist rule and those who limited it in the main to Stalinism. In both cases, however, the comparison with fascist systems was at least implicitly preserved.[8]

In the meantime, totalitarianism had been adopted in the 1950s as the fundamental prop of leading scholarly interpretations of Nazism, as in the classic pioneering works of Karl Dietrich Bracher. A political scientist himself, Bracher has pointed out the caution needed in developing a general theory of totalitarianism through constitutional or sociological categories resting on all too meagre empirical historical research. Such research was vital, in his view, to reveal the many varied forms of totalitarian rule, but would confirm the essential similarity in the techniques of rule of the Bolshevik/communist and Nazi/fascist systems. Bracher was unwilling to tie himself to the static, constitutive, and insufficiently differentiated features of the Friedrich model which could do scant justice to the 'revolutionary dynamic' which he saw as the 'core principle' distinguishing totalitarian from other forms of authoritarian rule. The decisive character of totalitarianism lies for him in the total claim to rule, the leadership principle, the exclusive ideology, and the fiction of identity of rulers and ruled. It represents a basic distinction between an 'open' and a

[6] Arendt, *Origins* (see ch. 1 note 31). See the remarks of Klaus Hildebrand, 'Stufen der Totalitarismus-Forschung', *PVS* 9 (1968), pp. 406-8; Martin Kitchen, *Fascism* (London, 1976), pp. 30-1; and Ayçoberry, pp. 130-3.
[7] Friedrich first advanced his model in his essay, 'The Unique Character of Totalitarian Society' in the volume he edited, *Totalitarianism* (Cambridge, Mass., 1954), and extended it in Friedrich and Brzezinski, *Totalitarian Dictatorship and Autocracy*. For criticism from within the 'totalitarianism' approach and a revised model, see Leonard Schapiro, *Totalitarianism* (London, 1973).
[8] See Schlangen, ch. 4.

'closed' understanding of politics.[9] The fundamental value of the totalitarianism concept resides therefore in its ability to recognize the primary distinction between democracy and dictatorship. Though Bracher sees that, as in all political and social theories which go beyond simple description, totalitarianism theories have their weaknesses, he claims that now as before, even after Hitler and Stalin, there is 'the phenomenon of totalitarian claims to rule and the tendency to the totalitarian . . . temptation' (which in this context he goes on to associate with the New Left among German intellectuals and also with the growth of terrorism of Left and Right in the Federal Republic in the 1970s).[10] In his view, the primary question of the totalitarian character of political systems cannot be shirked either in the interest of scholarly clarity and objectivity, or in view of the political and human consequences of such dictatorships and the tendencies towards totalitarianism in present-day society.

Though other eminent scholars have applied and continue to apply the concept of totalitarianism to characterize what they see as the essence of the Nazi system, it suffices here to summarize Bracher's use of the concept. Not only has he been at the pinnacle of scholarship on Nazism since the 1950s, but he has also consistently argued the case for totalitarianism within the framework of understanding different models of political domination and has more than any other historian been instrumental in the retention and even current revival of the totalitarianism concept in its application to Nazism. However, doubts must remain about Bracher's employment of a rather undifferentiated divide between 'open' and 'closed' understandings of politics as a key ordering principle for defining totalitarianism, about his lack of clear distinction between totalitarianism as a tendency and as a system of rule, about the arguable value of the concept of 'revolutionary dynamic' when applied to various societies which Bracher would regard as 'totalitarian', and, fundamentally, about the attribution of relatively superficial common characteristics to regimes revealing many significant differences of organization and aim.

We can turn now to a brief outline of opposed interpretations locating Nazism in the family of inter-war European fascisms, rejecting at the same time the comparison with Soviet Communism inherent in the totalitarianism approach.

Fascism

A new wave of interest in fascism as a phenomenon experienced in most countries of inter-war Europe was prompted in no small measure in the 1960s by the appearance of Ernst Nolte's highly influential book *Der Faschismus in seiner Epoche* in 1963.[11] Within five years several major international conferences had been held, numerous anthologies were in print containing studies of the nature and manifestation of fascist movements throughout Europe, and a con-

---

[9] For a succinct statement of his position of 'totalitarianism', see *Totalitarismus und Faschismus. Eine wissenschaftliche und politische Begriffskontroverse* (Munich/Vienna, 1980), pp. 10–17, 53–4, 69–70.
[10] Karl Dietrich Bracher, *Schlüsselwörter in der Geschichte* (Düsseldorf, 1978), pp. 109–10, 121–3.
[11] Ernst Nolte, *Der Faschismus in seiner Epoche* (Munich, 1963), Engl. trans., *The Three Faces of Fascism* (London, 1965, subsequent references to Mentor edn., New York/Toronto, 1969).

siderable scholarly literature had built up.[12] Scholarly interest in comparative fascism merged with, and was then in part overtaken by, political interest on the Left in the later 1960s during the period of the 'New Left' challenge to the values of contemporary liberal–bourgeois society. The political conditions of the 1960s spurred and steered, therefore, a revival of marxist theories of fascism derived from the writings of contemporary marxist analysts of the fascist phenomenon alongside the proliferation of non-marxist interpretations of fascism.[13] In the case of both marxist and non-marxist interpretations, it can generally be said that, as with totalitarianism, most of the strands of the debate reach back practically as far as the phenomenon of fascism itself.

Marxist Theories

The first serious attempt to explain fascism in theoretical terms was undertaken by the Comintern in the 1920s. The Comintern understanding, initially of Italian fascism, was founded on the notion of a close instrumental relationship between capitalism and fascism. Derived from the Leninist theory of imperialism, the theory held that the coming inevitable collapse of capitalism fostered an increased need on the part of the most reactionary and powerful groups within the now highly-concentrated finance capital to secure their imperialist aims by manipulating a mass movement capable of destroying the revolutionary working class and therefore of safeguarding in the short-term capitalist interests and profits to be achieved through expansion and war. Fascism was thus the necessary form and final stage of bourgeois–capitalist rule. According to this interpretation, therefore, politics was a direct function of economics and wholly subordinated to it; the fascist mass movements were a product of capitalist manipulation; fascist rule served the function of bolstering profit; fascist leaders were thereby the 'agents' of the capitalist ruling class. The key question to be asked was: to whose advantage did the system work?

[12] E.g. Eugene Weber, *Varieties of Fascism* (New York, 1964); 'International Fascism, 1920–1945', *JCH* 1 (1) (1966); Ernst Nolte, *Die faschistischen Bewegungen* (Munich, 1966); Francis L. Carsten, *The Rise of Fascism* (London, 1967); Stuart J. Woolf, ed., *European Fascism* (London, 1968), and *The Nature of Fascism* (London, 1968); Wolfgang Schieder, 'Faschismus', in C.D. Hernig, ed., *Sowjetsystem und demokratische Gesellschaft. Eine vergleichende Enzyklopädie* (7 vols., Freiburg/Basel/Vienna, 1966–72), vol. 2 (1968), columns 438–77; Renzo de Felice, *Interpretations of Fascism* (Cambridge, Mass., 1977, first Italian edn. 1969). For useful later anthologies and surveys of the literature, see Wolfgang Wippermann, *Faschismustheorien* (Darmstadt, 1972); Wolfgang Schieder, ed., *Faschismus als soziale Bewegung* (Hamburg, 1976); Hans-Ulrich Thamer and Wolfgang Wippermann, *Faschistische und neofaschistische Bewegungen* (Darmstadt, 1977); Walter Laqueur, ed., *Fascism. A Reader's Guide* (Harmondsworth, 1979); Stanley Payne, *Fascism: Comparison and Definition* (Madison, Wisconsin, 1980); Stein Ugelvik Larsen *et al.*, *Who were the Fascists? Social Roots of European Fascism* (Bergen, 1980); Wolfgang Wippermann, *Europäischer Faschismus im Vergleich, 1922–1982* (Frankfurt am Main, 1983) and Detlef Mühlberger, ed., *The Social Basis of European Fascist Movements* (London/Sydney, 1987). The most outstanding recent study is that of Roger Griffin, *The Nature of Fascism* (London, 1991).
[13] E.g. Ernst Nolte, ed., *Theorien über den Faschismus* (Cologne, 1967); Wolfgang Abendroth, ed., *Faschismus und Kapitalismus. Theorien über die sozialen Ursprünge und die Funktion des Faschismus* (Frankfurt am Main/Vienna, 1967); Reinhard Kühnl, ed., *Texte zur Faschismusdiskussion I. Positionen und Kontroversen* (Reinbek bei Hamburg, 1974); Reinhard Kühnl, *Formen bürgerlicher Herrschaft* (Reinbek bei Hamburg, 1971); Manfred Clemenz, *Gesellschaftliche Ursprünge des Faschismus* (Frankfurt am Main, 1972). A cross-section of 'New Left' work of the 1960s can be seen in *Das Argument* 1–6 (1964–70). For a sharp critique, see Heinrich August Winkler, *Revolution, Staat, Faschismus* (Göttingen, 1978), ch. 3.

And the answer left no doubt as to the intrinsic link between the fascist lackeys and the capitalist rulers. Though a short summary can do scant justice to the debates within the Comintern and to the varied glosses and interpretations which were advanced (the most far-sighted and nuanced by Clara Zetkin), it can be said that the view just described prevailed in essence to be encapsulated at the thirteenth plenary meeting of the Executive Committee of the Communist International in December 1933, and in its final form in the Dimitroff definition of 1935, mentioned in chapter 1. It remained the basis of Soviet and East German writing on Nazism down to the recent upheavals in eastern Europe.[14]

The contemporary dominance of the 'orthodox' Comintern thinking meant that 'nonconformist' marxist interpretations often received less attention than they merited at the time. The subtle interpretations, for example, of the KPD 'renegade' August Thalheimer, excluded from the Communist Party in 1928, and the Austrian theorist Otto Bauer received due recognition only during the revival of fascist studies in the 1960s and 1970s, though their influence on recent western marxist interpretations of fascism has generally been greater than the Comintern formulation.

Thalheimer, in a series of essays published in 1930 but gaining full recognition only in the late 1960s, and Bauer, in an essay printed in 1924 and elaborated upon in a chapter of a book written in 1936, both based their understanding of fascism on Marx's writings on Bonapartism, in particular his *Eighteenth Brumaire of Louis Bonaparte*, written immediately after the *coup d'état* of 2 December 1851. Though neither equated Bonapartism with fascism (which at the time of their original publications remained chiefly in its Italian manifestation), both saw in Marx's interpretation of the French *coup d'état* a significant pointer to understanding the mechanics of the fascist relation to the capitalist ruling class. Marx's work had rested on his assertion that the mutual neutralization of the social classes in the struggle for power in France had enabled Louis Bonaparte, supported by the lumpenproletariat and the mass of a-political peasant small-holders, to build the executive authority of the State into a relatively independent power. Applying Marx's analysis to fascism allowed Thalheimer and Bauer to distinguish between the social and political domination of the capitalist ruling class, to give weight to the autonomous importance of the fascist mass backing, to see fascism as only one of a number of possible ways out of the crisis of capitalism and by no means the equivalent of the final stage of capitalism *en route* to socialism, and, finally, to give weight to the relative autonomy of the fascist executive once in power. In each case, this interpretation brought them into direct conflict with the 'orthodox' Leninist line (though in his last writings in 1938 Bauer played down Bonapartism and came much closer to a Leninist analysis of imperialism). The crucial point was the dialectical relationship between the economic rule of the 'big bourgeoisie' and the political supremacy of the fascist 'ruling caste', financially supported by capitalists but not created by them. Though petty bourgeois in composition, the fascist party in power was bound, however, to become the instrument of the economic ruling class, especially its more warlike elements, but the inner contradictions within the system which would result in clashes of interest

---

[14] For a summary of GDR research by some of its foremost historians, see Eichholtz and Gossweiler, *Faschismusforschung* (see ch. 1 note 27).

between the fascist caste and the capitalist ruling class could only be soluble through war.[15]

While the Comintern theory remained, until the upheavals of 1989, operational in the GDR as the key to an understanding of fascism, variants of the Bonapartist approach (such as can also be seen in Trotsky's perceptive writings on fascism[16]) have greatly influenced the theoretical writings of western marxists since the 1960s. In recent years, however, writing on fascism on the Left has been significantly affected by a third major strain of marxist fascism interpretation, derived from Gramsci's work (in particular his concept of 'bourgeois hegemony') and articulated by Nicos Poulantzas, whose interpretation we will consider more closely in chapter 3.[17] The neo-Gramscian approach lays far more emphasis than other marxist interpretations on the conditions of *political* crisis, arising when the state can no longer organize the political unity of the dominant class and has lost popular legitimacy, and which make fascism attractive as a radical populist solution to the problem of restoring the dominant class's 'hegemony'. Marxist interpretations of fascism, briefly described here, will concern us in the following chapter when we deal with the relationship of politics to economics in the Nazi system of rule.

Non-Marxist Interpretations

While, as I have indicated, most recent marxist interpretations of fascism have adopted or built upon theories which were current in the 1920s and 1930s, early 'bourgeois' or non-marxist interpretations – few if any of them actually amounting to a *theory* of fascism – have generally been found seriously wanting by later scholarship. The 'moral crisis of European society' view, for instance, favoured by Croce, Meinecke, Ritter, and later Golo Mann, has had only the most indirect impact upon later non-marxist fascism interpretation. Wilhelm Reich's attempt to combine marxism and freudianism in interpreting fascism as a consequence of sexual repression, and Erich Fromm's collective psychology approach arguing for an 'escape from freedom' to take refuge in submission, have also provided little methodological impetus for current analysis of fascism. Only Talcott Parsons's approach through the 'anomie' of modern social structures and the conflict-laden coexistence of traditional, archaic value-systems and modern social processes can be said to have 'left an indelible imprint' on later non-marxist analyses of fascism linked to theories of

---

[15] On Thalheimer, Bauer, and 'Bonapartism', see esp. Gerhard Botz, 'Austro-Marxist Interpretations of Fascism', in 'Theories of Fascism', *JCH* 11 (4) (1976), pp. 129–56, esp. pp. 131–47; Jost Dülffer, 'Bonapartism, Fascism, and National Socialism', in *JCH* 11 (1976), pp. 109–28; and Hans-Gerd Jaschke, *Soziale Basis und soziale Funktion des Nationalsozialismus. Studien zur Bonapartismustheorie* (Opladen, 1982). See also Kitchen, ch. 7; Aycoberry, pp. 57–64; and Hildebrand, *Das Dritte Reich*, pp. 125–6. Winkler, *Revolution*, ch. 2 and pp. 83 ff. offers a critique. And for an excellent evaluation of interwar marxist analyses of fascism ('orthodox' and 'deviant'), together with a selection of the most important texts, see David Beetham, *Marxism in Face of Fascism* (Manchester, 1983).
[16] Leon Trotsky, *The Struggle against Fascism* (New York, 1971). Trotsky regarded the presidential cabinets of Brüning, von Papen, and Schleicher, not Fascism itself, as 'Bonapartism'. See Robert S. Wistrich, 'Leon Trotsky's Theory of Fascism', *JCH* 11 (1976), pp. 170–1.
[17] Nicos Poulantzas, *Fascism and Dictatorship* (London, 1974).

modernization.[18] Non-marxist scholarship on comparative fascism, since its revival in the 1960s, found its drive chiefly from three different directions: from the 'phenomenological' history of ideas approach emanating from Ernst Nolte's work; from a number of varied 'structural-modernization' approaches; and from 'sociological' interpretations of the social composition and class base of fascist movements and voters.

Nolte's self-proclaimed 'phenomenological method' seems to amount in practice to little more than taking the self-depiction of a phenomenon seriously – in this case the writings of fascist leaders. Biting critics have suggested that it turns out 'to be essentially Dilthey's good, old method of empathy', or 'little more than historicism in fancy dress'.[19] Nolte gives little serious consideration to the social foundations of fascism, since he finds socio-economic explanations of fascism inadequate. Rather, his analysis of the development of fascist ideas brings him to what he rather grandiosely calls a 'metapolitical' conception of fascism as a generic and autonomous force. In a somewhat mystical and mystifying conclusion, he sees fascism as 'practical and violent resistance to transcendence'. By 'transcendence' he understands a twofold process of mankind's quest for emancipation and progress (which he terms 'practical transcendence'), and of man's search beyond this world for salvation, 'reaching out of the mind beyond what exists and what can exist toward an absolute whole' – i.e. belief in God and an after-life (which he calls 'theoretical transcendence'). Fascism is in essence, therefore, anti-modernist; but in the emphasis on the notion of 'violent resistance to transcendence', Nolte distinguishes fascism from mere 'reaction' and sees it as a European movement which was both anti-traditional and anti-modern, which, in rejecting first and foremost its mirror image of communism, at the same time threatened also the existence of bourgeois society. Finally, in his stress on 'fascism in its epoch' (the original German title of his major work), Nolte is claiming that fascism was historically time-bound, that 'it would not be possible for the "same" sociological configuration in a different period and under other world conditions to produce an historically relevant phenomenon that can qualify as fascism, at least not . . . in the form of European national fascism'.[20]

Nolte's was an important book and, as mentioned earlier, stirred up interest in the problem of generic fascism more than any other single work of the 1960s. But it is difficult to see that either methodologically or in terms of its conclusions it has gained a wide following. Other writers on comparative fascism, also working from the self-image of the fascists, have argued that fascism was revolutionary rather than backward-looking, that it 'looks much like the Jacobinism of our time'.[21] Secondly, the omission of detailed analysis of the nature and dynamics of the socio-economic foundation of fascist movements is a significant limitation of Nolte's work. Finally, from a different perspective

---

[18] See Talcott Parsons, 'Democracy and Social Structure in Pre-Nazi Germany', and 'Some Sociological Aspects of the Fascist Movements', in his *Essays in Sociological Theory* (London/ Toronto, 1949). The quotation is from Geoff Eley, 'The Wilhelmine Right: How it Changed', in Richard J. Evans, ed., *Society and Politics in Wilhelmine Germany* (London, 1978), p. 115.

[19] Sauer, p. 414 (ch. 1 note 39); Kitchen, p. 40.

[20] Nolte, *Three Faces*, pp. 529, 537 ff., 566–7. The quotation is from Ernst Nolte, 'The Problem of Fascism in Recent Scholarship', in Henry A. Turner, ed., *Reappraisals of Fascism* (New York, 1975), p. 30.

[21] Weber, *Varieties*, p. 139.

it has been questioned whether Nolte has done more than describe similar manifestations of a type of political system which he calls 'fascism', but which showed vitally different degrees of intensity throughout Europe, in other words missing the point that the differences outweigh the similarities, which would call into question the very existence of the phenomenon itself.[22]

The second major non-marxist *group* of approaches (for they contain many varied nuances and differences of emphasis) is that linked to modernization theories, in which fascism is seen as one of a number of different paths along the route to modern society. In one variant of the modernization approach, which Klaus Hildebrand has dubbed the 'structural–functional theory', fascism is regarded as 'a special form of rule in societies which find themselves in a critical phase of the process of social transformation to industrial society and at the same time objectively or in the eyes of the ruling strata are threatened by the possibility of a communist upheaval'.[23] Fascism gains its chief impetus, in this view, from the resistance of residual 'elites to the egalitarian tendencies of industrial society. Other approaches see fascism as a form of developmental dictatorship (Gregor), as primarily a phenomenon encountered in agrarian societies in a particular phase of their transition to modernization (Organski), as a product of the road to modernism of an agrarian society which has encountered only 'revolutio.1 from above', resulting in revolutionary unrest – with temporary modernizing force – of a thoroughly reactionary class (the peasantry) which is doomed to extinction (Barrington Moore).[24]

The main problem of the 'structural–functionalist' approach seems to lie in its over-emphasis on the resistance of the ruling élites to change at the expense of the weight to be attached to the autonomous dynamism of the fascist mass movements themselves. Coupled with this is the difficulty of establishing which states afflicted by fascism were precisely in this process of transition to a pluralistic industrial society. At best this seems to apply to Italy and Germany, though the degree of the transition was so different in the two countries that doubts remain about the value of the 'model'.[25] The chief difficulty with those modernizing theories which place fascism chiefly in an agrarian context is that they seem scarcely to apply to the German case, where Nazism developed in a highly-industrialized society. Significantly, Organski – one of the most prominent exponents of this approach – leaves Germany out of his model, while Barrington Moore's stimulating and wide-ranging analysis of different patterns of modernizing development rooted in the varied nature of the power base of the landed élites greatly over-emphasizes the importance of feudal traditions to the success of fascism, correspondingly underrating significantly the relationship to the dynamics of a fully-fledged capitalist economy and bourgeois society. Such modernization approaches as concentrate specifically upon Germany (e.g. the works of Dahrendorf and Schoenbaum[26]) are not concerned with a theory

[22] Hildebrand, *Das Dritte Reich*, p. 136.
[23] Wolfgang J. Mommsen, 'Gesellschaftliche Bedingtheit und gesellschaftliche Relevanz historischer Aussagen', in Eberhard Jäckel and Ernst Weymar, eds., *Die Funktion der Geschichte in unserer Zeit* (Stuttgart, 1975), pp. 219–20; Hildebrand, *Das Dritte Reich*, p. 136.
[24] A.J. Gregor, *The Ideology of Fascism* (New York, 1969); A.F.K. Organski, 'Fascism and Modernization', in Woolf, ed., *The Nature of Fascism*, pp. 19–41; Barrington Moore Jr., *Social Origins of Dictatorship and Democracy* (London, 1967).
[25] Pointed out in Hildebrand, *Das Dritte Reich*, pp. 137–8.
[26] Ralf Dahrendorf, *Society and Democracy in Germany* (London, 1968); David Schoenbaum, *Hitler's Social Revolution* (London, 1966; all subsequent references to Anchor Books edn., New York, 1967).

of fascism, but rather with the modernizing impact (if largely unintended) of Nazism itself. These interpretations are evaluated in chapter 7.

A third influential non-marxist approach to fascism has been Seymour Lipset's 'sociological' interpretation of fascism as lower-middle-class radicalism – the 'extremism of the centre', as he dubbed it.[27] According to this view, fascism arose when mounting economic distress and a perceived threat both from big capital and organized labour forced middle-class strata which had previously supported centrist liberal parties to turn to the extreme Right. Such an interpretation has in recent years come under fire from various directions. First, it has been shown that the lower-middle-class vote in Germany before the rise of Nazism – and Lipset's argument was heavily based upon the German case – went to parties which in no sense could be regarded as 'liberal' or moderate centrist parties, but were distinctly rightist (authoritarian, nationalist, and often racist) in complexion. A vote for a fascist party was in fact the end of a long process of gradual rightwards shift in voting patterns.[28] Secondly, the Nazi Party received its main voter support in large cities – as has recently been demonstrated – from well-to-do districts representing the established upper bourgeoisie not the precariously placed or declining lower-middle-class social groups of the classic Lipset theory, while at the other end of the social scale the Nazis gained a higher level of backing from the working class (if not making serious inroads into 'organized' labour) than had been presumed.[29] Finally, it has been objected, exclusive concentration on the political behaviour of the lower middle class ignores completely both the role of the élites in bringing fascism to power and also the obvious subordination of lower-middle-class interests to those of big capitalism during the regime phase of fascism.[30]

It has not been my intention to attempt a full critique of the widely differing interpretations of fascism, but rather to illustrate the fact that, despite con-

[27] Seymour Martin Lipset, *Political Man. The Social Bases of Politics* (New York, 1960), ch. 5.
[28] See Heinrich August Winkler, 'Extremismus der Mitte? Sozialgeschichtliche Aspekte der nationalsozialistischen Machtergreifung', *VfZ* 20 (1972), pp. 175–91; and Thomas Childers, *The Nazi Voter. The Social Foundations of Fascism in Germany, 1919–1933* (Chapel Hill/London, 1983). On voter support for Nazism see now the sophisticated study of Jürgen Falter, *Hitlers Wähler* (Munich, 1991).
[29] For the big city vote, see Richard F. Hamilton, *Who voted for Hitler?* (Princeton, 1982). The wide social spectrum of Nazi support is emphasized by Childers, Jürgen W. Falter, 'Wer verhalf der NSDAP zum Sieg?', *APZ* (14 July 1979), pp. 3–21, and Heinrich August Winkler, 'Mittelstandsbewegung oder Volkspartei? Zur sozialen Basis der NSDAP', in Schieder, ed., *Faschismus als soziale Bewegung*, pp. 97–118. For the social structure of the party membership, see Michael Kater, *The Nazi Party. A Social Profile of Members and Leaders, 1919–1945* (Oxford, 1983). A good survey of the literature on the social composition of Nazi support, in particular the vexed questions of the nature and extent of working-class backing for Nazism and whether the SA had a more 'middle-class' or 'proletarian' character, can be found in Mathilde Jamin, *Zwischen den Klassen. Zur Sozialstruktur der SA-Führerschaft* (Wuppertal, 1984), pp. 11–45. New evidence on the structure of Party membership is presented by Detlef Mühlberger, *Hitler's Followers. Studies in the Sociology of the Nazi Movement* (London, 1991).
[30] Bernt Hagtvet and Reinhard Kühnl, 'Contemporary Approaches to Fascism: A Survey of Paradigms', in Larsen *et al.*, pp. 26–51, here p. 31. This is a perceptive analysis of the problems of comparative fascism. From a different perspective, see also Juan J. Linz, 'Some Notes towards a Comparative Study of Fascism in Sociological Historical Perspective', in Laqueur, pp. 13–78.

siderable advances in developing sophisticated typologies of fascist movements, there is no prospect in view of any theory of fascism which might win universal approval. No single marxist theory can command general acceptance even among marxist scholars, while some of the weaknesses and criticisms of 'bourgeois' interpretations have been indicated. Finally, as mentioned earlier, some leading scholars – whether favouring a 'totalitarianism' approach or not – question the whole basis of studies of comparative fascism, arguing that profound differences between the 'fascist' movements render any concept of generic fascism meaningless.

Following this brief description of the stages of development of the concepts of totalitarianism and fascism, we can now turn to consider critically whether either model type satisfactorily embraces the phenomenon of Nazism.

General Reflections on the Concepts of 'Totalitarianism' and 'Fascism'

Neither 'totalitarianism' nor 'fascism' is a 'clean' scholarly concept. Both terms have, from the beginning of their usage, served a double function: as an ideological instrument of negative political categorization, often serving in common parlance as little more than 'boo-words'; and as a heuristic scholarly device used in an attempt to order and classify political systems. It is as good as impossible to treat them as 'neutral' scholarly analytical tools, detached from political connotations. Scholarly debate about the use of the terms illustrates above all the closeness of the mesh of history, politics, and language.[31] This is reflected, too, in the lack of agreement about precise definitions as well as usages of the terms.

Furthermore, there is often less than clarity about the link between concept and theory. If 'theory' is taken to be a system of interrelated statements, deriving from and based upon each other, with general explanatory power, and 'concept' as an abstract linguistic short-cut, without independent standing and offering no systematic explanation, then it could be argued that in the case of totalitarianism Friedrich produced a conceptual definition, but one which does not provide a genuine theory of totalitarianism. In the case of fascism, most non-marxist approaches, as mentioned earlier, are essentially descriptive and rest on no clearly-defined theoretical premises, while marxist approaches derive from theoretical positions but the applied theory is not always based upon a clear conceptual definition and sometimes even upon what comes close to a tautological one.[32]

Though both 'fascism' and 'totalitarianism' approaches seek to provide typologies of political systems, these are of quite a different kind. The emphasis in fascism 'theories' is upon fascist *movements* – upon the conditions of growth, aims, and function of these movements as distinct from all other forms of political organization. (Though this is also true of the Comintern theory and

[handwritten margin note: Fascism – Conditions of growth, aims + function]

[31] See Karl Dietrich Bracher, 'Betrachtung: Terrorismus und Totalitarismus', in his *Schlüssel-wörter*, pp. 103–23 (a lecture given in 1977 at a CDU conference on the causes of terrorism), and the comments of Bracher and Martin Broszat in *Total. und Fasch.*, pp. 10–11, 32–3.
[32] Uwe Dietrich Adam, 'Anmerkungen zu methodischen Fragen in den Sozialwissenschaften: Das Beispiel Faschismus und Totalitarismus', *PVS* 16 (1975), pp. 55–88, here esp. pp. 75–6.

its later application, much more emphasis has generally here been attached to the nature of fascist dictatorship rather than to the 'movement' phase.) Totalitarianism models, on the other hand, are practically by definition largely uninterested in the pre-power phase, except in so far as it betrays 'totalitarian' ambitions. The focus is rather on *systems* and *techniques of rule*. Many questions, therefore, of vital importance to the analyst of fascist movements – regarding, for instance, the socio-economic 'causes' of fascism, the social composition of fascist parties, and the relationship of fascist movements to the existing 'ruling class' – are of little importance to the totalitarianism theorist. Significant concerns in the totalitarianism approach, on the other hand, such as the existence of a single monopoly party, plebiscitary legitimation of rule, or the dominance of an official ideology, are usually regarded as secondary by analysts of fascism, who stress rather the major differences in the aims, social base, and economic structures of fascist and communist regimes.

Both 'fascism' and 'totalitarianism' are concepts extending beyond single systems of rule to 'generic types'. As such, they both demand rigorous comparative method. Yet in practice, thorough comparative analysis has often been lacking, particularly so in the totalitarianism model, and both approaches have traditionally been top-heavy in their reliance on the case of Nazi Germany.[33] Valuable systematic comparative research has been undertaken in recent years into the structure of fascist movements,[34] but much comparative work remains to be done on the character of fascist institutions in power. From the totalitaianism perspective, research into Stalinist government and society has reached nowhere near the level of penetration of that into the Nazi Regime, and comparisons are in practice often highly superficial.

Despite the fact that the concepts are politically irreconcilable – protagonists of a general fascism concept rest their position upon the view that right-wing dictatorships are *fundamentally different* from left-wing dictatorships, while protagonists of a totalitarian approach begin with the premise that fascist and communist dictatorships are *basically similar* – prominent German scholars have recently claimed the indispensability of both concepts in analysing modern political structures and have argued that it is possible to apply both approaches in different ways in examining Nazism.[35] This seems to attract the difficulty of applying *comparative* concepts to a single phenomenon while leaving unresolved the problem of whether the comparative concept itself is a valid one. Nevertheless, that each of the concepts undeniably contains political overtones does not in itself disqualify them from having scholarly value and intellectual validity. Hence, there remains the need to test the explanatory value of each of the terms as vehicles for assessing the essential character of Nazism.

[33] Examples for fascism are the books of Clemenz (note 13), Richard Saage, *Faschismustheorien* (Munich, 1976), and Niels Kadritzke, *Faschismus und Krise* (Frankfurt am Main/New York, 1976), and, for totalitarianism, Hans Buchheim, *Totalitäre Herrschaft. Wesen und Merkmale* (Munich, 1962).
[34] There is an excellent summary of up-to-date findings in Larsen *et al.* (note 12).
[35] Kocka, 'Ursachen', pp. 14–15, and the comments of Kocka, Broszat, Schieder, and Nolte, in *Total. und Fasch.*, pp. 32–53.

Nazism as Totalitarianism?

Critics of the totalitarianism concept fall into two main categories: *(a)* those who reject categorically any deployment of a concept or theory of totalitarianism; and *(b)* those who are prepared to concede it some theoretical validity, but who regard its practical deployment as a tool of analysis as limited in potential. The arguments in favour of the second position are, in my view, more convincing.

(a) Categorical rejection of totalitarianism as a wholly worthless concept is usually pressed on the following grounds:[36]

(i) Totalitarianism is no more than a Cold War ideology, devised and deployed by western capitalist states in the 1940s and 1950s as an anti-communist instrument of political integration, and continuing to be used as such to the present day. Apart from the fact that, as we have seen, the concept and its application existed long before the Cold War, the undoubted and usually crude political use to which it was put in the Cold War of itself no more deprives totalitarianism of potential value as a scholarly analytical tool than the often equally crude political exploitation of the term 'fascism' robs theories of fascism of any validity.

(ii) The totalitarianism concept treats the form – the outward shape of the systems of rule – as content, as their essence. As a result, it fully ignores the completely different aims and intentions of Nazism and Bolshevism – aims which were wholly inhumane and negative in the former case and ultimately humane and positive in the latter case. The objection is not altogether convincing. As Adam has pointed out,[37] the argument is based upon a deduction from the future (neither verifiable nor falsifiable) to the present, a procedure which in strict logic is not permissible. There is also a presumption that form and content can be so dissociated from each other that a comment on the form says nothing about the content – a point rejected even by materialist dialectics. Furthermore, the emphasis upon the ultimate humanity of Bolshevism contrasted with the inhumanity of Nazism correlates a presumed idealistic intention of the one system with the known reality of the other, and shirks the question of possible actual similarities in techniques of domination between the Stalinist and Hitler regimes. The purely functional point that communist terror was 'positive' because it was 'directed towards a complete and radical change in society', whereas 'fascist (i.e. Nazi) terror reached its highest point with the destruction of the Jews' and 'made no attempt to alter human behaviour or build a genuinely new society'[38] is, apart from the debatable assertion in the last phrase, a cynical value judgement on the horrors of the Stalinist terror.

(b) Four substantial criticisms are raised by those who do not reject the totalitarianism model out of hand, but see its application as very limited:

(i) The concept of totalitarianism, however defined, can only unsatisfactorily grasp the peculiarities of the systems it attempts to classify. Broszat pointed out, for instance, in the introductory remarks to his masterly analysis of the

---

[36] Kitchen, ch. 2, comes close to this position.
[37] Adam, 'Anmerkungen', pp. 64–7.
[38] Kitchen, p. 31.

'Hitler State', the difficulty of locating the amorphous structurelessness of the Nazi system in any typology of rule.[39] The totalitarianism concept can, in fact, only speak in a generalized and limited fashion about the similarities of systems, which on closer inspection are so differently structured that comparisons are forced to remain highly superficial. Hans Mommsen has indicated, for example, how different the Nazi Party and the Soviet Communist Party were from each other in structure and function, and how little it says, therefore, simply to refer to both Nazi Germany and Soviet Russia (even confining the treatment to the Stalinist period) as 'one-party states'.[40] Equally significant were the major differences in the essential character of leadership in the two states, so that the roles of Hitler and Stalin can only with difficulty be typified as those of 'totalitarian dictators'. And the fundamental contrasts in the control of the Nazi and Soviet economies are an even more striking example of highly misleading generalizations emanating from the totalitarianism approach – in this instance about centralized 'totalitarian' economies.

(ii) The totalitarianism concept cannot cope adequately with change within the communist system. The extension of the concept to post-Stalinist USSR and other eastern-bloc states is forced to see the essence of totalitarianism as lying elsewhere than in the specific features of Stalinism usually taken to be comparable with Nazism (e.g. terror, leadership cult etc.). Still retaining the implicit (if not explicit) linkage with Nazism and other 'right-wing dictatorships', such attempts often rapidly widen into outright absurdity.

(iii) The decisive disadvantage of totalitarianism as a concept is that it says nothing about socio-economic conditions, functions, and political aims of a system, but is content to rely solely upon emphasis of techniques and overt forms of rule (exclusivity of ideology, tendency to comprehensive mobilization etc.).[41] Since one of the most obvious and striking differences between the Nazi and the Soviet systems lies in the socio-economic sphere, it has been pointed out that 'the value of an analysis which ignores the relations of production and the resulting social structure of the two systems is strictly limited'.[42]

(iv) The legitimacy of the totalitarianism concept rests upon the upholding of the values of western 'liberal democracy' and the distinction between 'open' and 'closed' government, between 'shared' and 'unified' power. There is, however, built into the totalitarianism concept an ambivalence between describing historically real systems of rule (Nazism, 'Stalinism') and being widened out into a 'tendency' which extends to so many modern dictatorships and even to sections of society within western democracies that the concept loses much of its analytical value.[43]

These criticisms are generally advanced by those who nevertheless would not wish altogether to discard the concept of totalitarianism. They claim – and I would agree with their argument – that it is in itself a wholly legitimate exercise, whatever essential differences existed in ideology and socio-economic struc-

[39] Martin Broszat, *Der Staat Hitlers* (Munich, 1969), p. 9, not included in the Engl. trans., *The Hitler State* (London, 1981).
[40] Hans Mommsen, in *Total. und Fasch.*, pp. 18–27.
[41] Jürgen Kocka, in *Total. und Fasch.*, pp. 39–44.
[42] Kitchen, p. 31.
[43] Martin Broszat, in *Total. und Fasch.*, pp. 32–8.

tures, to compare the forms and techniques of rule in Germany under Hitler and the Soviet Union under Stalin; and that a new scale and concept of the development of force in governmental systems, in attempted comprehensiveness of control and manipulation, in methods (based on modern technology) of dynamic plebiscitary mobilization of the population behind its rulers, and a radical intolerance of any focus of co-existing alternative loyalties or any form of institutional 'living space' except under the regime's own terms, corresponding therefore to the *attempted* politicization of all facets of social experience, can justifiably be seen in both systems. The spectrum of dissent ranging to 'resistance' in Nazi Germany (and *pari passu*, though so far little analysed, in Stalin's Russia) can in fact only be understood in the light of the relationship to the demands of a regime which made a 'total claim' on behaviour and manifestations of outward conformity, hence creating nonconformist and oppositional behaviour which even in other authoritarian systems would not have been politicized and turned, thereby, into political dissent.[44] If the redundant echoes of 'atomized mass society' theories can be dispensed with, then it may indeed be at the social rather than the institutional level that, if not the full-blown, politically loaded, concept of totalitarianism, then the more modest notion of the 'total claim' of a regime on its subjects could prove heuristically useful in a comparative analysis of behavioural patterns – acclamatory and oppositional – in quite differently structured societies and political systems.[45] Even the posing of an extreme 'total claim' might then be seen as symptomatic of the 'crisis management' of regimes in transitory, unstable periods rather than as lasting characteristics of rule.

Beyond this, it seems to me that depictions of Nazism as a 'totalitarian system' are best avoided, not simply because of the inescapable political colouring attached to the label 'totalitarianism', but because of the weighty conceptual problems which the term poses and which have been outlined above. There remains a final possibility of deploying the concept in a non-comparative sense, restricting its usage to Nazi/fascist systems alone and reverting to something like its earlier usage by Franz Neumann and others in distinguishing phases of development in the impact of a dynamic mass movement with 'total' claims upon the legislative and executive structures of the state. Broszat's analysis of the Nazi state, for instance, uses the adjective 'totalitarian' divorced from comparison with the USSR to distinguish the more radical phase of Nazi government after 1937–8 from the earlier merely 'authoritarian' phase.[46] Quite apart from the question of attaching distinctive labels to the periods of the Third Reich before and after 1937–8, and of ridding 'totalitarianism' of its usual comparative connotations with the USSR, it might be seriously doubted whether, in dealing with the Nazi state alone, the adjective 'totalitarian' is needed at all simply as a synonym for progressively radicalizing dynamism.

[44] See Ian Kershaw, *Popular Opinion and Political Dissent in the Third Reich. Bavaria 1933–1945* (Oxford, 1983), esp. pp. 374 ff.
[45] For a perceptive assessment of the impact of Nazism on German society, see Detlev Peukert, *Volksgenossen und Gemeinschaftsfremde* (Cologne, 1982). Engl. trans., *Inside Nazi Germany. Conformity and Opposition in Everday Life* (London, 1987).
[46] Broszat, *The Hitler State*, pp. ix–xiv, 346 ff. In his later work, Franz Neumann came to deploy the 'totalitarianism' concept in its conventional 'Cold War' usage. See his *The Democratic and the Authoritarian State* (New York, 1957).

Others, developing the same line of interpretation, find the term wholly redundant.[47]

All in all, the value of the totalitarianism concept seems extremely limited, and the disadvantages of its deployment greatly outweigh its possible advantages in attempting to characterize the essential nature of the Nazi Regime.

Nazism as Fascism or Unique Phenomenon?

Opponents of the use of a generic concept of fascism advance two principal and serious objections to the ranking of Nazism as fascism: firstly – an objection I find justified – that the concept is often extended in inflationary fashion to a wide variety of movements and regimes of wholly disparate character and significance; and secondly, but in my view less persuasive, that the concept is unable satisfactorily to embrace the singular characteristics of Nazism, and that the differences between Italian fascism and German National Socialism significantly outweigh whatever superficial similarities they might appear to possess.

(a) The first criticism pertains particularly, though not solely, to marxist interpretations of fascism. The intrinsic relationship between fascism and capitalism in the marxist–leninist version of fascism theory, for instance, extends the notion of 'fascist dictatorship' to cover numerous kinds of repressive regime, and no fundamental distinction is drawn between military dictatorships and mass-party dictatorships in terms of the essence of rule. Since, according to this view, the mass base of a fascist party is a manipulated product of the ruling capitalist class without any autonomous force, the importance of the mass movement (which most non-marxist analysts would regard as a significant difference between military authoritarian regimes and fascist regimes) recedes. Hence, GDR scholars classed such disparate regimes as existed in Poland, Bulgaria, and Hungary in the inter-war period, in Portugal under Salazar and Caetano and Spain under Franco, in Greece under the Colonels, Argentina under the Generals, Chile under Pinochet, and other South American dictatorships, as 'fascist' alongside 'Hitler Fascism'.[48] Decisive for GDR historians was not the outward form of the dictatorship, but its essence as the weapon of the most aggressive elements of finance capital. Nevertheless, GDR scholarship did come to distinguish very clearly between two basic types of fascist dictatorship; the *normal* form – usually a military dictatorship – in countries with relatively unadvanced capitalist economies; and the *exceptional* form – mass-party fascism – of which only the two examples of Italy and Germany have so far been experienced, both arising in highly unusual conditions within the framework of a complete national crisis.[49] Consideration of the

---

[47] E.g. Hans Mommsen, in *Total. und Fasch.*, p. 65, where he states: 'The totalitarianism theory is the myth which stands in the way of any *real* social historical explanation [of Nazism]', particularly because of its teleological tendency to take the end product for granted before examining the conditions of its growth.

[48] E.g. Manfred Weißbecker, 'Der Faschismus in der Gegenwart', in Eichholtz and Gossweiler, pp. 217 ff.; Kurt Gossweiler, *Faschismus und antifaschistischer Kampf* (Antifaschistische Arbeitshefte, Röderberg Verlag, Frankfurt am Main, 1978), pp. 18–23.

[49] Kurt Gossweiler, *Kapital, Reichswehr und NSDAP, 1919–1924* (East Berlin, 1982), ch. 1 provides a thoughtful discussion.

relationship between capitalism and Nazism, on which this theory rests, will have to wait until the following chapter. It suffices here to say that, however unconvincing the underlying principles are, GDR interpretations compared very favourably with the writings of parts of the 'New Left' in the Federal Republic, where the concept of fascism was extended to any form of 'repressive' government which serves to uphold the domination of economic power-groups, thus allowing western capitalist systems – and the Federal Republic in particular – to be dubbed 'fascist' or at least 'fascistoid' or 'proto-fascist'.[50] In such cases, where the fascism concept is widened in hopelessly nebulous fashion, it seems perfectly correct to speak of a trivialization of the horror of Nazism.

(b) The second, related, criticism claims that no theory or concept of generic fascism can possibly do justice to the peculiarities and unique characteristics of Nazism. While movements calling themselves fascist or national socialist existed in most European countries outside the Soviet Union in the inter-war period, it is widely accepted that fully-fledged, self-sustaining fascist dictatorships deriving their impetus from mass parties consolidated power only in Italy and Germany (leaving aside puppet or quisling governments of the war years). A comparison of fascism in all its stages can accordingly be made only for the systems in these two countries.[51] Yet in the eyes of some leading authorities, the differences between the two regimes were so profound that the term 'fascism' should be reserved for the Italian system under Mussolini, while Nazism should be called 'National Socialism' and regarded as a unique phenomenon (though, interestingly enough, falling in terms of techniques of rule within the category of 'totalitarian systems'). Since, in this view, the generic concept of fascism does not even apply to the two leading species within the genus, it had better be discarded altogether. The central differences emphasized in this argument focus on the dynamic nature of the Nazi race ideology, which had no exact parallel in Italian fascism; on the discrepancy between the Nazi elevation of the *Volk* over the state, contrasted with Italian fascist étatism; on the anti-modern, archaic aims and ideology of Nazism compared with the modernizing tendencies of Italian fascism; on the totality of the Nazi conquest of state and society as against the far more limited penetration of the established order by the Italian fascists; and, not least, on the contrast between a relatively 'traditional' imperialistic policy on the part of Italy, and a qualitatively different drive for racial domination, eventually of the whole world, by the Nazi regime. And since this last and most crucial distinction is, according to such interpretations, attributable directly to Hitler himself, it is claimed that 'the case of Hitler' was unique, and cannot be subjected to the generalizations of comparative fascism, not even to a comparison limited to Italy and Germany.[52]

[50] See the theoretical remarks of Adam, 'Anmerkungen', pp. 70–6; and Winkler, *Revolution*, pp. 108 ff.
[51] See the comments of Schieder, in *Total. und Fasch.*, pp. 45–9. MacGregor Knox, 'Conquest, Foreign and Domestic, in Fascist Italy and Nazi Germany', *JMH* 56 (1984), pp. 1–57 provides an interesting comparative essay on the Mussolini and Hitler regimes.
[52] Hildebrand, *Das Dritte Reich*, pp. 139–42; Hillgruber, *Endlich genug?*, pp. 17, 38, 42; Bracher, *Zeitg. Kontrov.*, chs. 1–4, and in *Total. und Fasch.*, pp. 14–17; Henry A. Turner, 'Fascism and Modernization', in Turner, *Reappraisals*, pp. 132–3; see also De Felice, p. ix (introductory comments of Charles F. Delzell) and pp. 10–12, 180.

These criticisms cannot be lightly passed over. Indeed, examination of two central issues – the relationship between capitalism and Nazism, and the personal role of Hitler in the Nazi system – form the direct subject of later chapters. There is space here only for a number of general observations about the criticisms of the generic fascism approach, related to the alternative possibility of emphasizing the uniqueness of Nazism.

A number of the supposed major differences between Nazism and Italian fascism are open to debate. This would apply, for instance, to the stress on the 'backward-looking' nature of Nazism in distinction to the 'modernizing' pressures of fascism in Italy. Research has called such a distinction increasingly into question, as chapter 7 indicates.[53] Quite apart from such qualification, the uniqueness of specific features of Nazism would not of itself prevent the location of Nazism in a wider genus of political systems. It might well be claimed that Nazism and Italian fascism were separate species within the same genus, without any implicit assumption that the two species ought to be well-nigh identical. Ernst Nolte has stated that the differences could easily be reconciled by employing a term such as 'radical fascism' for Nazism.[54] Winkler has indicated that for him Nazism was 'also but not only "German fascism" ',[55] while Linz regarded it as a 'distinctive branch grafted on the fascist tree'.[56] Jürgen Kocka, in a recent subtle essay on the causes of Nazism, again sees no incompatibility between the unique features of National Socialism in Germany and its attribution to a broader class of generic fascism, indispensable for putting the Nazi phenomenon in a wider than purely national perspective and understanding the social and political contexts in which such a movement could arise and take power.[57] Such approaches rightly stress the significant similarities between Nazism and the many movements (above all the Italian one) which called themselves fascist. Such similarities included: extreme chauvinistic nationalism with pronounced imperialistic, expansionist tendencies; an anti-socialist, anti-marxist thrust aimed at the destruction of working-class organizations and their marxist political philosophy; the basis in a mass party drawing from all sectors of society, though with pronounced support in the middle class and proving attractive to the peasantry and to various uprooted or highly unstable sectors of the population; fixation on a charismatic, plebiscitarily legitimized leader; extreme intolerance towards all oppositional and presumed oppositional groups, expressed through vicious terror, open violence, and ruthless repression; glorification of militarism and war, heightened by the backlash to the comprehensive socio-political crisis in Europe arising from the First World War; dependence upon an 'alliance' with existing élites – industrial,

---

[53] For modern traits in Nazism, see e.g. Peukert, pp. 42–7; Tim W. Mason, 'Zur Entstehung des Gesetzes zur Ordnung der nationalen Arbeit, vom 20. Januar 1934: Ein Versuch über das Verhältnis "archaischer" und "moderner" Momente in der neuesten deutschen Geschichte', in Hans Mommsen *et al.*, eds., *Industrielles System und politische Entwicklung in der Weimarer Republik* (Düsseldorf, 1974), pp. 322–51; Horst Matzerath and Heinrich Volkmann, 'Modernisierungstheorie und Nationalsozialismus', in Jürgen Kocka, ed., *Theorien in der Praxis des Historikers* (Göttingen, 1977), pp. 95–7; Hans-Dieter Schäfer, *Das gespaltene Bewußtsein. Deutsche Kultur und Lebenswirklichkeit 1933–1945* (Munich/Vienna, 1981), pp. 114–62; Martin Broszat, 'Zur Struktur der NS-Massenbewegung', *VfZ* 31 (1983), pp. 52–76.
[54] Nolte, in *Total. und Fasch.*, pp. 77–8, and in *Three Faces*, pp. 529, 569–77.
[55] Winkler, *Revolution*, p. 66.
[56] Linz (note 30), p. 24.
[57] Kocka, 'Ursachen', esp. p. 15.

agrarian, military, and bureaucratic – for their political breakthrough; and at least an initial function – despite a populist-revolutionary, anti-establishment rhetoric – in the stabilization or restoration of social order and capitalist structures.[58]

The establishment of fundamental generic characteristics linking Nazism to movements in other parts of Europe allows further consideration on a comparative basis of the reasons why such movements were able to become a real political danger and gain power in Italy and Germany, whereas in other European countries they mainly remained an unpleasant, but transitory irritant. Among other things, one would undoubtedly have to lay stress on features prominent, though in different strengths, in both Italy and Germany before the First World War and massively accentuated through the traumatic consequences of the War itself. Common to both countries were the powerful imperialist–expansionist strains pronounced among the ruling élites and bolstered by the widespread extreme chauvinism in the bourgeois classes of these new states – self-perceived 'have-not nations'; the co-existence and conflict of highly modern strands of development and powerful remnants of archaic social structures and value-systems in societies simultaneously undergoing the processes of national integration, transition to a bourgeois constitutional state, and rapid industrialization;[59] and finally, but not least, deeply fractured political systems, whose splintered parliamentary structures reflected deep social and political cleavages, fostering the feeling that a strong, but 'populist', leadership was necessary to impose unity 'from above' – in the first instance by crushing those standing in the way of unity, primarily 'the marxist Left'. The different scale of the social and political conflict spheres in Italy and Germany helps explain the different level of radicalization in the two countries when beset by different, though related, comprehensive crises of the political system – directly unleashed by the War in the Italian case, unfolding, after a long period of political instability, during the world economic crisis in Germany.

It is within this perspective, rather than divorced from it in an emphasis upon Nazism as an altogether unique phenomenon, that the peculiarities of the German radical variant of fascism can be brought out by analysis of the specific features of the German political culture and its relationship to socio-economic structures. There need be no contradiction, therefore, between acceptance of Nazism as (the most extreme manifestation of) fascism and recognition of its own unique characteristics within this category, which can only properly be comprehended within the framework of German national development.

Such an argument would not, however, satisfy Bracher, Hildebrand, Hillgruber, and others, who would argue that Nazism was not only in form but

---

[58] See Kocka, 'Ursachen', p. 15, and in *Total. und Fasch.*, pp. 39, 44. See also Winkler, *Revolution*, p. 66. Recent analyses by the British scholars Roger Griffin and Roger Eatwell, though applying different definitions, have no difficulty in including Nazism as an integral part of their comparative studies of fascism. See Roger Griffin, *The Nature of Fascism* (London, 1991) and Roger Eatwell, 'Towards a New Model of Generic Fascism', *Journal of Theoretical Politics* 4 (1992), pp. 161–94. In particular, Griffin's emphasis on 'palingenetic ultranationalism' – extreme populist nationalism focused upon national 'rebirth' and the eradication of presumed national decadence – as the core of fascist ideology, self-evidently embraces Nazism.
[59] The importance of this simultaneous three-fold transition is stressed by Schieder, in *Total. und Fasch.*, pp. 45–9.

in essence a uniquely German phenomenon, and that this essence or uniqueness was located in the person and ideology of Adolf Hitler. This personalization of the essence of Nazism is, in fact, at the crux of the debate over the historical place and characterization of Nazism. The major differences do not lie in explaining Nazism's origins and the circumstances of its rise to power. Bracher has tended to emphasize the specific features of German–Austrian ideological development in order to lay full weight on the racial–*völkisch* dimension of Nazi ideology; Hillgruber and Hildebrand have stressed the particular constellation of German power-politics and the overwhelming continuities between 1871 and 1933 (only to be broken thereafter) intrinsic to the Prussian–German State.[60] These are important strands of an overall explanation of Nazism and, despite differences of emphasis, are generally compatible with those works – for example, by Wehler, Kocka, Puhle, and Winkler[61] – which look rather to Germany's specific socio-economic structures as the focal point of their explanations. Yet this latter group have no hesitation in accepting Nazism, for all its singularities, as a form of fascism; while the former group deny this categorization and insist that it was *sui generis*. The breaking-point is clearly 'the case of Hitler': whether Nazism can be set aside from fascism in Italy and elsewhere because it was *in its essence* 'Hitlerism'. According to the latter approach, not the causes of Nazism's rise but the character of the dictatorship itself is decisive. And here, the differences between Italian fascism and Nazism, whose rule rested on the implementation of the ideas and policies of the monocratic dictator, Hitler, were fundamental.[62]

This 'Hitler-centrism' is itself an understandable over-reaction against some crude left-wing interpretations which reduced Hitler to a mere cipher. However, irreplaceable though Hitler undoubtedly was in the Nazi movement, the equation Nazism = Hitlerism unnecessarily restricts the vision and distorts the focus in explaining the origins of Nazism; deflects away from rather than orientates towards consideration of the political manifestations in other European countries which shared (and continue to share today) important affinities and common characteristics with Nazism; and finally – as I hope to argue in later chapters – provides in itself a quite unsatisfactory explanation of the dynamic radicalization of politics within the Third Reich itself.

This evaluation of the concepts of totalitarianism and fascism in relation to Nazism's alleged uniqueness as a phenomenon has suggested the following conclusions:

[60] See Bracher, 'The Role of Hitler', in Laqueur, pp. 209-10, fully developed in Karl Dietrich Bracher, *The German Dictatorship* (Harmondsworth, 1973), esp. ch. 1; Andreas Hillgruber, 'Kontinuität und Diskontinuität in der deutschen Außenpolitik von Bismarck bis Hitler', in his *Großmachtpolitik und Militarismus im 20. Jahrhundert* (Düsseldorf, 1974), pp. 11-36, and *Endlich genug?*, pp. 48 ff.; Klaus Hildebrand, 'Hitlers Ort in der Geschichte der preußisch-deutschen Nationalstaates', *HZ* 217 (1973), pp. 584-632, and his *Foreign Policy*, (see ch. 1 note 17), esp. Introdn. and Concl.
[61] E.g., Wehler, *Kaiserreich* (see ch. 1 note 16); Jürgen Kocka, *Angestellte zwischen Faschismus und Demokratie* (Göttingen, 1977); Hans-Jürgen Puhle, *Von der Agrarkrise zum Präfaschismus* (Wiesbaden, 1972); Heinrich August Winkler, *Mittelstand, Demokratie und Nationalsozialismus* (Cologne, 1972).
[62] See Bracher, *Zeitg. Kontrov.*, pp. 30, 88-9, 99; Hillgruber, *Endlich genug?*, pp. 40-2; and Klaus Hildebrand, 'Nationalsozialismus oder Hitlerismus?', in Michael Bosch, ed., *Persönlichkeit und Struktur in der Geschichte* (Düsseldorf, 1977), pp. 55-61, here esp. pp. 56-7.

(1) The concept of fascism is more satisfactory and applicable than that of totalitarianism in explaining the character of Nazism, the circumstances of its growth, the nature of its rule, and its place in a European context in the inter-war period. The similarities with other brands of fascism are profound, not peripheral. Nazism's features place the phenomenon squarely within the European-wide context of radical anti-socialist national–integrationist movements, which also rejected the forms though not the economic substance of bourgeois society, derived from the era of open imperialist conflict and emerged to prominence in the upheavals following the First World War.

(2) This is not incompatible with the retention of the concept of totalitarianism, though this latter concept is much less usable and its value is strictly limited. Nazism undoubtedly did have a 'total' (or 'totalitarian') claim, which had consequences both for its mechanics of rule and for the behaviour – acclamatory and oppositional – of its subjects. Consequences for the mechanics of rule were reflected especially in new forms of plebiscitary mass mobilization through new technologies of rule combined with an exclusive dynamic ideology and monopolistic demands on society. On the basis of these features, it is legitimate to compare the forms of rule in Germany under Hitler and the Soviet Union under Stalin, even if, for the reasons adduced earlier, this comparison is doomed from the outset to be superficial and unsatisfactory. Moreover, 'totalitarianism' according to our analysis, if to be used at all, would have to be restricted to passing phases of extreme instability reflected in the paranoid sense of insecurity of the regimes, rather than being seen as a lasting structure of rule. From a long-range perspective, the entire period of the Third Reich and the bulk of Stalin's rule could be said to fall within such a categorization. This would be a reason additional to those mentioned earlier to exclude the application of the comparative totalitarianism concept to post-Stalinist communist systems, where it rapidly approaches futility if not outright absurdity.

(3) The peculiar features which distinguish Nazism from other leading manifestations of fascism are only to be fully comprehended within the structures and conditions of German socio-economic and ideological–political developments in the industrial–bourgeois era. The person, ideology, and function of Hitler have to be located in and related to these structures. Without question, Hitler played personally a vital part both in the rise of Nazism and in the character of Nazi rule. But the significance of his role can only be assessed by relating his input to the conditions which produced and shaped him, and which he could not autonomously control even at the height of his power. Nazism was in many respects indeed a unique phenomenon.[63] But its uniqueness cannot – except in a superficial sense – be solely attributed to the uniqueness of its leader.

[63] In their *The Racial State, Germany 1933–1945* (Cambridge, 1991) Michael Burleigh and Wolfgang Wippermann emphasise the 'specific and singular character' of Nazism (p. 306). I agree with their interpretation of the Third Reich 'as a singular regime without precedent or parallel'. In order to uphold this claim, nevertheless, systematic comparison of the regime with other modern state systems and not simply a description – however compelling – of Nazi race policies is necessary. To contend that theories based upon totalitarianism or global theories of fascism are 'poor heuristic devices' (p. 307) for understanding Nazism is, in my view, therefore, going too far. The extent to which Burleigh and Wippermann stress the quest for racial purification as the core of Nazism is, in my opinion, valid. But this does not rule out as conclusively as they appear to presume other valid perspectives and questions about Nazism's comparability with other forms of fascism and/or totalitarianism. (For a comment on their criticism of the application of modernization theory to Nazism, see ch. 7 note 61.)

3

Politics and Economics in the Nazi State

The question of the relationship between Nazism and the dominant economic forces in Germany has remained one of the most contentious issues of debate among scholars since the theoretical deliberations of the Comintern in the 1920s and 1930s. It is a debate in which preconceived theoretical (and ideological) positions are often at their most apparent. With the double development of the opening up of major archival sources and the revival of marxist scholarship in the West during the 1960s, the debates began for the first time seriously to preoccupy non-marxist historians. The enormous improvement since then in the level of empirical knowledge on the Nazi economy has been accompanied by new levels of sophistication in interpretation, though the central areas of concern and the focal points of conflicting interpretation have changed relatively little in the meantime.

One major issue revolves around the extent to which the Nazi rise to power was a product of the character of German capitalism and of the machinations and political aims of the leaders of German industry. This issue, relating to the pre-dictatorship phase, will not concern us here. It must suffice to point out that, for all the remaining scholarly divisions, a growing body of recent scholarship rejects both the crude instrumentalism of a view which sees Nazism as a movement 'reared' and controlled from the outset by capitalist interests, and the equally crass counter-argument denying any structural links between capitalism and the rise of Nazism. Such scholarship – both marxist and non-marxist – broadly accepts two structural connections between capitalism and the rise of Nazism. First, it is clear that there was an increasing readiness among powerful sectors of the industrial élite long before the Nazi political break-through to discard the Weimar Republic in favour of a more palatable authoritarian solution which would restore profitability in the first instance through repression of labour. Secondly, among an industrial sector in many ways split and disorientated by the economic crisis of the early 1930s, there was an increased willingness in the deepening recession even among sections of industry not especially well disposed towards the Nazis to tolerate at least a Nazi share in government in order to provide the political framework within which the capitalist system could reproduce itself.[1] Important for our concern in this

[1] For an excellent survey and evaluation of the literature on the relationship of capitalism and Nazism before 1933, see Dick Geary, 'The Industrial Élite and the Nazis in the Weimar Republic', in Peter D. Stachura, ed., *The Nazi Machtergreifung* (London, 1983), pp. 85-100. David Abraham's *The Collapse of the Weimar Republic. Political Economy and Crisis* (Princeton, 1981), provoked a storm about the author's use of evidence (see *CEH* 17 1984, pp. 159-293). Unfortunately, the 'cleaned-up' second edition (New York, 1986) has still given rise to major objection and criticism. See Peter Hayes, 'History in an Off Key: David Abraham's Second *Collapse*',

chapter is the very fact that the Nazis presented, as it were, the last hope rather than the first choice of much of industry in offering a form of State which would uphold capitalist interests. Together with the pervading and continuing divisions within the economic élites about strategies for recovery, this ruled out obvious alternatives, bound the industrial leadership even if initially only in a negative fashion to the Nazi State, and offered the new Nazi leaders some scope and potential for political initiatives.

This relates closely to the second major issue which has preoccupied scholars exploring the connections between capitalism and Nazism: the extent to which the politics of the Nazi regime between 1933 and 1945 were shaped and determined by economic considerations, notably the interests of German industry. Put in slightly different and more pointed fashion, this amounts to the question of how far the regime was able to acquire a degree of political autonomy amounting in practice to a primacy of ideological and political objectives over economic aims and interests. This question is our concern in this chapter.

Interpretations

Even in the GDR, where economic relations had of course from the beginning been central to analyses of 'Hitler Fascism', it was only from the 1960s that more detailed archival research provided the base for a more subtle and differentiated scholarship, the prime example of which was Dietrich Eichholtz's study of the German war economy, published in 1969.[2] This brought out far more strongly than had previously been the case the contradictions and conflicts within various monopoly capitalist 'groupings' and corresponded in some of its findings with new work on the Nazi economy by western scholars. The general tenor of the gradually emerging research in the West, mainly carried out by non-marxists, was to see a far closer structural relationship between German industry and the policies of the Nazi leadership than had formerly been accepted, and to reject rather primitive notions of a highly centralized State 'command economy' which had formed part and parcel of the 'totalitarianism' model. The American scholar Arthur Schweitzer, for instance, emphasized what he regarded as a 'coalition' between the Nazi leadership and business élites in a period of 'partial fascism' down to 1936, although – anticipating, if from a different theoretical position, the debate about the 'primacy of politics' which was to take place shortly afterwards – he saw the period of 'full fascism' after 1936 as one in which business became increasingly dependent on

---

Business History Review 61 (1987), pp. 472–92. Henry A. Turner, *German Big Business and the Rise of Hitler* (Oxford, 1985), provides by contrast a meticulously researched empirical study of relations between business leaders and Nazis. Especially useful for the role of big business in the immediate prelude to Hitler's takeover of power is Reinhard Neebe, *Großindustrie, Staat und NSDAP* (Göttingen, 1981). And for a masterly analysis of the entire economic crisis and its significance for Nazi economic policy after 1933, see Harold James, *The German Slump. Politics and Economics 1924–1936* (Oxford, 1986).

[2] Dietrich Eichholtz, *Geschichte der deutschen Kriegswirtschaft, 1933–1945* (East Berlin, 1969). The second volume appeared in 1984. For a survey of GDR historical writing, see Andreas Dorpalen, *German History in Marxist Perspective. The East German Approach* (Detroit, 1985). Chapter 8 examines the Nazi era.

the political and ideological goals of the Nazi leadership.[3] Dieter Petzina's analysis of the Four Year Plan demonstrated how far removed it was from a genuine 'planned economy' and how closely the political–ideological interests of the Nazi leadership coincided with the economic interests of what was emerging as the strongest sector of German big business, the great chemical combine of IG Farben.[4] And Alan Milward uncovered the underlying weaknesses of a war economy which had necessitated Blitzkrieg as the only feasible strategy and had been centralized and rationally administered only after coming under Speer's control from 1942.[5]

Scholarly debate on the character of the Nazi economy was given a sharp stimulus by the appearance in 1966 of the essay by the British marxist historian Tim Mason on the 'primacy of politics' in the Third Reich.[6] Mason's article was framed in terms of a challenge both to existing marxist–leninist orthodoxy and to the main thrust of 'liberal–bourgeois' approaches to Nazism. While the former denied the existence of an autonomous political realm in representing the political–ideological sphere as part of the superstructure of the socio-economic system, the latter had tended to treat the economy as more or less subjected along with everything else to the unquestioned political priorities and autonomy of a ruthless, ideologically-motivated dictatorship. Mason's conclusion, based on an analysis of economic relations in the Third Reich, was 'that both the domestic and foreign policy of the National Socialist government became, from 1936 onward, increasingly independent of the influence of the economic ruling classes, and even in some essential aspects ran contrary to their collective interests'. He went in fact so far as to accept that 'it became possible for the National Socialist state to assume a fully independent role, for the "primacy of politics" to assert itself'. This – from a marxist viewpoint – startling conclusion was qualified only to the extent that, in Mason's terms, this relationship in the Third Reich upturned the norm in capitalist states, and was 'unique in the history of modern bourgeois society and its governments'.[7]

Mason pointed to a number of different aspects of the economic development of Nazi Germany to support his thesis: the far-reaching exclusion of the representatives of industry from the direct decision-making processes; the extraordinarily rapid growth of the economic role of the State itself in providing orders for industry and thereby creating markets and acting as a determining factor in production; the transfer of capitalist competition from a struggle for markets to a struggle within an armaments-dominated economy for raw materials and labour – leading to the endangering of entire sectors of industry and extensive state intervention and regulation; the decline of economic interest groups in shaping state policy; and the inability of the leaders of the armaments economy to force through before 1942 the redistribution of the social product

---

[3] Arthur Schweitzer, *Big Business in the Third Reich* (Bloomington, Indiana, 1964).
[4] Dieter Petzina, *Autarkiepolitik im Dritten Reich. Der nationalsozialistische Vierjahresplan* (Stuttgart, 1968).
[5] Alan S. Milward, *The German Economy at War* (London, 1965).
[6] Tim Mason, 'Der Primat der Politik – Politik und Wirtschaft im Nationalsozialismus', *Das Argument* 8 (1966), pp. 473–94. All references below are to the English version, 'The Primacy of Politics – Politics and Economics in National Socialist Germany', in Henry A. Turner, ed., *Nazism and the Third Reich* (New York, 1972), pp. 175–200.
[7] Mason, 'Primacy', pp. 175–7.

in terms of sharp inroads into the standard of living, which they had been demanding since the beginning of the Third Reich. In Mason's view, these features of the Nazi political economy either came into being or were massively accelerated from 1936–7, so that it is permissible to speak of 'weighty structural changes in economy and society', and consequently of a significant increase in the autonomy of the State from that date.[8]

The classical marxist–leninist counter-thrust was not long in coming. It was provided by two leading GDR scholars, Dietrich Eichholtz and Kurt Gossweiler, after Mason had parried without too much difficulty an attack by another GDR historian, Eberhard Czichon, which contained empirical weaknesses and theoretical crudity and was premised on a number of basic misunderstandings of Mason's argument.[9] Eichholtz and Gossweiler argued that Mason's interpretation removed fascism from the realms of historical explicability, reducing it to the level of an historical accident, adding that if Mason were right it would amount to 'a complete refutation of marxist social analysis' – an over-dramatized allegation which would seem to rest on a misreading of Marx and Engels. Their own approach began by attempting a justification of the Comintern definition of fascism (despite the admitted need for greater precision and refinement), followed this with a summary tribute to Lenin's theory of imperialism and its relation to fascism, and repeated the marxist–leninist theory of state monopoly capitalism. This lengthy theoretical exposition was then followed by a relatively short 'empirical' section, centring upon the changes in 1936 and aiming to show that alterations in the political course of the Third Reich were intrinsically related to developments in the dominant factions of state monopoly capitalism. It was not enough, they argued, to see finance capital simply as the beneficiary rather than the 'inspiration and initiator' of fascist policy; rather, analysis of the changing structure of state monopoly capitalism disproved Mason's thesis and demonstrated that capital was far from relinquishing its power to the State after 1936. Instead, the Nazi State provided the ground for an intensified struggle within monopoly capitalism – a struggle reaching its peak during the war, which itself was the direct product of the aims and wishes of the most reactionary, chauvinist, and imperialist sections of finance capital.[10]

Did then the Nazi regime follow the interests of 'big business' in pursuing policies which culminated in war and genocide, or was it 'its own boss'? The primacy of politics or economics in the Third Reich, polarized in the debate between Mason and his GDR antagonists, has remained a central area of controversy in interpreting the Nazi dictatorship. Scholarly interpretations continue to be deeply divided – on political–ideological as well as historical–philosophical grounds.

In the dominant 'liberal–bourgeois' historiography there is little doubt about the nature of the relationship. Economic issues claim little space, for instance,

[8] Tim Mason, 'Primat der Industrie? – Eine Erwiderung', *Das Argument* 10 (1968), p. 199. Despite its marxist intonation, Mason's argument clearly shared much common ground with the approach of 'liberal bourgeois' historians, who, not unnaturally, welcomed this advocacy of the primacy of politics over economics by a marxist writer.
[9] Eberhard Czichon, 'Der Primat der Industrie im Kartell der nationalsozialistischen Macht', *Das Argument* 10 (1968), pp. 168–92; Dietrich Eichholtz and Kurt Gossweiler, 'Noch einmal: Politik und Wirtschaft 1933–1945', *Das Argument* 10 (1968), pp. 210–27.
[10] Eichholtz and Gossweiler, 'Noch einmal', pp. 220–7.

in Karl Dietrich Bracher's *The German Dictatorship*, and the 'primacy of politics' question was dispatched within a single paragraph:

> The very fact that a capitalist economy could be led into war in so non-economic a fashion and mobilized fully only during the war itself (after 1941–2) proves the absolute primacy of the political goals. Here, too, Hitler was anything but an instrument of the capitalists. Their cooperation followed the same pattern found in the governmental and cultural policies: the cooperating experts and economists were instruments and objects, not originators, of this policy. Economic efficiency and primacy of politics, not capitalist, middle-class, or socialist doctrines, determined the course.[11]

In similar vein, Ernst Nolte has written of the industrialists being 'completely eliminated as a major political factor'[12] and Klaus Hildebrand of 'economy in the service of politics',[13] while Andreas Hillgruber, in a brief recapitulation of differing approaches to the history of Nazism did not even regard the economy as one of his selected areas of debate.[14] Rather more cautiously, Karl Dietrich Erdmann, in a widely-read textbook, comments: 'Scholarship – apart from soviet marxist historical writing – is agreed that a determining industrial influence on the foreign and war policy decisions of Hitler cannot be proven in the sources'.[15] Finally, a most uncompromising statement of the position can be found in a recent survey of research on the Nazi economic recovery by the English historian Richard Overy, who writes: 'Over all the internal divisions within industry stood the authority and interests of the Nazi movement itself. Industry was subordinate to the requirements of the party. Control over the whole economy passed into the state's hands during the political crisis of 1936–7 and the establishment of the Four Year Plan.'[16]

Such assertive 'primacy of politics' arguments, it might be argued, posit a far clearer distinction between the sphere of politics and that of economics than in fact exists. They further imply a clarity of purpose and intent and a decisive command-role of Hitler and the Nazi leadership which might again be subject to qualification. Finally, they level the attack at an instrumentalist 'primacy of economics' argument which would be defended by even few marxist historians today.

Most western marxist approaches to the relationship of economics and politics in the Third Reich, whatever their differences of emphasis, tend to take their starting-point either from a type of 'Bonapartist' interpretation as originally advanced, for example, by August Thalheimer, or from an adaptation of Gramsci's emphasis on the state as a form of bourgeois 'hegemony'.

Mason's original 'primacy of politics' article was itself, even if not explicitly stated, closely related to Bonapartist notions of the growth of autonomy of the executive from the economic ruling class, and his position – or variants of

[11] Bracher, *The German Dictatorship* (see ch. 2 note 60), p. 416.

[12] Ernst Nolte, 'Big Business and German Politics: A Comment', *AHR* 75 (1969–70), p. 76.

[13] Hildebrand, *Das Dritte Reich*, pp. 160–1.

[14] Hillgruber, *Endlich genug?*, pp. 28–32 offers only a four-page disapproving summary of marxist/bonapartist interpretations of the 'social and economic aspects of the Third Reich'.

[15] Karl Dietrich Erdmann, *Deutschland unter der Herrschaft des Nationalsozialismus 1933–1939* (Gebhardt Handbuch der Geschichte, Band 20, Munich, 1980), pp. 141–2.

[16] Richard J. Overy, *The Nazi Economic Recovery 1932–1938* (Studies in Economic and Social History, London, 1982), p. 58.

it – has been followed by a number of leading marxist authorities on fascism. Reinhard Kühnl, for instance, accepted that 'the fascist state had to . . . possess a certain autonomy and freedom of decision as regards the economic power groups. It could not be the organ of execution of the ruling economic power groups in their entirety, for these had no common will; but it could also not be the instrument of a single economic fraction, because otherwise a stabilization of the entire system would not have been possible'. Hence there existed a 'partial independence of the political power' from the dominant economic interests. He concluded: 'That the freedom of decision of this executive is limited by the principles of the capitalist social order, remains undisputed. Even so, it seems legitimate to speak of at least a partial autonomy of the fascist executive from its allies, that is the socially dominant upper bourgeoisie'.[17] Another prominent West German marxist historian, Eike Hennig, adopted a not dissimilar position. He spoke of a 'division of labour' of 'political power' and 'economic domination' under Nazism, and commented favourably upon Mason's thesis and upon 'Bonapartist' interpretations.[18] Alfred Sohn-Rethel, who in the early years of the Third Reich was in a unique position as a marxist 'insider' at the hub of German industrial interest representation, writes of the 'subsumption' of industrial interests under 'the fascist state dictatorship of the party' and of the 'political imprisonment of the bourgeoisie in its fascist dictatorship'.[19] He makes it clear, in his analysis of the Nazi economy published decades after its initial formulation, that this is no subjection of the capitalist class or of 'big business' in the way that the 'totalitarianism' approach of 'liberal' historians would have it. Rather, the Nazi executive and the capitalist class were bound to each other inexorably by the rules of capital itself – by the need for an exceptional form of exploitation in order to revitalize capitalism and extricate it from its great crisis. The Nazi executive's monopoly of power derived from its ability to safeguard the objective interests of the bourgeoisie by maximizing its profits in these conditions of extreme crisis of capitalism. This was done by turning away from the international market economy to a more 'absolute' form of capitalist accumulation based directly upon the power of the State – upon outright repression, 'plunder', and ultimately war. Once embarked on this road, there was no going back. The process was irreversible, and the economic élites were bound to it – they were 'all in the same boat' as Schacht put it. Nazi political domination was therefore anchored in the crisis position of the capitalist bourgeoisie. But at the same time, this political domination remained dependent on the dynamic of the 'absolute' form of capitalist exploitation which had been unleashed, and therefore on the continued economic dominance of big capital.[20]

A rather different marxist approach to the relationship of capitalism and the Nazi State is advanced by Nicos Poulantzas, in a theoretical work drawing for illustration upon the historical reality of fascism in Italy and Germany,

---

[17] Kühnl, *Formen* (see ch. 2 note 13), pp. 123, 141. Kühnl's points here could, of course, be argued for any capitalist state.

[18] Eike Hennig, *Thesen zur deutschen Sozial- und Wirtschaftsgeschichte 1933 bis 1938* (Frankfurt am Main, 1973), pp. 126–8, 248–9.

[19] Alfred Sohn-Rethel, *Ökonomie und Klassenstruktur des deutschen Faschismus* (Frankfurt am Main, 1973), pp. 110–11, Engl. trans., *The Economy and Class Structure of German Fascism* (2nd edn., London, 1987).

[20] Sohn-Rethel, pp. 90 ff., 173 ff. The quotation from Schacht is cited on p. 174.

and owing more to Gramsci than any other marxist thinker.[21] Central to Poulantzas's interpretation is the notion of fascism as the most extreme form of 'exceptional capitalist state' – others being military dictatorship and Bonapartist regimes. The reason why fascism should be the type of 'exceptional capitalist state' to emerge was determined by the specific nature of the class struggle, relations of production, and the particular form of political crisis. Poulantzas rejected as unsatisfactory theories of fascism not only the Comintern version of fascism as the direct agent of monopoly capital, and the interpretation (which he attributes to 'social democratic circles') of fascism as 'the political dictatorship of the petty bourgeoisie', but also Bonapartist conceptions based on a notion of class equilibrium. According to Poulantzas, Bonapartist views rest upon a misinterpretation of Marx's formulation of the 'opposition of State and Society' and the 'independence' of the State in relation to civil society and have led marxist theoreticians 'to attribute to the fascist State a *type* and *degree* of relative autonomy which it does not in fact possess, and in the end makes them unable to define correctly the relations between fascism and big capital. . . . This relative autonomy of the State, taken to the limit, would even mean breaking the tie between the State and the hegemonic fraction; hence completely false descriptions of fascism using the war economy – openly and for a long period – against the interests of big capital and in declared opposition to it' – a misinterpretation he associates with Mason, in company with the 'élitist' theories of Schweitzer and Neumann.[22]

Though rejected by Poulantzas in connection with Bonapartist approaches, the notion of 'relative autonomy' is in fact central to his own interpretation. Fascism – i.e. the fascist party and the fascist state – has in his view a 'relative autonomy' both from the unstable power bloc of the politically dominant classes and from 'the fraction of big monopoly capital' whose dominance within the power bloc fascism has (re-)established. Fascism's relative autonomy derives on the one hand from the internal contradictions within the power alliance and on the other from the contradictions between the dominant and dominated classes. Fascism's 'complex relationship' with the 'dominated classes' is in fact 'precisely what makes fascism indispensable to mediate a reestablishment of political domination and hegemony'. In other words, whereas in Bonapartist theory the State derives from an equilibrium between the two main social forces, without thereby becoming a neutral mediator in the class struggle, the fascist state according to Poulantzas 'never ceases to organize political domination', possesses a far smaller 'margin for manoeuvre', and serves the objective function not of increasing its own independence from capital and the creation of a primacy of politics over economics, but of reestablishing the domination of the ruling fraction of monopoly capital. In Poulantzas's writings (not just on fascism), the political sphere – state power – always enjoys a relative autonomy from the economic sphere – capital – and this relative autonomy is extended to an exceptional degree under fascism. But it lasts in this exceptional degree only for a short period before the dominance of big monopoly capital is re-established.[23]

---

[21] Poulantzas (see ch. 2 note 17). Jane Caplan, 'Theories of Fascism: Nicos Poulantzas as Historian', *HWJ* 3 (1977), pp. 83–100, offers an excellent, penetrating critique.
[22] Poulantzas, pp. 84–5 and note 17.
[23] Poulantzas, pp. 85–6; and see Caplan, pp. 86–8.

Common to all the variant marxist theories summarized here is an acceptance of some degree of autonomy of the Nazi State from the power of even the most dominant capitalist forces. The level of autonomy posited is at its maximum in the Mason approach, where it amounts to a *primacy* of politics over economics; it is at its minimum in Poulantzas's interpretation, where it lasts only for a very short time in order to reassert the dominant position of monopoly capital. These differing marxist views are at least in agreement, therefore, that the suggestion of an *identity* between Nazism and capitalism in which the Nazi State apparatus functions as the executive instrument of the ruling class of the most extreme sections of monopoly capital is simplistic and mistaken. In fact, even GDR historians softened the earlier rigid instrumentalist line, though there was no retreat from the notion that 'in the last instance' the economic base – the interests of the monopoly bourgeoisie – determines the political course of action.

The question, therefore, which each of these marxist interpretations poses is: what weight can be attached to the concept of 'relative autonomy' as an explanatory factor in understanding the unfolding of Nazi policy and the relationship of Nazism and capitalism? Subsumed within this are a number of other problems posed by marxist analyses, some of a more empirical nature. Do marxist interpretations, for instance, accord sufficient importance to Nazi ideological aims? Are they in danger, even granting the 'relative autonomy' of the State, of grossly underestimating the 'Hitler factor' – not only Hitler's actual executive role (however it is defined) but also his functional position as integrating element and charismatic focus of mass plebiscitary support? Given the latter, do marxist analyses tend to exaggerate the undoubted importance of the big capital bloc and underrate correspondingly other power blocs – in particular the army leadership, the party with its mass base, and the rapidly developing power-centre in the SS-police apparatus? Do they pay sufficient attention to the changing chronology of relations between Nazism and the industrial élite, and to the complexities of decision-making processes in the Third Reich? (Poulantzas's historical treatment of the dictatorship period in Germany contains, for example, some serious empirical flaws which vitiate his periodization and gravely endanger his theoretical conclusions.[24]) As regards decision-making processes, do marxist analyses clearly separate direction, influence, and execution – an important distinction, not least in economic policy-making – and do they tend to assume that partial identity of aim equals influence? Finally, even accepting that exceptional forms of capitalism (Sohn-Rethel) existed under an exceptional form of capitalist state (Poulantzas), do marxist theories underrate or ignore the extent to which Nazism was prompting the growth of economic organization which had little to do with classical capitalism and, in the eyes of some authorities,[25] was moving in the direction of post-capitalist economics?

The evaluation and interpretation which follows attempts to take account of some of these critical questions alongside the problems posed by 'liberal' approaches to the 'primacy of politics'.

[24] See Caplan, pp. 87 ff.
[25] E.g. Winkler, *Revolution* (see ch. 2 note 13), pp. 100, 154 note 90; Saage, *Faschismustheorien* (see ch. 2 note 33), pp. 72–3; Gert Schäfer, 'Ökonomische Bedingungen des Faschismus', *Blätter für deutsche und internationale Politik* 15 (1970), pp. 1260 ff.: Alan S. Milward, 'Fascism and the Economy', in Laqueur (see ch. 2 note 12), pp. 435, 443–4.

Evaluation

A starting-point of analysis is to question whether the polarization into 'primacy of politics' or 'primacy of economics' does not amount to an extreme oversimplification of a complex structural interrelationship between the policies of the Nazi State and the interests of German capital. The reduction to alternatives of 'politics' and 'economics' both impermissibly narrows the concept of 'politics' and operates on a crude and misleading dichotomy between 'state' and 'society'. The tenor of recent work on the Nazi economy has been to suggest instead that the closely interwoven aims and interests of the Nazi leadership and of German capital influenced and affected each other, making it difficult to separate a specifically 'political' and specifically 'economic' sphere, and therefore to distinguish a clear 'primacy'. In William Carr's words, 'ideological, strategic, and economic factors are too closely intermeshed in a country's foreign policy to permit of a clinical separation',[26] while Hans-Erich Volkmann outrightly rejects the question of 'primacy' as now a redundant one.[27] Volkmann prefers to speak of 'a far-reaching congruity of interest' between the State and major industry, of a (partial) 'identity of interest of the economy and National Socialism', of such a close interlinkage of politics and economics in the Nazi State that one can describe it as a 'coercive identity'. He refers further to the 'interweaving' of the political–economic substructure, and to a 'mutual dependence of political leadership and industry' also during the War itself. Nor, in his view, did the 'common cause' which Germany's economic élites entered into with the Nazis from the turn of the year 1932–3 onwards develop into a 'primacy of politics' after 1936.[28] Rather, the State and the leading sectors of industry merged even more closely than before, so that before and especially during the war initiative, responsibility, and administrative control over the functioning of the economy – and with this extensive influence over political and military decisions inextricably bound up with the economy – passed to private industry. There developed, therefore, according to this interpretation, an increasing blurring of the boundaries between state economic administration and the sphere of the private economy. Volkmann argues, in distinction to the line of GDR historians, that the Nazi regime was not put into power by German capital in order to extend Germany's economy through territorial expansion at the behest of German industry. But, nevertheless, once in power the Nazis had no need to subject the economy also to its political demands. Rather, 'the leading German economic circles placed themselves in the service of the power political intentions of the German fascist government in order in this way to attain a closed economic area, largely independent of world economic vicissitudes, in which a high measure of autarky could be effected'.[29]

Such an argument is plausible and carries conviction. Nevertheless, as

---

[26] William Carr, *Arms, Autarky, and Aggression* (2nd edn., London, 1979), p. 65.
[27] Hans-Erich Volkmann, 'Politik, Wirtschaft und Aufrüstung unter dem Nationalsozialismus', in Manfred Funke, ed., *Hitler, Deutschland und die Mächte* (Düsseldorf, 1978), pp. 279, 289.
[28] Volkmann, 'Politik, Wirtschaft und Aufrüstung', pp. 273, 279–80, 289.
[29] Volkmann, 'Politik, Wirtschaft und Aufrüstung', pp. 290–1; Hans-Erich Volkmann, 'Zum Verhältnis von Großwirtschaft und NS-Regime im Zweiten Weltkrieg', in Karl Dietrich Bracher *et al.*, eds., *Nationalsozialistische Diktatur 1933–1945. Eine Bilanz* (Bonn, 1983), pp. 480–508.

Volkmann's hint of the economy functioning 'at the service' of the political intentions of the regime appears tacitly to admit, the acceptance of inter-dependence and affinity of interest still leaves open the explanation for the peculiar thrust, dynamic, and character of Nazi policy. Unquestionably, the alliance between the Nazi leadership and the industrial–military complex, cemented by the rearmament and expansionist programme, lasted into the final phase of the Third Reich as each of the partners found itself increasingly bound to the logic of the development they had set in train. But it might still be claimed that the weighting within this 'alliance' tipped gradually but inexorably towards the Nazi leadership, so that at the crucial junctures of development in the Third Reich the political and ideological demands of the Nazi leaders came to play an increasingly dominant role in determining policy. In fact, the ultimately self-destructive irrational momentum of the Nazi regime seems only explicable on this premiss: the faster the regime careered madly out of control and towards the abyss, the greater was the scope for political–ideological initiatives out of sequence with and in the end directly negating the potential of the socio-economic system to reproduce itself.

In order to understand this process, the position and role of 'big business' has to be located with the context of the complex and changing multi-dimensional ('polycratic') power-structures in the Third Reich. Fundamental to this is the need to break away both from the 'totalitarianism' model of a cen-tralized command economy and monolithic state in the hands of Hitler and a clique of Nazi leaders, and from the alternative, almost equally monolithic, model of the Nazi State as the direct representative and most aggressive form of rule of finance capital. Far more illuminating as an interpretative concept is the notion, first formulated by Franz Neumann and more recently expanded and developed by Peter Hüttenberger, of the Nazi regime as an unwritten 'pact' (or 'alliance') between different but interdependent blocs in a 'power-cartel'.[30] This cartel was initially a triad composed of the Nazi bloc (comprising the various component parts of the Nazi movement), 'big business' (including large landowners), and the army. From around 1936 it could be said to have acquired a fourth grouping as the Nazi bloc itself fell into two main subdivisions around the party organization proper and the increasingly powerful SS–Police–SD complex.[31] Though the blocs in the 'power-cartel' remained intact – and their interdependence sustained – until the end of the Third Reich, their relationship to each other and weighting within the 'cartel' altered during the course of the dictatorship. Broadly speaking, the change took place in the direction of an extension of the power of the Nazi bloc, and in particular of the SS–Police–SD complex, with a corresponding weakening – though never to the point of insignificance or complete submission – of the relative positions within the 'cartel' of 'big business' and the armed forces leadership.

The 'pact' of 1933 was based upon the mutual interests but not complete identity of the Nazi bloc, 'big business', and the army. The bond of alliance bet-ween Nazism and the army provided for a free hand for the new Nazi rulers in radically reorganizing Germany's internal political order in return for an acknowledgement of the Reichswehr as 'the most important institution in the

---

[30] Neumann (see ch. 2 note 5); Peter Hüttenberger, 'Nationalsozialistische Polykratie', *GG* 2 (1976), pp. 417–42.
[31] Huttenberger, pp. 423 ff., 432 ff.

State' together with the pledging of a comprehensive programme of rearmament which met goals held dear by the army throughout Weimar.[32] Massive rearmament came to be the main catalyst which assured the dynamic fusion of interests of army, industry, and Nazi leadership.[33] Initially, German 'big business', divided in itself and with partially contradictory economic aims, was far from uniformly or wholly enthusiastic about giving total priority to rearmament.[34] However, the crushing of the Left, the free hand accorded to industry, the reordering of industrial relations, and in general the new political climate, formed the basis of a positive relationship between the Nazi government and 'big business' – a relationship which became cemented by the stimulus to the economy through the work creation programme and then in growing measure by the massive profits to be derived from the armaments boom.

Though forming the most dynamic element within the 'power cartel', the Nazi bloc – possessing direct control neither over economic production nor over military power – was in a relatively weak position during the early years of the dictatorship. The strength of Nazism's 'partners' was reflected in the pressures prompting the destruction in June 1934 of the threat posed by the SA to the established order. Furthermore, the serious economic difficulties which confronted the regime in mid 1934, aggravated by economic repercussions abroad caused by the anti-Jewish measures, and coupled with a still precarious diplomatic position, meant that the regime's room for manoeuvre in this period was closely circumscribed by economic as well as strictly political factors.

In these conditions, the relative strength of the 'bargaining position' of 'big business' within the 'power cartel' was assured. It was reflected in the position of Hjalmar Schacht, President of the Reichsbank and from 1934 as Economics Minister one of the most powerful men in the Nazi State. However, Schacht's key position in controlling foreign trade and exchange – and therefore the raw material imports so essential for the armaments industries – was an obvious source of potentially serious conflict, since it meant intervention in an area – armaments policy – which was absolutely central to the interests not only of Hitler and the Nazi leadership but also of the armed forces and important and influential sectors of industry (in particular the electro-chemicals lobby centred upon IG-Farben).[35] Schacht was gradually coming, therefore, to represent only one – and as it transpired not the most powerful – wing of industry concerned with improving Germany's international trading position, and losing at the same time the support of the increasingly strong industrial grouping which backed and stood most to gain from autarkic policies. At first imperceptibly, but inexorably, Schacht's power was on the wane. And by the time the immanent tension in the Nazi economy between the demands of rearmament and the demands of consumption broke into full crisis in the spring and summer of 1936, the power relations within the original 'cartel' had already therefore begun to alter shape. The clash within 'big business' between those supporting

---

[32] See Wilhelm Deist, *The Wehrmacht and German Rearmament* (London, 1981), pp. 21 ff.

[33] See Dieter Petzina, 'Hauptprobleme der deutschen Wirtschaftspolitik', *VfZ* 15 (1967), p. 50, and the contribution by Hans-Erich Volkmann to Wilhelm Deist *et al.*, *Das Deutsche Reich und der Zweite Weltkrieg*, vol.1 (Stuttgart, 1979), pp. 208 ff.

[34] See Michael Geyer, 'Etudes in Political History: Reichswehr, NSDAP, and the Seizure of Power', in Stachura (see note 1), p. 114.

[35] See Hüttenberger, p. 433. On IG-Farben, see now the authoritative study by Peter Hayes, *Industry and Ideology: IG Farben in the Nazi Era* (Cambridge, 1987).

the Schacht line and those pushing for accelerated autarkic policies – with obvious corollaries for both domestic and especially foreign policy – can be said to have weakened (at least temporarily) the position of industry as a whole. Meanwhile, the position of the Nazi leadership, and of Hitler in particular, was immeasurably stronger than it had been in 1933, and a successful mastering of the crisis contained the potential for a further strengthening of the Nazi bloc within the overall power constellation in the Third Reich.[36]

The resolution of the immediate crisis – though it stored up future massive economic problems for the regime – was the introduction of the Four Year Plan, announced at the Party rally in September 1936 and setting Germany on an accelerated rearmament and autarkic policy as preparation for war. It was a decision in which politics and economics, ideology and material interest, were inextricably intermeshed.

Hitler's secret memorandum justifying the Plan – which was significantly given only to Göring, Blomberg, and (much later) Speer but not to Schacht – reads like the clearest demonstration of a 'primacy of politics', emphasizing that 'the nation does not live for the economy', but rather that 'the economy, economic leaders, and theories . . . all owe unqualified service in this struggle for the self-assertion of our nation'.[37] However, it has been rightly noted that Hitler's intervention 'should not be seen primarily as a capricious meddling in economic matters by a restless dictator'.[38] Rather, Hitler's memorandum came at the end of a process in which the dominant economic position had been grasped by the chemicals giant IG-Farben, which had forged an axis in particular with the Air Ministry and with the party through the key figure of Göring. IG-Farben had provided the technical details for the Four Year Plan, and their top management came to be completely merged with State officials in the running of the Plan. It would also be mistaken to imagine that industry was irredeemably split as a result of the introduction of the Plan. Heavy industry suffered more a temporary setback than the permanent defeat which Mason posited.[39] The threat posed by the setting up in 1937 of the state-owned steel corporation, the Reichswerke-Hermann-Göring, in the teeth of opposition from Germany's steel barons, can be exaggerated. The high production costs of the state concern in fact held steel prices high; and far from indicating an onslaught on private ownership, it coincided with a major 're-privatization' wave, including the return to private hands of the mammoth United Steelworks. Finally, the blockage in iron production which the state concern had been established to by-pass was over before its production had got under way.[40]

Research has done much, therefore, to qualify the notion that the Four Year Plan marked a sharp divide in the influence of industry and the breakthrough to a decisive 'primacy of politics'. At the same time, it is still significant that

---

[36] Hüttenberger, pp. 433–5.
[37] 'Denkschrift Hitlers über die Aufgaben eines Vierjahresplans', *VfZ* 3 (1955), pp. 204–10, here p. 206.
[38] Carr, *Arms, Autarky, and Aggression*, p. vi.
[39] Mason, 'Primacy', p. 185. Hüttenberger (p. 434) rightly points out that the autarky conflict did not result in a split in the political position of 'big business'.
[40] George W. F. Hallgarten and Joachim Radkau, *Deutsche Industrie und Politik von Bismarck bis in die Gegenwart* (Reinbek bei Hamburg, 1981), pp. 255–8; see also Petzina, *Autarkiepolitik*, pp. 104 ff.

the economic reorientation in 1936 was carried out initially against the wishes of important sectors of the once mighty heavy industry, and that as a result of the Four Year Plan and the replacement of Schacht by Göring as the dominant figure in the economy, the constraints of what might be regarded as the previous 'economic establishment' on the Nazi leadership diminished sharply. Moreover, the foundation of the Reichswerke-Hermann-Göring in 1937, if marking no long-term threat to private industry, did register the fact, as Petzina pointed out, 'that private industrial interests were not automatically identical with the interests of the regime, and that in a case of conflict the regime would not shy away from effecting its aims against the resistance of sections of heavy industry'.[41] As Milward put it, 'nothing could have more clearly demonstrated that, however sympathetic to the business world and however dependent on it, the Nazi government had its own interests which it was prepared to pursue'.[42]

With the successful mastery of the crisis of 1936, the Nazi leadership gained an enhanced position of strength within the 'power cartel' which brought with it an increased priority to and scope for ideological considerations in the formulation of policy. This was particularly the case in the spheres of foreign policy, where the traditional authority of the Foreign Office had diminished, and strategic–military planning, where the *Wehrmacht*'s influence had also waned. By early 1938, in fact, the SS–Police–SD bloc was powerful enough to weaken the *Wehrmacht*'s position still further by instigating the Blomberg–Fritsch affair, a symbolic turning-point in the army's transition from a power to a mere functional élite.[43] Certainly, the influence of leading business circles on German foreign policy in the later 1930s, as indeed earlier, has often been underestimated.[44] Clearly, too, German expansion into Austria and Czechoslovakia was both a logical and necessary step economically as well as strategically. German firms profited hugely from this expansion, as did some major concerns from the 'aryanization' of the economy in 1938. Ideological, strategic, and economic interests still went hand in hand. But the impetus was increasingly shifting towards a high-risk policy in which the inbuilt and unstoppable momentum of the arms race harnessed to the ideological expansionism of the Nazi leadership shaped the contours within which economic interest operated.

In the wake of the forced rearmament policy from 1936 onwards, Germany's economic problems – chronic shortages of foreign exchange, raw materials, and labour, strains, blockages, over-heating, balance of payments difficulties, inflationary tendencies – mounted alarmingly. Expansionism as the only solution to Germany's otherwise gloomy economic prospects was a central theme of Hitler's monologue to the leaders of the armed forces in November

[41] Petzina, *Autarkiepolitik*, p. 105. For the economic development of the Reichswerke-Hermann-Göring, see Richard J. Overy, 'Göring's "Multi-National Empire" ', in Alice Teichova and P.L. Cottrell, eds., *International Business and Central Europe, 1918-1939* (Leicester, 1983), pp. 269–98. And for the circumstances of their foundation, see Overy's more recent article, 'Heavy Industry and the State in Nazi Germany: The Reichswerke Crisis', *European History Quarterly* 15 (1985), pp. 313–40.
[42] Milward, 'Fascism and the Economy', p. 434.
[43] See Hüttenberger, p. 435; and Klaus-Jürgen Müller, *Armee, Politik und Gesellschaft in Deutschland 1933-1945* (Paderborn, 1979), pp. 39–47, Engl. trans., *Army, Politics, and Society in Germany, 1933-1945* (Manchester, 1984).
[44] See Hallgarten and Radkau, Part II, chs. 3–4.

1937.[45] Hitler repeated his remarks on the threatening economic pressures in a speech to the armed forces' commanders in August 1939, days before the attack on Poland, when he stated that for Germany it was easy to make decisions: 'We have nothing to lose; we have everything to gain'. Because of our restrictions, our economic situation is such that we can only hold out for a few more years. Göring can confirm this. We have no other choice, we must act'.[46] The dire prognoses of Germany's economic future without expansion were coming from all sides of industry, agriculture, and from the *Wehrmacht*'s Economic Inspectorate. Strong though the evidence is for this mounting economic crisis, it is weak in suggesting that economic pressures played the decisive role in affecting either the timing or the reasons for the outbreak of war. Strategic considerations took first rank, while the increasingly critical economic situation, itself deriving in no small measure from the political–ideological premises of the regime, appears to have played chiefly the role of confirming Hitler in the view that his original diagnosis of Germany's plight was correct, and that time was running out.[47] Certainly, the most aggressive, expansionist noises were emanating from 'big business' circles at this time – prominent, though by no means isolated, the imperialist demands of IG-Farben boss Karl Krauch. And obviously, expansion fed expansion in economic as well as political–military terms. But compared with Austria and Czechoslovakia, as Radkau points out, the attack on Poland 'had relatively little to do with the main lines of interest of the concerns' and 'generally the east was for capital much less attractive than, say, the south-east'.[48] This did not of course hinder in any way German firms from profiting massively from the ruthless exploitation of conquered Poland.

Economic determinants continued during the war itself to be inseparably interlocked with ideological and military–strategic factors in shaping the character and pattern of German aggression. And the chronic problems of availability and allocation of raw materials and labour meant a voice for the leaders of the dominant war industries which could not be ignored in the shaping of policy decisions. Given the particular development of German capitalism during the Third Reich, especially since 1936, the imperialist war of plunder was

[45] Noakes and Pridham, vol. 3, pp. 680–7. *IMT*, 25, pp. 402–13, Doc. 386-PS. On the 'Hossbach memorandum', see now Jonathan Wright and Paul Stafford, 'Hitler, Britain and the Hoßbach Memorandum', MGM 42 (1987), pp. 77–123 (abbreviated version in *History Today* (March 1988), pp. 11–17).

[46] *IMT*, 26, pp. 338 ff., here p. 340, Doc. 798-PS.

[47] Carr, *Arms, Autarky, and Aggression*, p. 65. The evidence for the economic crisis is summarized in Timothy W. Mason, 'Innere Krise und Angriffskrieg 1938/1939', in F. Forstmeier and H.-E. Volkmann, eds., *Wirtschaft und Rüstung am Vorabend des Zweiten Weltkrieges* (Düsseldorf, 1975), pp. 158–88. For criticism and qualification of Mason's emphasis upon the internal crisis as the decisive factor in the timing of the war, see Ludolf Herbst, 'Die Krise des nationalsozialistischen Regimes am Vorabend des Zweiten Weltkrieges und die forcierte Aufrüstung. Eine Kritik', *VfZ* 26 (1978), pp. 347–92; Heinrich August Winkler, 'Vom Mythos der Volksgemeinschaft', *AfS* 17 (1977), pp. 488–9; Jost Dülffer, 'Der Beginn des Krieges 1939: Hitler, die innere Krise und das Mächtesystem', *GG* 2 (1976), pp. 443–70; Milward, 'Fascism and the Economy', p. 437; Richard J. Overy, 'Hitler's War and the German Economy: A Reinterpretation', *EcHR* 35 (1982), pp. 272–91; and Overy's more recent article, 'Germany, "Domestic Crisis" and War in 1939', *Past and Present* 116 (1987), pp. 138–68, which unleashed a fierce reply from Tim Mason and further contributions to the 'debate' by Richard Overy and David Kaiser in *Past and Present* 122 (1989), pp. 200–40.

[48] Hallgarten and Radkau, pp. 302–3, 366–8.

a logical necessity – increasingly the only option available;[49] German industry was structurally implicated in the policy decisions which culminated in destruction and inhumanity on a scale unprecedented in Europe.

It is necessary, however, to distinguish between the economy as a structural determinant in helping to frame the course and character of aggression, and the specific needs and perceived interests of particular groups within the economy. Much emphasis on 'the primacy of politics' concentrates rather simplistically and misleadingly on merely the question of whether decisions in the Third Reich were taken directly in the interests of German capitalists. This line of argument remains in essence little more than a superficial attack on naïve versions of the instrumentalist 'agent theory' – of the Nazi leadership as the puppets of 'big business'. Reality was somewhat more complex, as the decision to invade the Soviet Union illustrates.

In this decision, too, ideological motivation can hardly be separated as an autonomous factor from questions of military–strategic and economic necessity. It is too simple to look no further than Hitler's ideological obsession – important though this was – in explaining the reasons for the invasion of the Soviet Union in 1941. Unquestionably, the ideological hatred of 'Jewish Bolshevism' which had been pumped into Germans for years under the Nazi regime provided for the horrifically brutal character of the 'war of annihilation' in the east. But – a point to which we shall return in a later chapter – strategic considerations revolving around the unfinished war in the West and especially the prospects of combating the USA also played a crucial role in the thinking of Hitler and the Nazi and military leadership about the Soviet Union in 1940–1. Last, but certainly not least, there was the economic dimension. The German dependence upon raw materials from the Soviet Union, and the critical threat to grain and above all else to oil supplies posed by Soviet expansion in east and south-east Europe following the Nazi–Soviet Pact of 1939 meant that the entire German war effort was endangered if the Soviet Union remained unconquered. The possibility of the Soviet air-force destroying the vital Rumanian oil-fields, providing more than half of German supplies, was decisive. As Hitler told his generals in January 1941, 'in the era of air power Russia can turn the Rumanian oil fields into an expanse of smoking debris . . . and the life of the Axis depends on those oil fields'.[50]

This obvious importance of the economic dimension to decision-making on military–strategic questions is, however, not necessarily synonymous with the perceived needs of German industrialists. Joachim Radkau, a left-wing West German historian, argues on the basis of a detailed study of available sources, that contrary to expectation there is little evidence of complete identity of interest between Nazism and 'big business' in the preparation of the attack on the Soviet Union: 'Disregarding ideological anti-communism, in general no hostility towards Soviet Russia can be recognized from the practical wishes and recommendations of business – often indeed a striving for improvement in

---

[49] For emphasis upon the Blitzkrieg as the only possible strategy available to Germany, see Alan S. Milward, 'Der Einfluß ökonomischer und nicht-ökonomischer Faktoren auf die Strategie des Blitzkriegs', in Forstmeier and Volkmann, pp. 189–201, here esp. pp. 200–1. The conception of a 'Blitzkrieg economy' is wholly rejected by Overy, 'Hitler's War', and more fully in ' "Blitzkriegswirtschaft"?' *VfZ* 36 (1988), pp. 379–435.

[50] Cited in Norman Rich, *Hitler's War Aims* (2 vols., London, 1973–4), vol. 1, p. 207. See also Hallgarten and Radkau, p. 309.

relations. Business [*die Wirtschaft*] played a much clearer role in advancing the Stalin–Hitler Pact than in preparing the attack on the Soviet Union'. Trade with Russia – not least for heavy industry – had been important in the 1920s and early 1930s; the evidence which Radkau assembles – though it does not speak wholly in unison – suggests that some prominent sections of industry were placing their hopes in a revival of economic links rather than in the ideologically motivated smashing of the Soviet Union, and that many industrialists were not enamoured with the investment risks and likely benefits to be gained in the newly-conquered 'Lebensraum'.[51] Again, however, such views did not limit in any way whatsoever the readiness to exploit in the most barbarous fashion the human as well as the material resources of the conquered territories. Furthermore, such views were out of step with the unstoppable momentum – economic as well as military – of the Nazi war. The dominant economic forces in Germany were completely at one with this war effort. The collaboration of the rest was assured by the fact that there was no escape from the course of events which they themselves had helped initiate and had fostered: they were committed to flourish or perish with the Nazi regime.

The ace in the hand of proponents of a 'primacy of politics' approach is always taken to be the extermination of the Jews – on the face of it, the most blatant refutation of the view that the interests of 'big business' were behind Nazi policy. Indeed, the Ministry for the Occupied Eastern Territories already expressly stated in autumn 1941 that 'economic considerations are to be regarded as fundamentally irrelevant in the settlement of the [Jewish] problem'.[52] And, as Mason pointed out in his 'primacy of politics' essay, 'among the first Polish Jews who were gassed in the extermination camps were thousands of skilled metal workers from Polish armament factories'.[53]

The deployment of scarce transport facilities to ferry human cargo across Europe for instant extermination at a time when German industry was desperate for manpower – even though some Jewish labour did continue to be used almost to the end of the war – was hardly compatible with 'rational' economic interest. However, as we shall bring out more fully in a later chapter, it would be a distortion to remove the 'Final Solution' from the material as well as the ideological context of the complex development which led to Auschwitz. 'Big business' was largely indifferent to early anti-Jewish measures in the Nazi State, except where German foreign trade was adversely affected by negative responses abroad. Such criticisms on economic grounds of the anti-Jewish 'boycott movement' and of wild terror actions against Jews were voiced, for instance, by the Economics Minister Schacht in 1935.[54] Under the growing pressure of the armaments economy, however, 'big business' had a direct interest in the acquisition of Jewish capital and keenly promoted the 'aryanization' of Jewish concerns in late 1937 and 1938.[55] Moreover, the expanding power and autonomy within the overall power structure of the regime of the

[51] Hallgarten and Radkau, pp. 383 ff. See also Winkler, *Revolution*, pp. 99, 153–4 note 89.

[52] Cit. Hans Buchheim *et al.*, *Anatomie des SS-Staates* (Olten/Freiburg, 1965), vol. 2, p. 377.

[53] Mason, 'Primacy', p. 195.

[54] Uwe Dietrich Adam, *Judenpolitik im Dritten Reich* (Düsseldorf, 1972), pp. 123–4; Karl A. Schleunes, *The Twisted Road to Auschwitz. Nazi Policy toward German Jews, 1933–1939* (Urbana/Chicago/London, 1970), pp. 153 ff.

[55] Schleunes, pp. 159 ff.; Helmut Genschel, *Die Verdrängung der Juden aus der Wirtschaft im Dritten Reich* (Göttingen, 1966), pp. 222 ff.

SS–Police–SD complex, which by the end of 1938 had gained control over the implementation of anti-Jewish policy, meant that anti-Jewish measures now acquired a rapidly increasing momentum of their own. With the massive extension of the 'Jewish Question' in the Occupied Territories and the administratively insoluble character of the 'problem', the inner dynamic of a course of development which could by now only logically end in physical extermination could not be checked. In any case, there was still at this stage no contradiction between the relative autonomy of the SS apparatus within the regime and the interests of German capital. Germany's major industrial concerns were more than willing to take advantage of the concentration of Jewish labour in the Polish ghettos, with a free hand for total exploitation at absolutely minimal cost. Whatever 'wastage' took place was bearable in the period of expansion, when abundant slave labour to satisfy the needs of the whole German economy seemed close at hand.[56] By the time the course of the war – and with it the prospects and interests of German industry – had changed dramatically, wholesale physical extermination of the Jews, which had gradually crystallized as the solution to a growing administrative nightmare arising from the 'problem' the Nazi rulers had created for themselves, was in full swing and unstoppable.

The extermination of the Jews was, therefore, *ultimately* a 'policy' which contradicted economic rationality. But it emerged as the final stage in a process which for long was compatible with, even where not directly in the interests of, German capital. The 'Final Solution' became a possibility through the conditions of war and brutal conquest. The obsession with the 'Jewish Question' chiefly belonged to the Nazi bloc within the 'power cartel' of the Third Reich. However, the other power élites showed no hesitation in helping to implement anti-Jewish measures and to turn ideological obsession into policy decisions. Above all, *all* sections of the 'power cartel' worked to bring about the barbarous war of conquest which made genocide an attainable reality rather than a lunatic vision.

German industry's direct implication and collaboration in the Nazi plunder, exploitation, destruction, and mass murder in the Occupied Territories continued to the end. Whereas certain groups within the armed forces and the old aristocracy underwent a development from initial reserve to outright antipathy towards the Nazi regime, culminating in their involvement in the plot against Hitler on 20 July 1944, industrial leaders were notably missing from resistance circles. Yet by the last year of the war, it was becoming increasingly apparent to 'big business' that the complete abyss of destruction which was looming was the contradiction of any 'rational' economic policy. Even so, the divorce between the radical nihilism of the Nazi bloc and the material interests of German industry only became total during the last phase of the war, in the wild lashings of the regime in its death-throes. A symbolically decisive moment, as Alan Milward points out, occurred in January 1944 'when the Führer supported [Labour Plenipotentiary] Sauckel's impossible plans to deport a further million workers from France during that year against the advice of Speer and the Ministry of War Production to organize more war production in the occupied

---

[56] Kurt Pätzold, 'Von der Vertreibung zum Genozid. Zu den Ursachen, Triebkräften und Bedingungen der antijüdischen Politik des faschistischen deutschen Imperialismus', in Eichholtz and Gossweiler, *Faschismusforschung* (see ch. 1 note 27), pp. 181–208, here pp. 206–8.

territories. From that moment the position of the Ministry of War Production and of the businessmen who ran it became increasingly weaker than that of the more radically fascist parts of the administration. The business circles which had sought to control the movement in 1933 now had their most pessimistic fears fulfilled; they had themselves become the plaything of a political revolution'.[57]

Until the last stages of the war, the benefits of the Third Reich to all those sections of industry and finance connected with armaments production were colossal. Undistributed profits of limited liability companies were four times higher in 1939 than they had been in 1928.[58] The monopoly concerns were the greatest single winners – and in prime place the chemicals giant IG-Farben, whose annual net profit, which had stagnated between 1933 and 1935, doubled in 1936 from RM 70 to 140 million, rocketed to RM 300 million by 1940, and doubtless reached stratospheric heights – though these are undocumented – thereafter.[59] The mammoth profits of the major concerns were no incidental by-product of Nazism, whose philosophy was closely tied in with provision of a free hand for private industry and eulogization of the entrepreneurial spirit.[60] Private industry was indispensable to the rearmament effort, and this gave its representatives a very considerable bargaining power, which they did not hesitate to use to their advantage throughout the Third Reich. However, it is important to recall the distinction between the initiation, execution, and exploitation of policy. I have argued here that while major capitalist enterprise could massively increase its profits through Nazi policy, control over the execution of policy moved unmistakably towards the specifically 'Nazi bloc' in the 'power cartel'. And as the groups in the 'Nazi bloc' gained the upper hand in policy execution, so also the initiation of policy in crucial areas with a direct bearing on the economy shifted inexorably away from 'big business', though coming only at a late stage to stand diametrically in opposition to the prime capitalist interest in its own reproduction. By then, the level of intervention by the Nazi State in both labour and capital markets, coupled with the autarkic exclusion of the new German *imperium* from world markets, had certainly promoted a capitalism quite differently structured from that analysed by Marx.[61] However, speculation about the future nature and role of capitalism in a victorious Nazi 'new order' seems vacuous. Ultimately, the madly escalating nihilistic dynamic of Nazism was incompatible with the lasting construction and reproduction of any economic order.

[57] Milward, 'Fascism and the Economy', pp. 434–5. The growing gulf between the interests of the regime and those of a crucial industry, that of coal, are well demonstrated in John R. Gillingham. *Industry and Politics in the Third Reich* (London, 1985).
[58] Dietmar Petzina, *Die deutsche Wirtschaft in der Zwischenkriegszeit* (Wiesbaden, 1977), p. 141; Milward, 'Fascism and the Economy', p. 435.
[59] Hallgarten and Radkau, p. 262.
[60] See Hallgarten and Radkau, pp. 227 ff., 269 ff.
[61] Milward goes so far as to claim ('Fascism and the Economy', p. 435) that fascist regimes did not preserve capitalism, but 'changed the rules of the game so that a new system was emerging'. But in an important study of the ways in which economic planning in the Nazi state (especially in the Economics and Armaments Ministries) was evolving under the impact of total war, Ludolf Herbst shows how ideals of technocratic efficiency, liberated from the dead hand of state bureaucracy, were being developed as models of a post-war order which would rest upon German industrial strength. See Ludolf Herbst, *Der Totale Krieg und die Ordnung der Wirtschaft* (Stuttgart, 1982).

In the preceding analysis, I have attempted to break away from what are in my view oversimplistic alternative interpretations – the 'primacy of politics' or 'primacy of economics' – of the complex relationship of Nazism and 'big business' in the Third Reich. To insist that 'in the last instance' economic factors determine, seems indeed – to say the least – an inadequate explanation of the growing paramountcy of Nazism's radical nihilism over 'rational' economic interest. At the same time, the classic 'liberal' interpretation of the 'primacy of politics', posited implicitly or explicitly on notions of the 'totalitarian' control over an economy 'in the service' of a single-minded dictatorship, is scarcely more convincing in its simplification of the power structure of the Third Reich and its inbuilt overemphasis upon the personality and ideology of Hitler. This, however, and the contrasting interpretation offered here, based upon an understanding of the 'polycratic' character of the 'power cartel' in the Third Reich, raises a new set of questions revolving around the place and function of Hitler in the government of Nazi Germany. The next chapters focus upon this central problem of interpretation.

4

Hitler: 'Master in the Third Reich' or 'Weak Dictator'?

Locating Hitler's role and function within the Nazi system of rule is less straightforward than initially it may seem. Indeed, it has become a central problem of interpretation in a debate between leading historians of the Third Reich – a debate which, it has been said, sometimes resembles in its complexities the theological wrangles of the Middle Ages,[1] and which certainly contains a degree of rancour extending beyond the conventional disagreements of historians. The unusually heated and sometimes bitter tone of the debate[2] reflects in some ways the three dimensions – historical–philosophical, political–ideological, and moral – of writing on Nazism (especially in West Germany) which were outlined in chapter 1. Above all, the moral issue – the feeling that the evil of the central figure of the Third Reich is not being adequately portrayed, that Hitler was underestimated by contemporaries and is now being trivialized by some historians – lies at the root of the conflict and determines the character of the debate. The moral issue is itself indissoluble from questions about historical method and philosophy – how to write the history of Nazism – which in turn are inseparable from political and ideological value-judgements also relating to present-day society.

The key issue in historical–philosophical terms is the role of the individual in shaping the course of historical development, as against the limitations on the individual's freedom of action imposed by impersonal 'structural determinants'. In the present case, this focuses upon the question of whether the terrible events of the Third Reich are chiefly to be explained through the personality, ideology, and will of Hitler, or whether the Dictator himself was not at least in part a (willing) 'prisoner' of forces, of which he was the instru-

---

[1] John Fox, 'Adolf Hitler: The Continuing Debate', *International Affairs* 55 (1979), p. 261.
[2] See the rancorous exchanges prompted by Klaus Hildebrand, 'Nationalsozialismus ohne Hitler?', *GWU* 31 (1980), pp. 289-305: 'Externus', 'Hildebrands Lied – oder: Wie die GWU ihre Leser informiert', *Geschichtsdidaktik* 5 (1980), pp. 325-7; K.D. Erdmann, 'Antwort an einen Dunkelmann: Wie informiert GWU ihre Leser?', *GWU* 32 (1981), pp. 197-8; Klaus Hildebrand, 'Noch einmal: Zur Interpretation des Nationalsozialismus', *GWU* 32 (1981), pp. 199-204; 'Externus', 'Die GWU und ihr Frontberichterstatter: Fortsetzung eines "Gedankenaustausches" ', *Geschichtsdidaktik* 6 (1981), pp. 233-8; Wolfgang J. Mommsen, 'Die "reine Wahrheit" über das nationalsozialistische Herrschaftssystem?', *GWU* 32 (1981), pp. 738-41; and Klaus Hildebrand, 'Die verfolgende Unschuld', *GWU* 32 (1981), p. 742. The tone is upheld in Hofer's later essay (see ch. 1 note 2). Hildebrand's original piece was a one-sided report on a conference staged by the German Historical Institute, London, at Windsor in 1979, which highlighted the huge divides in interpretation of the Third Reich, especially among West German historians. The conference papers published in Hirschfeld and Kettenacker (see ch. 1 note 23), some considerably revised, scarcely convey the acrid debates which accompanied some of them during the conference.

ment rather than the creator, and whose dynamic swept him too along in its momentum. The historiographical positions are graphically polarized in the frequently-cited comment of the American historian Norman Rich, that 'the point cannot be stressed too strongly: Hitler was master in the Third Reich',[3] and in the diametrically opposed interpretation of Hans Mommsen, of a Hitler 'unwilling to take decisions, frequently uncertain, exclusively concerned with upholding his prestige and personal authority, influenced in the strongest fashion by his current entourage, in some respects a weak dictator'.[4] Before attempting to evaluate these interpretations, it is necessary to outline the contours of the debate in the light of the recent historiography on Hitler and the structure of the Nazi State.[5]

Personality, Structure, and 'the Hitler Factor'

Studies founded upon the centrality of Hitler's personality, ideas, and strength of will to any explanation of Nazism take as their starting-point the premise that, since the Third Reich rose and fell with Hitler and was dominated by him throughout, 'National Socialism can indeed be called Hitlerism'.[6] Behind such an interpretation is in general a philosophy which stresses the 'intentionality' of the central actors in the historical drama, according full weight to the freedom of action of the individual and the uniqueness of his action. This type of thinking obviously characterizes biographies of Hitler, as well as the more recent vogue of 'psycho-historical' studies. It also, however, underlies some outstanding non-biographical studies of Nazism.

The 1970s saw the appearance of a number of Hitler biographies – amid the outpouring of mainly worthless products of the so-called 'Hitler Wave', indicating a macabre fascination with the bizarre personality of the Nazi leader.[7] Some of the findings of the new biographies themselves seemed to add little more than antiquarian detail to existing knowledge about Hitler, though the best of them, by Joachim Fest – significantly the first major biography of Hitler by a German – went a long way towards replacing Bullock's dated classic of the 1950s.[8] Even so, as perceptive critics pointed out amid the paeans of

---

[3] Rich (ch. 3 note 50), vol. 1, p. 11.
[4] Hans Mommsen, 'Nationalsozialismus', in *Sowjetsystem und demokratische Gesellschaft* (ch. 2 note 12), vol. 4 (Freiburg, 1971), column 702. Mommsen appears to have first made this heuristic point in his *Beamtentum im Dritten Reich* (Stuttgart, 1966), p. 98 note 26, where he stated that Hitler was 'in all questions which needed the adoption of a fundamental and definitive position, a weak dictator'.
[5] For historiographical surveys, see Wolf-Rüdiger Hartmann, 'Adolf Hitler: Möglichkeiten seiner Deutung', *AfS* 15 (1975), pp. 521–35; Andreas Hillgruber, 'Tendenzen, Ergebnisse und Perspektiven der gegenwärtigen Hitler-Forschung', *HZ* 226 (1978), pp. 600–21; Wolfgang Michalka, 'Wege der Hitler-Forschung', *Quaderni di storia* 8 (1978), pp. 157–90 and 10 (1979), pp. 123–51; William Carr, 'Historians and the Hitler Phenomenon', *German Life and Letters* 34 (1981), pp. 260–72; and, comprehensively, Schreiber (see ch. 1 note 6).
[6] Bracher, 'The Role of Hitler' (ch. 2 note 3), p. 198.
[7] For a devastating critique of 'Hitler-Wave' publications, see Eberhard Jäckel, 'Rückblick auf die sog. Hitler-Welle', *GWU* 28 (1977), pp. 695–710.
[8] Joachim C. Fest, *Hitler. Eine Biographie* (Berlin, 1973), Engl. trans., *Hitler* (London, 1974). Alan Bullock, *Hitler. A Study in Tyranny* (orig. edn., London, 1952; revised edn., Harmondsworth, 1962).

praise, Fest's stylistic study revealed some of the inbuilt weaknesses of the biographical approach – in particular when the subject of study was such a 'non-person' as Hitler.[9] Fest's work is rather unbalanced in coverage, for instance, in devoting undue attention to Hitler's early years; it ignores or plays down socio-economic issues; it is excessively concerned with the historically futile question of whether Hitler can be attributed with qualities of 'negative greatness'; and generally, it shows a far less sure touch when relating Hitler to the broader developments of German society and politics than when dealing with his personality. The inability of the biographical approach to avoid the extreme personalization of complex issues, reducing them to questions of Hitler's personality and ideology, characterizes, too, the widely-read and highly influential piece of quality journalism by Sebastian Haffner, which treats Nazism solely in terms of Hitler's 'achievements', 'successes', 'errors', and so on.[10]

The apogée of 'Hitler-centrism' is reached in the psycho-historical approach characterizing a number of new studies in the 1970s and coming close to explaining the war and the extermination of the Jews through Hitler's neurotic psychopathy, oedipal complex, monorchism, disturbed adolescence, and psychic traumas (allegedly fitting into the collective psychology of the German people).[11] Even if the findings were less dependent on conjecture and speculation, it is difficult to see how this approach could help greatly in explaining how such a person could become ruler of Germany and how his ideological paranoia came to be implemented as government policy by non-paranoids and non-psychopaths in a sophisticated, modern bureaucratic system. Wehler's sarcasm – and he is one of the few historians to have seriously tested the applicability of psycho-analysis to historical method – seems not misplaced: 'Does our understanding of National Socialist politics really depend on whether Hitler had only one testicle? . . . Perhaps the Führer had three, which made things difficult for him – who knows? . . . Even if Hitler could be regarded irrefutably as a sado-masochist, which scientific interest does that further? . . . Does the "final solution of the Jewish question" thus become more easily understandable or the "twisted road to Auschwitz" become the one-way street of a psychopath in power?'[12]

The most important studies which take the centrality of Hitler's person and ideology as their interpretative focus are of immeasurably higher quality, and are not biographically orientated at all. Unlike most of the biographies (excluding Bullock and Fest), the wide range of works by Bracher, Hillgruber, Hildebrand, and Jäckel – to name the leading figures – has made a major

[9] See Hermann Graml, 'Probleme einer Hitler-Biographie. Kritische Bemerkungen zu Joachim C. Fest', VfZ 22 (1974), pp. 76–92. Bracher's doubts about the biographical approach are expressed in his 'The Role of Hitler', pp. 194–7. Some of Graml's criticisms would apply to the recent biography by Marlis Steinert, Hitler (Paris 1991) which, though solid, informative, reliable, and abreast of the latest literature, does not altogether satisfactorily accomplish its aim of blending social, ideological, and cultural analysis into a biographical study. Alan Bullock's second work on the German dictator, Hitler and Stalin. Parallel Lives (London, 1991) does not quite match the brilliance of his first pioneering work.

[10] Sebastian Haffner, Anmerkungen zu Hitler (Munich, 1978), Engl. trans., The Meaning of Hitler (London, 1979).

[11] Among the leading productions are Robert Waite, Adolf Hitler. The Psychopathic God (New York, 1977), and Rudolf Binion, Hitler among the Germans (New York, 1976).

[12] Hans-Ulrich Wehler, 'Psychoanalysis and History', Social Research 47 (1980), p. 531.

contribution to an understanding of Nazism. What links their individually different approaches together is the notion that Hitler had a 'programme' (though not a crude blueprint for action), which in all essentials he held to consistently from the early 1920s down to his suicide in the Berlin bunker in 1945. His own actions were directed by his ideological obsessions; and the Third Reich was directed by Hitler; therefore the Führer's ideology became implemented as government policy. Roughly summarized, this is the basis of the 'programmatist' type of interpretation.

The conception of Hitler as a man fanatically pursuing defined objectives with relentless consistency (though with tactical flexibility) – replacing as late as the 1960s the view that he was little more than a power-grabbing, unprincipled opportunist – produced in sophisticated works such as those of Andreas Hillgruber a picture of the 'programmatist' Hitler bending German foreign policy to his determined will to accomplish long-term but clear-cut ideological goals.[13] This picture depended in turn upon a corresponding perception of Hitler's role in domestic policy as the supreme machiavellian working, with whatever tactical adroitness, to a preordained concept and pushing in a perniciously logical and internally rational series of steps towards total power in order to implement his ideological aims as government practice. The development of this interpretation of Hitler owed most to the work of Karl Dietrich Bracher.

For Bracher, a political scientist, the key question was how liberal democracy disintegrated and made way for 'totalitarian' dictatorship.[14] His exposition of the workings of the German 'totalitarian' dictatorship, emerging in a flow of centrally important studies from the mid 1950s onwards, attributed a pivotal role to Hitler and emphasized the motivating force of Hitler's ideology.[15] In an interesting bridge to the later 'structuralist' emphasis on the 'institutional anarchy' of the Third Reich, Bracher was already writing in 1956 that 'the antagonism between rival agencies was resolved solely in the omnipotent key position of the Führer', which 'derived precisely from the complex coexistence and opposition of the power groups and from conflicting personal ties'.[16] The emphasis on the Führer's actual omnipotence, however, distinguishes Bracher's position clearly from that of the later 'structuralists'. Moreover, the title of Bracher's essay – 'stages of totalitarian *Gleichschaltung*' – reflected the stress he placed upon the essentially planned, regulated, and 'rational' progression to preconceived goals, an argument he consistently reformulated in his major works. By a different route, Bracher had developed an interpretation of Hitler which clearly married with the 'programmatist' approach to foreign policy, and also with the Hitler-centrism of the best biographies.

Bracher affirmed his position in an interpretative essay in the mid 1970s addressed to the problem of 'the place of the individual within the historico-political process'..[17] He argues vehemently that Hitler was fatally underesti-

---

[13] See esp. Hillgruber, *Hitlers Strategie* (cf. ch. 1 note 17 this edition).
[14] This was the central problem tackled in Bracher, *Auflösung* (see ch. 1 note 32 this edition).
[15] The arguments are fully developed in two main monographs, *Machtergreifung* (see ch. 1 note 32 this edition) and *The German Dictatorship* (ch. 2 note 60 this edition).
[16] Karl Dietrich Bracher, 'The Stages of Totalitarian Integration (*Gleichschaltung*)', in Hajo Holborn, ed., *Republic to Reich. The Making of the Nazi Revolution* (New York, 1973), p. 128. This is the Engl. trans. of 'Stufen totalitärer Machtergreifung', *VfZ* 4 (1956), pp. 30–42.
[17] Bracher, 'The Role of Hitler', pp. 193–212.

mated in his own time, and that new patterns of research which reject 'totalitarianism' as a concept and view Nazism instead as a German variant of fascism are in danger of repeating the underestimation. Hitler, in his assessment, was a uniquely German phenomenon: the most radical expression of the ideas of extreme German nationalism and a genuine revolutionary, even if the changes he ultimately wrought were the opposite of those he had intended. Nazism cannot therefore be divorced from the person of Hitler, and, consequently, it is legitimate to call it 'Hitlerism': 'It was indeed Hitler's *Weltanschauung* and nothing else that mattered in the end, as is seen from the terrible consequences of his racist anti-semitism in the planned murder of the Jews'.[18]

This interpretation is advanced in its most uncompromising form in the work of Eberhard Jäckel and Klaus Hildebrand. In Jäckel's opinion, the Nazi regime can be dubbed an *'Alleinherrschaft'* – literally 'sole rule' – which he takes as meaning 'that the essential political decisions were taken by a single individual, in this case by Hitler'.[19] And implicit, if not stated in so many words, is the notion that these decisions followed logically from Hitler's *Weltanschauung*, which Jäckel analysed in a detailed study with the subtitle (in the English version): 'A Blueprint for Power'.[20] Hildebrand, too, though accepting that Nazism cannot be reduced solely to the personality of the Führer, insists upon the absolute centrality of the 'Hitler factor' to the course of development of the Third Reich, especially in the spheres of foreign and race policy, and argues forcefully for the *monocratic* rather than polycratic nature of Nazi rule. For Hildebrand, too, Nazism is ultimately Hitlerism.[21]

The contrasting approach, variously described as 'structuralist', 'functionalist', or (more disparagingly) 'revisionist', offers a fundamentally different interpretation of the Third Reich – concentrating, as the epithets suggest, more on the 'structures' of Nazi rule, the 'functional' nature of policy decisions, and 'revising' what is taken for an unjustifiable over emphasis of the personal role of Hitler in 'orthodox' historiography. In essence, all 'structuralist' interpretations go back to the masterly analyses published by Ernst Fraenkel and Franz Neumann in the 1940s.[22] It was only during the course of the 1960s, however, that the challenge to notions of the 'monolithic' 'totalitarian' State together with the theoretical influence of the newly developing 'structural history' and, derived from political science, of systems analysis, gradually affected writing on the Third Reich.

By the end of the 1960s a number of penetrating studies had laid bare the 'leadership chaos' of Nazi Germany, and established the base of what grew into the notion of 'polycratic' rule – a multidimensional power-structure, in which

---

[18] Bracher, 'The Role of Hitler', p. 201.

[19] Eberhard Jäckel, 'Wie kam Hitler an die Macht?', in Karl Dietrich Erdmann and Hagen Schulze, eds., *Weimar. Selbstpreisgabe einer Demokratie. Eine Bilanz heute* (Düsseldorf, 1980), p. 305. And see Jäckel's *Hitler in History* (Hanover/London, 1987), pp. 28–30.

[20] Eberhard Jäckel, *Hitlers Weltanschauung. Entwurf einer Herrschaft* (Tübingen, 1969), Engl. trans., *Hitler's Weltanschauung. A Blueprint for Power* (Middletown, Connect., 1972).

[21] Of Klaus Hildebrand's numerous essays, see esp. 'Nationalsozialismus oder Hitlerismus?' (see ch. 2 note 62), and 'Monokratie oder Polykratie?' (ch. 1 note 23).

[22] Ernst Fraenkel, *The Dual State* (New York, 1941); Neumann, *Behemoth* (see ch. 2 note 5). For theoretical comments on 'structuralist' approaches, see Jürgen Kocka, 'Struktur und Persönlichkeit als methodologisches Problem der Geschichtswissenschaft', in Bosch (see ch. 2 note 62), pp. 152–69.

Hitler's own authority was only one element (if a very important one).[23] Important works on, for example, the civil service, Party–State relations, the Gauleiter and their provincial power enclaves, the Rosenberg agency, the economy, and policy-implementation at regional and local level (in a study suggestively entitled *The Limits of Hitler's Power*), all helped to revise understanding of how Nazi rule operated in practice.[24]

Unquestionably the outstanding general analysis of the internal structure of the Nazi regime was Martin Broszat's *The Hitler State*, first published in German in 1969.[25] In a strict sense the title was a misnomer, since Broszat broke away from a personality-based, Hitler-centred treatment of Nazism to explore the causal connections between the development of the internal power-structure and the progressive radicalization of the Nazi regime, culminating in European-wide destruction on an unprecedented scale and genocide. In another sense, however, the title was apt. It reflects the antagonisms of a form of absolute leadership which could not be reconciled with the normal practice and organization of government. In Broszat's view – and here he differs from Bracher and others who accept the chaotic governmental structure of the Third Reich as a consequence of Hitler's skilful deployment of a skilful 'divide and rule' strategy – the administrative chaos was not consciously devised, but nor was it pure chance. Rather, it was the inevitable result of the form of authority exerted by Hitler, of his unwillingness and inability to regulate systematically the relationship between Party and State and to create an ordered system of authoritarian government. There was an uneasy 'power-sharing' in the early years of the dictatorship between the conservative 'authoritarian' forces in State and society and the largely negative 'totalitarian' forces of the Nazi mass movement, which having attained power indeed sought to take over as many spheres as possible, but otherwise had no clear ideas what to do with it – apart from attacking the Jews, the Left, and other 'enemies of the State' and minorities which did not fit into the 'national community'. This allowed Hitler's own authority to detach itself from both Party and from State and to develop a wide sphere of autonomy – expressed, however, in haphazard, piecemeal, and inconsistent fashion. The demise of collective, centralized government (the Cabinet never met again after 1938) promoted the disintegration of government into a proliferation of departments with ministries working largely independently of each other. Alongside ministries and Party offices were vital power-bases which crossed Party–State boundaries and derived their authority solely from a mandate of the Führer. The Four Year Plan and especially the SS–Police empire were the most important of these. The overall structure of government was reduced thereby to a shambles of constantly shifting power-bases and warring factions – but a shambles which unleashed immense energy and contained

[23] See esp. Hüttenberger, 'Polykratie' (see ch. 3 note 30). The term seems to have been coined originally by Carl Schmitt, one of the prime legal theorists of the Third Reich, and to have been first employed in a major analysis of Nazi government structure in 1960 by Gerhard Schulz in Bracher *et al.*, *Machtergreifung*, though within the context of a 'totalitarianism' interpretation.
[24] Hans Mommsen, *Beamtentum*; Peter Diehl-Thiele, *Partei und Staat im Dritten Reich* (Munich, 1969); Peter Hüttenberger, *Die Gauleiter* (Stuttgart, 1969); Reinhard Bollmus, *Das Amt Rosenberg und seine Gegner. Studien zum Machtkampf im nationalsozialistischen Herrschaftssystem* (Stuttgart, 1970); Petzina, *Autarkiepolitik* (see ch. 3 note 4); Edward N. Peterson, *The Limits of Hitler's Power* (Princeton, 1969).
[25] See ch. 2 note 39 for full reference.

its own inbuilt destructive momentum. In Broszat's interpretation, the Darwinian rivalry immanent to the system and the ill-co-ordinated attempts of the fractured government machine to 'interpret' the will of the Führer – to bureaucratize charismatic authority and channel vague ideological imperatives into coded law and practices of conduct – led inexorably to an accelerating decline into aggression, lawlessness, and criminal brutality.

Hitler's ideological obsessions were by no means ignored in this analysis. But the emphasis was shifted to the functional pressures within the various, and competing, components of the governmental 'system', aligned to chiliastic but in essence of necessity destructive goals, which could be transmuted into reality through the growing decay and collapse of coherent, 'rational' governmental control and policy planning. This posed a challenge to notions of a planned, consistent, systematic pursuit of clear objectives, which had underlain 'totalitarianism' theories and 'Hitlerism' approaches.[26] Hitler is seen by Broszat as tending more to *sanction* pressures operating from different forces within the regime rather than creating policy: the symbolic Führer authority is more important than the direct governing will of the person Hitler. The fixed points of Hitler's personal *Weltanschauung* served, therefore, largely a functional role.[27] They had so little to do with divisive day-to-day social and political issues that they could be resorted to as 'directions for action' (*Aktionsrichtungen*) and advanced as ultimate, long-range goals. Furthermore, 'Hitler was all the more compelled to keep coming back to them and to keep the Movement going, as other Party ideas of a new order proved illusory'. In this sense, Hitler's fixations with anti-semitism, anti-bolshevism, and with *Lebensraum* might be said to have had, at least in the early years of the Third Reich, a largely symbolic function, serving in Broszat's phrase chiefly as 'ideological metaphors'. In this rather complex argument, Hitler is certainly accorded a vital role in shaping the course of the Third Reich, but not in so simple and straightforward a fashion as the ideological 'intentionalists' would have it.

The most uncompromising exposition of the implications of the 'structuralist' approach for a reassessment of Hitler's position in the power-constellation of the Third Reich has been consistently advanced by Hans Mommsen, in a stream of important essays from the mid 1960s to the present time.[28] Mommsen's interpretation, showing many similarities to that of Broszat though generally expressed in bolder and more combative language, has developed into the dialectical counterpoint of Hildebrand's 'monocratic' 'Hitlerist' line of argument.[29] In a direct clash with Hildebrand in 1976,[30]

---

[26] See Broszat, *Der Staat Hitlers*, p. 9.

[27] The argument is fully expounded in Martin Broszat, 'Soziale Motivation und Führer-Bindung des Nationalsozialismus', *VfZ* 18 (1970), pp. 392–409, here esp. pp. 403–8.

[28] See e.g. Hans Mommsen, *Beamtentum*, esp. pp. 13–19; 'Nationalsozialismus' (see note 4, this chapter), columns 695–702; 'Ausnahmezustand als Herrschaftstechnik des NS-Regimes', in Funke (see ch. 3 note 27), pp. 30–45; 'Nationalsozialismus oder Hitlerismus?', in Bosch (see ch. 2 note 62), pp. 62–7; 'National Socialism: Continuity and Change', in Laqueur (see ch. 2 note 12), pp. 151–92; 'Hitlers Stellung im nationalsozialistischen Herrschaftssystem', in Hirschfeld and Kettenacker (see ch. 1 note 23), pp. 43–72; and his short text for the Deutsches Institut für Fernstudien an der Universität Tübingen, *Adolf Hitler als 'Führer' der Nation* (Tübingen, 1984). Some of his most important and influential essays have now been made available in English translation in Hans Mommsen *From Weimar to Auschwitz* (Oxford, 1991).

[29] Compare their contributions in Hirschfeld and Kettenacker.

[30] In Bosch, pp. 62–71, following Hildebrand's contribution, pp. 55–61.

Mommsen rejected 'personalistic' interpretations of Nazism as raising more questions than they answer and offering a retrospective over-rationalization of Hitler's 'policy'. Rather than operating on the basis of the concrete political calculations and compromise which are the essence of 'normal politics', Hitler's limited number of fanatically held but vague ideological fixations were (in Mommsen's view) incapable of offering a platform for rational decision-making. Hitler remained first and foremost a propagandist, with an eye to the *presentation* of an image and the exploitation of the opportune moment. His ideological statements ought therefore to be seen more as propaganda than as 'firm statements of intent'. Domestic policy is impossible to deduce from Hitler's ideological premises. Such a deduction would be hazardous, too, in the sphere of racial policy, where the 'Final Solution' cannot simply be reduced to the implementation of Hitler's intentions and has to be seen as the product of the complex structure of decision-making processes and the cumulative radicalization of the Third Reich. Even in foreign policy, which Mommsen elsewhere incorporated in his model,[31] there was little or no consistent planning to be seen. Rather than being based upon rational calculation, foreign policy was largely an outward projection of domestic policy – a spiralling radicalization in which the regime lurched from crisis to crisis, burning its boats in a series of *ad hoc* responses to recurrent emergencies, and creating a diminishing sense of reality in the pursuit of extravagant objectives.

Two issues lie at the forefront of Mommsen's concern: the absence of clear planning and consistent direction from Hitler; and the complicity of the German élites in Nazi policy. Both are directly related to the collapse of ordered government into self-destructive, self-generating disintegratory impulses. In a recent, particularly clear statement of his interpretation, Mommsen summed up: 'Hitler's role as a driving force, which with the same inner compulsion drove on to self-destruction, should not be underestimated. On the other hand, it must also be recognized that the Dictator was only the extreme exponent of a chain of antihumanitarian impulses set free by the lapse of all institutional, legal, and moral barriers, and, once set in motion, regenerating themselves in magnified form'. Moreover, since Hitler was by no means always the protagonist of the most radical solution – for example, in Church or economic policy where there was a danger of provoking unrest – it is far too easy 'to emphasize as the final cause of the criminal climax and terroristic hubris of National Socialist policy the determining influence of Hitler'. And if the most horrific crimes cannot be explained solely or even largely with reference to Hitler's personality, ideology, and will, then the role and complicity of the dominant élites that helped Hitler into power and sustained him when there, co-operating in and benefiting in good measure from the Nazi 'restoration of social order', must be the subject of special concern. Historical assessment of the Third Reich cannot, therefore, be reduced to the uniqueness of 'the Hitler phenomenon', but must instead tackle the more difficult but still relevant problems of the conditions and structures which allowed such barbarity to emerge

[31] See Hans Mommsen, 'Hitlers Stellung', pp. 57–61, 69–70; 'Ausnahmezustand', p. 45; 'National Socialism: Continuity and Change', pp. 177–9; his review of Hans-Adolf Jacobsen, *Nationalsozialistische Außenpolitik* (Frankfurt am Main/Berlin, 1968), *MGM* (1970), Heft 1, pp. 180–5; and *Adolf Hitler*, pp. 91–109.

and expand in a civilized and sophisticated industrial society.[32] The implications for wider interpretations and their attendant historical philosophies, and their underlying political standpoints, are clear.

What has come to be labelled the 'intentionalist' approach – i.e. deducing the development of the Third Reich from Hitler's ideological intentions – has an immediate and obvious appeal. Seldom has a politician stuck with such fanatical consistency to an ideological fixation as Hitler appears to have done in the period extending from his entry into politics to his suicide in the bunker. That the quest for *Lebensraum* and the extermination of the Jews, far from remaining the wild ravings of a lunatic-fringe beer-hall rabble-rouser, became horrific reality and were implemented as government policy by a regime led by Hitler, seems to point conclusively to the validity of the 'intentionalist' argument. However, for all its superficial attractiveness, such an argument contains a number of potentially serious flaws, as Tim Mason has pointed out. Methodologically, argued Mason, a concentration on Hitler's intentions short-circuits all fundamental questions of the character of social, economic, and political agencies of change. Underlying the approach is the dubious assumption that historical development can be explained by recourse to intuitive understanding of the motives and intentions of leading actors in the drama. Subsequent events are then rationalized in necessarily teleological fashion by their relation to such intentions, which function, therefore, both as cause and as sufficient explanation. In addition, there are major problems – simply in terms of the availability and quality of the sources – in attempting to reconstruct Hitler's reasons for decisions and the processes which led to decisions being made. The evidence is not always plain and consistent, and can be read in different ways. The 'Hitlerist' case has to be demonstrated, not merely asserted. Even its moral implications are not altogether clear. Since Hitler was by definition unique and unrepeatable, and his actions and intentions both a premise and a conclusion, whatever moral warning might be drawn from a study of Nazism is limited in its application.[33]

The 'structuralist' argument seems inherently more difficult to express, as the convoluted language sometimes employed by its exponents appears to betray. Notions of Hitler as weak and indecisive, of antisemitism and *Lebensraum* as 'ideological metaphors', of Nazism being bent on upholding rather than revolutionizing the social order, and of foreign policy as a device of domestic policy, do not carry instant conviction. There appears to be some strength in the argument that the 'structuralists' might have a point in the realm of domestic policy, where Hitler showed little active interest, but that in anti-Jewish and foreign policy it is a different story. And that, rather than collapsing under the weight of its own internal contradictions, administrative chaos, and self-destructive dynamic, Nazi Germany was only defeated by the assembled might of the Allies, also seems to speak against the 'structuralist' argument. Finally, the counter-factual rhetorical question of what the course of the German government might have been without Hitler in charge appears to clinch the case for stressing rather than de-emphasizing Hitler's importance.

[32] Hans Mommsen, 'Hitlers Stellung', pp. 66–7, 71.
[33] Mason, 'Intention and Explanation' (see ch. 1 note 25), pp. 29–35.

The 'structuralists' do not, however, ignore or play down Hitler's importance. They merely seek to locate this importance within the framework of
numerous additional pressures built into the governmental system. They start
from the premise that the processes of cumulative and progressive radicalization in the Third Reich were so complex in themselves that it would be
impossible to explain them without widening the focus away from Hitler's
personality and ideology, and without considering the Führer less in personality terms than in his functional role within a multi-dimensional (polycratic)
system of rule. The 'structuralist' argument is less easily disposed of than the
'intentionalists' often claim. A full assessment of these polarized interpretations must, however, extend over three inter-related but separate areas: the
character of Hitler's rule and the internal power structure of the Nazi State;
the implementation of anti-Jewish policy, in particular the process of decision-
making which initiated the 'Final Solution'; and the regime's foreign policy
and expansionist ambitions. Central to all three areas is the question of how
decisions were reached in the Third Reich. The last two areas, lying at the heart
of Hitler's *Weltanschauung*, are dealt with in subsequent chapters. The first
area forms the subject of the evaluation which follows.

Hitler's Power: an Evaluation

An examination of Hitler's power, whether he is to be seen as 'master in the
Third Reich' or 'a weak dictator', must begin with some conception of what,
potentially, might comprise his 'strength' and 'weakness' within the overall
power constellation in the Third Reich. At least three categories of possible
weakness appear to be distinguishable:

(i) It might be argued that Hitler was 'weak' in the sense that he regularly
shirked making decisions, and was compelled to do so in order to protect his
own image and prestige, dependent upon the Führer remaining outside factional politics and unassociated with mistaken or unpopular decisions. This
would mean that the chaotic centrifugal tendencies in the Third Reich were
'structurally' conditioned and not simply or mainly a consequence of Hitler's
ideological or personal predilections, or of a machiavellian 'divide and rule'
strategy.

(ii) Hitler could be regarded as 'weak' if it could be shown that his decisions
were ignored, watered-down, or otherwise not properly implemented by his
subordinates.

(iii) It might be claimed that Hitler was 'weak' in that his scope for action, his
manoeuvrability, was preconditioned and limited by factors outside his control
but immanent to the 'system', such as the demands of the economy or fear of
social unrest.

The following analysis attempts to relate these categories to an assessment of
Hitler's rule and the internal power structure of the Third Reich.

Historians are in no fundamental disagreement over the fact that the government of Nazi Germany was chaotic in structure. It is, of course, easy to exag-

gerate the 'ordered' character of any modern governmental system. However, it seems clear that the fragmentation and lack of co-ordination in the internal administration of the Third Reich existed to such an extreme degree that the overlapping, conflicting, and sometimes outrightly contradictory spheres of authority can be aptly depicted as 'chaotic'. The question is, what significance should be attached to this 'chaos'?

The 'intentionalist' type of approach sees in the confused lines of authority in the Third Reich a reflection of a calculated policy of 'divide and rule', testimony therefore of Hitler's pivotal role, his real power, and his preconceived planning of the take-over, consolidation, and wielding of total power with a view to carrying out his long-term objectives.[34] The opposed 'structuralist' line of interpretation regards the fragmented machinery of government rather as the inevitable product of Hitler's 'charismatic' form of leadership. This preconditioned rejection of the institutional and bureaucratic norms necessary for the 'rational' government of a modern state in favour of dependence on personal loyalty as the basis of authority – a transmission of the ethos of the Nazi Party since its early days to the task of running a sophisticated, modern government machine.[35] 'Charismatic' leadership also predetermined an essentially propagandistic preoccupation with avoiding any harmful inroads into the prestige and image of the Führer, hence the need to refrain from interference in internal conflicts and to remain aloof from day-to-day decision-making and association with possibly unpopular policy options.[36] In contrast to conceptions of a 'monocratic' dictatorship relentlessly pursuing its fixed goals with remorseless zeal and energy, this interpretation emphasizes the lack of efficiency, fragmentation of decision-making, absence of clear, rational 'middle-range' policies, and diminishing sense of reality – all promoting the immanent instability of the political system, the inevitable selection of negative goals, and cumulative radicalization.[37] Hitler's personal scope for action was limited, moreover, by the continued existence of other, real – if fluctuating – centres of power.[38]

Evidence of a machiavellian 'divide and rule' strategy – a claim which Hitler's former press chief Otto Dietrich made in his post-war memoirs[39] – is usually found in the deliberate blurring of lines of command and creation of a duplication or triplication of office. An example is the way in which Hitler

[34] See e.g. Bracher, 'Stages', and Diehl-Thiele, p. ix (where he speaks of a 'permanent improvisation within the framework of a principled divide-and-rule tactic').
[35] For the impact on the NSDAP of Hitler's 'charismatic' leadership, see Joseph Nyomarkay, *Charisma and Factionalism within the Nazi Party* (Minneapolis, 1967), and Wolfgang Horn, *Führerideologie und Parteiorganisation in der NSDAP (1919–1933)* (Düsseldorf, 1972).
[36] I try to tackle the making and impact of Hitler's popular image in my study *The Hitler Myth'. Image and Reality in the Third Reich* (Oxford, 1987), and more briefly in 'The Führer Image and Political Integration: The Popular Conception of Hitler in Bavaria during the Third Reich', in Hirschfeld and Kettenacker, pp. 133–63.
[37] See Hans Mommsen, 'National Socialism: Continuity and Change', pp. 176–8; Broszat, 'Soziale Motivation'; and the valuable essay of Jane Caplan, 'Bureaucracy, Politics, and the National Socialist State', in Peter D. Stachura, ed., *The Shaping of the Nazi State* (London, 1978), pp. 234–56.
[38] See Hüttenberger, 'Polykratie', and this edition ch. 3, for shifts within the 'power-constellation' of the Third Reich.
[39] Otto Dietrich, *Zwölf Jahre mit Hitler* (Cologne/Munich, 1955), pp. 129 ff. See Noakes and Pridham, vol. 2, p. 205.

broke up the unified control over the Party's organization which Gregor Strasser had built up. Following Strasser's resignation in December 1932, Hitler himself took over the formal leadership of the Party's 'Political Organization', strengthened the position of the Gauleiter at the expense of the Reich Leadership, and divided power at the centre between Robert Ley, who eventually adopted Strasser's old title as 'Reich Organization Leader' but with diminished power, and Rudolf Heß, given the title of 'Deputy Führer' in April 1933, with the right to decide in Hitler's name in all questions relating to the Party leadership.[40]

Another example is the refusal of Hitler to back the attempts of Wilhelm Frick, Reich Minister of the Interior, to instigate a rational system of centralized state control through far-reaching plans for 'Reich Reform'. In the early years of the Third Reich, Frick struggled to establish authority over the Reich Governors, most of whom were at the same time Gauleiter of the Party. The Reich Governors had been put in as Reich delegates in the Länder in April 1933, bearing a loose mandate to ensure the execution of the Reich Chancellor's policy through the Länder governments.[41] By January 1934 it looked as if Frick was on the way to success. The 'Law for the Reconstruction of the Reich', signed by Hitler, placed the Reich Governors under the administrative supervision of the Reich Minister of the Interior. (In abolishing Länder sovereignty, the law technically did away with the logic of having Reich Governors at all, but, typically, they remained in existence.) Following massive protests by the Reich Governors about their right of appeal to the Führer, Frick had to be content with a gloss by Hitler which in practice completely undermined Frick's authority. It was now stated that, although generally subordinated to Frick, 'an exception must be made for those cases which are concerned with questions of special political importance. In the view of the Reich Chancellor, such a regulation is consistent with his position of leadership'.[42] Frick's patiently devised schemes for Reich reform, aimed at introducing a centralized and rational system of authority, based on a Reich Constitution instead of the Enabling Act, went much the same way and were finally abandoned in the middle of the war, as were plans to introduce a senate to assist the Führer and to elect his successor.[43]

Whether one can read into these and other examples a systematic 'divide and rule' strategy is debatable. Hitler, in fact, promoted the construction of some huge power-bases. In the example mentioned above, Robert Ley was given control over the mammoth Labour Front to add to his authority over questions of party organization. But even this mini-empire was insignificant compared with the massive accretions of power which came to Göring and Himmler, with Hitler's active support. Nor was there much sign of anxiety on Hitler's part about Martin Bormann's accumulation of power in the war years. And the greatest threat to Hitler in the early phase of the Dictatorship, Ernst Röhm and the SA leadership, was eliminated only after Hitler had bowed to intense pres-

---

[40] Diehl-Thiele, pp. 204–6.
[41] 'Zweites Gesetz zur Gleichschaltung der Länder mit dem Reich,' *Reichsgesetzblatt* (1933/I, p. 173; and see Diehl-Thiele, pp. 37–60.
[42] Martin Broszat, *Der Staat Hitlers* (Munich, 1969), p. 153; see Diehl-Thiele, pp. 61–73.
[43] See Peterson, pp. 102–25; Broszat, *Hitler State*, pp. 286–8; and Hans Mommsen, 'National Socialism: Continuity and Change', p. 169.

sure from the army and had been pushed into it by Göring and Himmler.

What does seem clear is that Hitler was hypersensitive towards any attempt to impose the slightest institutional or legal restriction upon his authority, which had to be completely untrammelled, theoretically absolute, and contained within his own person. 'Constitutional Law in the Third Reich', stated Hans Frank, head of the Nazi Lawyers Association, in 1938, 'is the legal formulation of the historic will of the Führer, but the historic will of the Führer is not the fulfilment of legal preconditions for his activity'.[44] Hitler was correspondingly distrustful of all forms of institutional loyalty and authority – of army officers, civil servants, lawyers and judges, of Church leaders, and of cabinet ministers (whom he was unwilling to see even gathering informally between the increasingly infrequent cabinet sessions).[45]

The corollary of Hitler's extreme distrust of institutional bonds was his reliance on personal loyalty as the principle of government and administration. He appears to have had no consuming distrust of power-bases deriving from his own Führer-authority and held by his own chosen paladins – hence, his ultimate despair in the bunker at the final stab-in-the-back by Himmler, his 'loyal Heinrich'.[46] The appeal to personal loyalty had been Hitler's hallmark, especially in moments of crisis, since the early years of the Party.[47] The loyalty principle, a feature of *Party* management before 1933 in binding leaders as well as ordinary members to the person of the Führer, was carried over after 1933 into the practice of governing the Reich. In this sense, Robert Koehl's depiction of the Third Reich as less a totalitarian state than a neo-feudal empire has some meaning as an analogy.[48] In fact, however, the bonds of personal loyalty – a pure element of 'charismatic' rule – did not replace but were rather superimposed upon complex bureaucratic structures. The result was not complete destruction as much as parasitic corrosion. The avoidance of institutional restraints and the free rein given to the power ambitions of loyal paladins offered clear potential for the unfolding of dynamic, but unchannelled, energies – energies, moreover, which were inevitably destructive of rational government order.

As numerous studies have shown, the bonds of loyalty between Hitler and the Gauleiter, his trusted regional chieftains, vitiated any semblance of ordered government in the provinces.[49] Hitler invariably sided with his Gauleiter (or, rather, with the strongest Gauleiter) in any dispute with central authority or government ministries, protecting their interests and at the same time

[44] Hans Frank, *Im Angesicht des Galgens* (Munich-Gräfelfing, 1953), pp. 466–7.
[45] Lothar Gruchmann, 'Die "Reichsregierung" im Führerstaat. Stellung und Funktion des Kabinetts im nationalsozialistischen Herrschaftssystem', in G. Doecker and W. Steffani, *Klassenjustiz und Pluralismus. Festschrift für Ernst Fraenkel zum 75. Geburtstag* (Hannover, 1973), p. 202. For Hitler's attitude towards the 'establishment' élites, see Michael Kater, 'Hitler in a Social Context', *CEH* 14 (1981), pp. 251 ff.
[46] H.R. Trevor-Roper, *The Last Days of Hitler* (Pan Books, London, 1972 edn.), p. 202.
[47] See Nyomarkay, Horn (see note 35, this chapter) and Dietrich Orlow, *The History of the Nazi Party. Vol. 1: 1919–1933* (Newton Abbot, 1971).
[48] Robert Koehl, 'Feudal Aspects of National Socialism', in Turner, *Nazism and the Third Reich* (see ch. 3 note 6), pp. 151–74.
[49] E.g. Diehl-Thiele; Hüttenberger, *Gauleiter*; Peterson; Jochen Klenner, *Verhältnis von Partei und Staat 1933–1945. Dargestellt am Beispiel Bayerns* (Munich, 1974); and Jeremy Noakes, 'Oberbürgermeister and Gauleiter. City Government between Party and State', in Hirschfeld and Kettenacker, pp. 194–227, esp. pp. 207 ff.

securing himself a powerful body of support, loyal to him and to no one else. In Rauschning's judgement Hitler 'never ran counter to the opinion of his Gauleiter. . . . Each one of these men was in his power, but together they held him in theirs. . . . They resisted with robust unanimity every attempt to set limits to their rights of sovereignty. Hitler was at all times dependent on them – and not on them alone'.[50] As we saw, Frick's attempts to gain control over the Reich Governors foundered on Hitler's support for the objections of the Gauleiter. The mighty Himmler encountered the same problem in his dealings with the Gauleiter after he had been made Reich Minister of the Interior in 1943.[51]

At the level of central government, too, Hitler's ideological predisposition to let rivals fight it out and then side with the winner – an instinctive application of social darwinistic precepts – together with his ready recourse in a crisis to the establishment of new agencies, bypassing or cutting through existing institutions, with plenipotentiary powers directly commissioned by the Führer and dependent on his authority alone, militated strongly against the setting of rational policy priorities. The consequence was the inevitable disintegration of central government – reflected by the increasing infrequency of cabinet meetings down to their complete cessation in early 1938 – and the dissolution of government into a multiplicity of competing and non-co-ordinated ministries, party offices, and hybrid agencies all claiming to interpret the Führer's will. Hand in hand with this development went the growing autonomy of the Führer-authority itself, detaching itself and isolating itself from any framework of corporate government and correspondingly subject to increasing delusions of grandeur and diminishing sense of reality.[52]

The chaotic nature of government in the Third Reich was also markedly enhanced by Hitler's non-bureaucratic and idiosyncratic style of rule. His eccentric 'working' hours, his aversion to putting anything down on paper, his lengthy absences from Berlin, his inaccessibility even for important ministers, his impatience with the complexities of intricate problems, and his tendency to seize impulsively upon random strands of information or half-baked judgements from cronies and court favourites – all meant that ordered government in any conventional understanding of the term was a complete impossibility. 'Ministerial skill', it was pointed out after the war, 'consisted in making the most of a favourable hour or minute when Hitler made a decision, this often taking the form of a remark thrown out casually, which then went its way as an "Order of the Führer" '.[53]

It would be misleading to conclude from this comment that, however eccentrically arrived at, a steady stream of decisions flowed downwards from Hitler's lofty pinnacle. Rather, he was frequently reluctant to decide in domestic affairs and generally unwilling to resolve disputes by coming down on one side or the

[50] Cited in Peterson, p. 7, and see also pp. 14–15, 18–19. Rauschning was, however, spoiling a good point by exaggerating when he added that the secret of Hitler's leadership 'lay in knowing in advance what course the majority of his Gauleiter would decide upon and in being the first to declare for that course': Hermann Rauschning, *Hitler Speaks* (London, 1939), pp. 214–15.

[51] Diehl-Thiele, pp. 197–200 (and note 70).

[52] See Broszat, *Hitler State*, chs. 8–9.

[53] Ernst von Weizsäcker, *Erinnerungen* (Munich etc., 1950), pp. 201–2. See also Fritz Wiedemann, *Der Mann, der Feldherr werden wollte* (Velbert, 1964), pp. 68 ff.

other, much preferring parties to a dispute to sort it out themselves.[54] It would be too simple to attribute this, and the governmental disorder in the Third Reich in general, solely to Hitler's personal quirks and eccentric style. Certainly, he was languid, lethargic, and uninterested in what he regarded as trivial matters of administrative detail beneath his level of concern. But it does seem clear that the protection of his own position and prestige was an important factor in predetermining his unwillingness to intervene in problem areas and to let things ride as long as possible – by which time a solution almost invariably suggested itself, and the contours of support had been clarified, the opposition (if any) already isolated. Thus, cabinet meetings in the early years of the Dictatorship were in no sense a forum for genuine debate preceding a policy decision. Hitler hated chairing the meetings in which he might potentially be forced into retreat on a given issue. Consequently, he 'reserved the right to decide when a difference of opinion could be brought before the cabinet. This way it came less and less to discussion. Each minister presented his draft, on which agreement had already been reached and Lammers [the head of the Reich Chancellory] recorded that all were agreed'.[55] Even so, the cabinet meetings were allowed to atrophy into non-existence. As regards legislation, this remained the usual procedure: draft laws were circulated to all ministers concerned, difficulties and disputes ironed out, and Hitler's sanction given only after all parties concerned had already resolved their differences. In 1943 Bormann reiterated the procedure that 'all orders and decrees must be given to all involved before their declaration; the Führer is to be approached only after *all* involved have taken a clear position'.[56] Effectively, it was the transfer to the complex business of state administration of the Party's basic 'principle of letting things develop until the strongest has won' – hardly a foundation for 'rational' decision-making. In any case, already by the mid 1930s influence on important affairs of state had passed to the shifting personnel of Hitler's most trusted cronies and government ministers were left to read in the press about what had taken place.[57]

Distant rather than immediate leadership in everyday affairs, and hesitancy about deciding before the situation had all but resolved itself were not simply a reflection of Hitler's style of rule, but were necessary components of his 'charismatic' Führer-authority, helping to maintain both in the ruling circle and among the people themselves the myth of Hitler's unerringly correct judgement and his independence from factional disputes – from 'normal politics'. The soaring popularity of Hitler, contrasted with the massive unpopularity of the Party and of so many aspects of the daily experience of Nazism, can only be attributed to the image of a Führer who seemed to stand aloof from political infighting and the grey daily reality of the Third Reich.[58] To an extent, Hitler

[54] See Peterson, pp. 4 ff.

[55] Lutz Graf Schwerin von Krosigk, *Es geschah in Deutschland* (Tübingen/Stuttgart, 1951), p. 203. See also Peterson, p. 31.

[56] Cited in Peterson, p. 39.

[57] Krosigk, p. 203; Gruchmann, pp. 193–4. An important reassessment of Hitler's role in decision-making during the war is now provided by Dieter Rebentisch's study, *Führerstaat und Verwaltung in Zweiten Weltkrieg* (Stuttgart, 1989).

[58] I attempt to argue this in my essay 'Alltägliches und Außeralltägliches: ihre Bedeutung für die Volksmeinung', in Detlev Peukert and Jürgen Reulecke, eds., *Die Reihen fast geschlossen Beiträge zur Geschichte des Alltags unterm Nationalsozialismus* (Wuppertal, 1981), pp. 273–92, esp. pp. 285 ff.

had to live up to his image. This, too, conditioned a leadership style of aloof-ness, non-interference, 'moderation' in sensitive areas (such as the 'Church struggle'), and tendency always to side with 'the big battalions'.[59] The need to produce ever greater feats of achievement to bind the masses closer to him and to prevent a sagging of the regime's 'vitality' into stagnation, disenchantment, and likely collapse was one further weighty factor which militated against the establishment of a 'state of normalcy' in the Third Reich, promoting instead the radical but essentially negative dynamism which had formed the basis of the social integration of the Nazi movement, but which could hardly end but in destruction.

The governmental chaos of the Third Reich seems better explained if the notion of a systematic strategy of 'divide and rule' is left aside – even though Hitler's conscious protection of his authority against any potential attempt to limit it institutionally is evident. Though the chaotic structure of government was for the most part not a deliberate creation, it would seem unsatisfactory evidence for the view that Hitler was 'in some respects a weak dictator'.[60] Indeed, the notion of 'weakness' seems misplaced in this case. If Hitler had *wanted* a different governmental structure, but been prevented from attaining it; or if he had *intended* to make decisions, but found himself unable to do so, then there would have been some conflict between 'intention' and 'structure' and it might have been possible to conclude that Hitler was 'weak'. Since there is no evidence for either point, but rather every indication that Hitler was con-tent, indeed wanted, to keep out of wrangles among his subordinates, had little interest in participating in the legislative process – especially in areas of peripheral concern – except where his own authority was directly invoked, and actively furthered rather than tried to hinder the government chaos on occa-sion, then one would have to accept that was no incompatibility in this area bet-ween 'intention' and 'structure', and thus reject the conclusion that, because of the 'structural' restrictions imposed on his dictatorship, Hitler was 'weak'.

Our second criterion of weakness was whether Hitler made decisions which were then ignored, by-passed, or inadequately implemented by subordinates.

Certainly, Hitler's inclination towards often impulsive, verbal agreement to proposals casually presented to him at opportune moments by his underlings, who thereafter interpreted his spontaneous remarks as sufficient sanction and an 'unalterable decision', did lead to occasional embarrassment. One such instance occurred in October 1934, when Labour Front boss Robert Ley per-suaded Hitler to sign a decree enhancing the authority of the Labour Front at the expense of both employers and the Trustees of Labour. Ley had not taken his proposal to either the Ministry of Labour or the Ministry of Economics, and both, together with Hess on behalf of the Party, protested so vehemently that Hitler – unwilling to antagonize Schacht and the industrial leadership – was obliged to yield to the pressure. Characteristically, the decree was not revoked – which would have been a slight on the Führer's prestige – but remained a dead letter, simply ignored by all with Hitler's tacit approval, even though Ley con-tinued to refer to it in an attempt to extend his own power.[61] Difficulties were

---

[59] Peterson, p. 7.
[60] See note 4 this chapter.
[61] The decree is printed in Walther Hofer, ed., *Der Nationalsozialismus. Dokumente 1933–1945* (Frankfurt am Main, 1982), p. 87.

also caused in early 1935 by Hitler's agreement to a proposal by Labour Minister Seldte to a unified wage structure for building workers (replacing the regionally weighted system in operation). The objections of the Gauleiter – most prominently of Gauleiter Kaufmann of Hamburg – about the effects of the necessary reduction of wages in certain areas on worker morale, carried weight with Hitler, however, and he typically ordered a further indefinite period of deliberation before the proposed wage revision should come into force – meaning that the matter was shelved and forgotten.[62] Examples can also be found, especially in the early years of Nazi rule, where Hitler had to bow to economic pressure, and where unpalatable decisions were forced upon him – such as in 1933, when he had to accept the provision of financial support for ailing Jewish department stores in order to prevent staff redundancies adding to the total unemployed.[63] On the other hand, in one of the few divisive issues to come before the Reich Cabinet, and one on which he himself felt strongly, Hitler pushed through the Sterilization Law of July 1933 despite the voiced objections of Vice Chancellor Von Papen (who on this occasion was articulating the views of the Catholic lobby).[64]

It would be rash to claim, on the evidence for the implementation of Führer directives, that Hitler was a 'weak dictator'. The 'limits of Hitler's power' which Peterson advanced are arguably 'limits' only when juxtaposed with a wholly idealistic notion of 'total power'. Moreover, Peterson provided no convincing example of a directive held by Hitler to be of central importance that was ignored or blocked by his subordinates or others. More important for the workings of Nazi government than whether Hitler can be regarded as a 'strong' or 'weak' dictator, is the fact that he produced so *few* directives in the sphere of domestic politics. It becomes difficult, therefore, to establish precisely what his aims were in the domestic sphere, other than the elimination of 'enemies of the State' and the psychological as well as material mobilization for the war which he felt was inevitable within a very short time.[65] This aim was compatible with social developments diametrically opposite to those which had been preached by Nazi ideologues.[66] It is nevertheless in the area of mobilization of the German people for war, the central task of domestic policy, that it has been claimed that Hitler's real 'weakness' was to be found.

Tim Mason, above all, argued that Hitler's scope for action – especially in the central period of the Third Reich between 1936 and 1941 – was seriously restricted by tensions built into the Nazi economy and not subject to control by the Führer's 'will' or 'intention'.[67] The key determinant, in his view, of the

[62] BAK, R43II/541, fos. 36–95 and R43II/552, fos. 25–50. See also Timothy W. Mason, *Sozialpolitik im Dritten Reich* (Opladen, 1977), pp. 158–9.
[63] Peterson, p. 48.
[64] Gruchmann, p. 191.
[65] See Peterson, p. 432. Rebentisch's recent study (see note 57 this chapter) has shown that Hitler's involvement in domestic issues during the war was greater than had been assumed. His sporadic interventions, usually prompted by his subordinates or even on occasion by articles in the press, scarcely amounted, however, to a coherent set of directives for a clear formulation of policy.
[66] See Schoenbaum (see ch. 2 note 26), p. 285.
[67] Timothy W. Mason, 'The Legacy of 1918 for National Socialism', in Anthony Nicholls and Erich Matthias, eds., *German Democracy and the Triumph of Hitler* (London, 1971), pp. 215–39; *Sozialpolitik*, esp. chs. 1 and 6; 'Innere Krise und Angriffskrieg' (see ch. 3 note 47); and 'Labour in the Third Reich', *Past and Present* 33 (1966), pp. 112–41.

Nazi leadership's thought and action in the domestic sphere was the lesson drawn from the Revolution of 1918 of the dangers of working-class unrest. Hitler in particular was extraordinarily sensitive towards discontent among workers, aware that psychological motivation alone was extremely short-lived, and consequently that material sacrifices must be kept to a minimum. Hence, according to Mason, the Third Reich amounted to a huge social imperialist gamble, in which the material satisfaction of the masses could only be brought about through successful foreign expansion, but where the accomplishment of that expansion was significantly impaired by the unwillingness of the regime to impose even short-term reductions in living standards necessary for the effective functioning of an armaments-led economy. As a consequence, the regime developed no consistent social policy and was in essentially a weak position when faced with the logic of the economic class struggle and the need to square the circle of paying for armaments without drastic reductions in consumer spending. Hitler's own role was one of increasing helpless apathy and inactivity, a product of 'anxious insecurity' and growing pessimism. Strength of will alone, argued Mason, could not suffice to combat the class antagonism. The industrial opposition of German workers, exploiting their bargaining position in a time of acute labour shortage even without the help of trade unions, assisted in promoting a major economic crisis, which developed into a general crisis for the regime and necessarily affected the timing of war, determining that on economic grounds – and to preserve the social peace and protect the regime's own threatened position – war had to come sooner rather than later. Moreover, the war itself had to be conducted without major sacrifices of a material kind by the German people. Hence, mobilization of the population was half-hearted and incomplete (compared, for example, with Britain), and production for the war economy was hampered.[68] The weakness of the regime, therefore, went to the very heart of its ethos – war – and limited its potential to such an extent that it could be argued that the regime's destruction was not simply a matter of external defeat, but was implicit in its essence – was 'structurally determined' by its internal contradictions.

There is no shortage of evidence to illustrate Hitler's acute sensitivity towards any sign of threat to 'social peace'. Speer recorded in his memoirs Hitler's private admissions of anxiety about loss of popularity giving rise to domestic crises.[69] Worries about the social unrest which might ensue from rapidly rising prices in 1934 prompted Hitler to restore the office of Reich Commissar for Price Surveillance and to maintain it purely for propaganda purposes long after its head, Carl Goerdeler, had requested its dissolution on the grounds that there was nothing effective for it to do.[70] During mounting consumer problems and worrying reports of growing tension in industrial areas in 1935-6, Hitler was even prepared – temporarily – to forego imports for armament production in

---

[68] See Timothy W. Mason, *Arbeiterklasse und Volksgemeinschaft* (Opladen, 1975), esp. ch. 21, and also Milward, *German Economy at War* (above ch. 3 note 5). The counter-position is that of Overy (refs. in ch. 3 notes 47 and 49). For labour and the war economy see the rather different interpretations of Wolfgang Werner, *'Bleib übrig!' Deutsche Arbeiter in der nationalsozialistischen Kriegswirtschaft* (Düsseldorf, 1983), and Stephen Salter, *The Mobilization of German Labour* (unpubl. D. Phil. thesis, Oxford, 1983).

[69] Albert Speer, *Erinnerungen* (Frankfurt am Main, 1969), p. 229.

[70] See BAK, R43II/315a, esp. fols. 188–240.

order to prevent the socially undesirable consequences of food rationing.[71] In 1938, despite desperate pleas from the Ministry of Food and Agriculture, Hitler categorically refused to raise food prices because of the damaging effect on living standards and worker morale.[72] In the first months of the war, the regime retreated on its plans for labour mobilization in the wake of worker protest at the impact on wages, working conditions, and living standards.[73] And the unwillingness of the regime to push through the comprehensive mobilization of women for the war effort probably has to be located not simply in Hitler's views on the role of women but in Nazi fears of the possible repercussions on morale and work discipline.[74]

The far-reaching conclusions which Mason draws from such evidence about the 'weakness' of Hitler and the regime have, however, been subjected to searching criticism from quite different directions, and Mason's overall thesis has found little general acceptance. It has been argued, for instance, that, whatever objective problems existed in the economy in 1938–9, the Nazi leadership – and in particular Hitler – exhibited no *consciousness* of a general political crisis of the system forcing a need for imminent war as the only way out.[75] In addition, it might be claimed that Mason has exaggerated the political significance and even the scale of industrial unrest, labelling as worker opposition to the system what was not specific to Nazism, but a feature (as in England during the war) of capitalist economies in periods of full employment.[76] The interpretation of a political crisis of the Nazi system in 1938–9 provoked by industrial opposition is, on these grounds, therefore highly dubious. As regards the timing of the war, it has been forcefully argued that, important though the domestic situation was, the decisive factor was the international balance of power and in particular the comparative armaments position of Germany's rival powers. The compulsion to act was not, therefore, conditioned by fear of internal unrest, but by the state of the arms race which Germany had unleashed.[77] While West German critics maintained that Mason underrated Hitler's 'politically autonomous aims', and that Hitler's 'decisions for war arose from political motives alone',[78] GDR historians claimed that in underestimating the aggressive imperialist aims, intentions, and policies of monopoly capital, Mason was elevating Hitler to the level of 'the only decisive acting force'.[79] Both sets of critics shared, therefore, from wholly opposed perspectives, the unease that Mason's attribution of weakness to Hitler and the Nazi regime leads to an interpretation in which the *intentions* of the regime are underplayed and it is mistakenly seen as stumbling into war from a position of weakness and without clear direction.[80]

[71] BAK, Zsg 101/28, fol. 331 ('Vertrauliche Informationen' for the press, 7 Nov. 1935).
[72] BAK, R43II/194, fol. 103.
[73] Mason, *Arbeiterklasse*, ch. 21; *Sozialpolitik*, pp. 295 ff.
[74] See Dörte Winkler, *Frauenarbeit im Dritten Reich* (Hamburg, 1977); Tim Mason, 'Women in Nazi Germany', *HWJ* 1 (spring, 1976), pp. 74–113 and esp. *HWJ* 2 (autumn, 1976), pp. 5–32; and Salter, *Mobilization*.
[75] See Herbst (see ch. 3 note 47).
[76] See Winkler, 'Vom Mythos der Volksgemeinschaft' (see ch. 3 note 47).
[77] See Dülffer, 'Der Beginn des Krieges' (see ch. 3 note 47).
[78] Hilderand, *Das Dritte Reich*, p. 159.
[79] Lotte Zumpe, review of Mason, *Arbeiterklasse*, in *Jahrbuch für Wirtschaftsgeschichte* (1979), Heft 4, p. 175.
[80] From the GDR side, the point was forcefully made by Kurt Gossweiler in a review of Mason, *Arbeiterklasse*, in *Deutsche Literaturzeitung* 99 (1978), Heft 7/8, p. 538.

These are weighty criticisms, even if at times they appear somewhat to distort the claims of Mason, who, for example, stressed that the primary *cause* of the war must be sought in the racial and anti-communist aims of the Nazi leadership and the economic imperialism of German industry, not in the crisis of the Nazi system.[81] They point, however, to the need to look for a synthesis of 'intention' and 'structure', rather than seeing them as polarized opposites. It seems, indeed, clear that Hitler's intentions and the socio-economic 'structural determinants' of Nazi rule were not antagonistic poles, pushing in opposite directions, but acted in a dialectical relationship which pushed in the same direction. Consequently, it is as good as impossible to separate as a causal factor 'intention' from the impersonal conditions which shape the framework within which intentions can become 'operational'. At the same time, it seems important to recognize that an 'intention' is not an autonomous force, but is affected in its implementation by circumstances which it may itself have been instrumental in creating, but which have developed a momentum of their own. In the present case, Hitler and the Nazi leadership (actively supported by prominent sections of the economic and military élites) unquestionably intended to wage the war which, in their view, would solve Germany's problems. But the war only gradually adopted concrete shape and form, and then by no means wholly in the way Hitler had envisaged it. As late as autumn 1935, Hitler's directive to ministers and army leaders, according to Goebbels's report, was as vague as: 're-arm and get ready. Europe is on the move again. If we're clever, we'll be the winners'.[82] The absolute priority accorded to rearmament, a *political* decision made at the very outset of the Third Reich, was at the root of the unresolvable tension in the economy between provision for armaments production and consumption. From 1936 onwards the die was cast and there could be no retreat if the regime were to survive. The course was set, and, despite preparations for a long war expected to commence around the mid 1940s, in practice for the only possible sort of war Germany could fight – a Blitzkrieg – in the nearer rather than more distant future. The economic problems intensified rapidly and enormously in 1937–9. Hitler could do little about them, though the impression to be gleaned from the sources is that he had little interest in doing anything, and fatalistically regarded the situation as only soluble after final victory in the war which he had always forecast as inevitable. By this date, Hitler was in any case more preoccupied with strategic questions and foreign affairs. The rapidly accelerating momentum of the worsening international situation confirmed Hitler's fears that time was running against Germany, that the only hope of success lay in gaining the advantage through an early strike. Diplomatic, strategic, and economic factors were by this time so intermeshed that it is impossible to single out one or the other as the sole determinant.[83] Together, they meant that by 1939 Hitler got the war which he had intended – but, from his 'programmatic' standpoint against the 'wrong' enemy (Britain) and at the best available but by no means ideal

[81] Mason, 'Innere Krise', p. 186. Rather, in his view, the crisis explains and determined the sort of war which Germany could conduct.
[82] Elke Fröhlich, ed., *Die Tagebücher von Joseph Goebbels. Sämtliche Fragmente* (4 vols., Munich/New York/London/Paris, 1987), vol. 2, p. 529, entry of 19 Oct. 1935.
[83] See Carr, *Arms, Autarky, and Aggression* (see ch. 3 note 26), p. 65.

juncture for Germany. Once in the war, a string of Blitzkrieg successes concealed for a while the underlying weaknesses of the German war economy which the Nazis were unable fully to mobilize and which only began to operate to some extent efficiently when the nation had its back to the wall.

Hitler's 'intentions' are indispensable to explaining the course of development in the Third Reich. But they are by no means an adequate explanation in themselves. The conditions in which Hitler's 'will' could be implemented as government 'policy' were only in small measure fashioned by Hitler himself, and, moreover, made the ultimate failure of his aims and the destruction of the Third Reich almost inevitable. The fact that little of what happened in domestic politics before at least the middle of the war can be said to have run counter to or contradicted Hitler's 'will' and 'intention' makes it difficult to conceive of him as a 'weak dictator' – however useful the concept might have proved heuristically. On the other hand, the implementation of Hitler's 'will' is not such a straightforward matter and foregone conclusion as the 'intentionalists' want to have it. If not a 'weak dictator', Hitler was not 'master in the Third Reich' in the implied meaning of omnipotence.

'Intention' and 'structure' are both essential elements of an explanation of the Third Reich, and need synthesis rather than to be set in opposition to each other. Hitler's 'intentions' seem above all important in shaping a climate in which the unleashed dynamic turned them into a self-fulfilling prophecy. The Third Reich provides a classic demonstration of Marx's dictum, cited by Mason: 'Men do make their own history, but they do not make it as they please, nor under conditions of their own choosing, but rather under circumstances which they find before them, under given and imposed conditions'.[84]

In the two following chapters we need to ask what relevance such conclusions have for anti-Jewish and foreign policy – areas in which Hitler's own ideological obsessions were more obvious than in the domestic arena.

---

[84] Cited in Mason, 'Intention and Explanation', p. 37. See Karl Marx, The *Eighteenth Brumaire of Louis Bonaparte* (Progress Publishers, Moscow, 1954 edn.), p. 10.

5

Hitler and the Holocaust

Explaining the Holocaust stretches the historian to the limits in his central task of providing rational explanation of complex historical developments. Simply to pose the question of how a highly cultured and economically advanced modern state could 'carry out the systematic murder of a whole people for no reason other than that they were Jews' suggests a scale of irrationality scarcely susceptible to historical understanding.[1] The very name 'the Holocaust', which acquired its specific application to the extermination of the Jews only in the late 1950s and early 1960s, when it came to be adopted (initially by Jewish writers) in preference to the accurately descriptive term 'genocide', has been taken to imply an almost sacred uniqueness of terrible events exemplifying absolute evil, a specifically Jewish fate standing in effect outside the normal historical process – 'a mysterious event, an upside-down miracle, so to speak, an event of religious significance in the sense that it is not man-made as that term is normally understood'.[2]

The 'mystification' and religious–cultural eschatology which has come for some writers to be incorporated in the term 'the Holocaust' has not made the task of Jewish historians an easy one in a subject understandably and justifiably 'charged with passion and moral judgement'.[3] Given the highly emotive nature of the problem, non-Jewish historians face arguably even greater difficulties in attempting to find the language sensitive and appropriate to the horror of Auschwitz. The sensitivity of the problem is such that over-heated reaction and counter-reaction easily spring from a misplaced or misunderstood word or sentence.

The perspective of non-Jewish historians is, however, inevitably different from that of Jewish historians. And if we are to 'learn' from the Holocaust, then – with all recognition of its 'historical' uniqueness in the sense that close parallels have not *so far* existed – it seems essential to accept that parallels *could* potentially occur in the future, and among peoples other than Germans and Jews. The wider problem alters in essence, therefore, from an attempt to 'explain' the Holocaust specifically through Jewish history or even German-

---

[1] Lucy Dawidowicz, *The War against the Jews 1933–45* (Harmondsworth, 1977), p. 17. For the following remarks, see Geoff Eley, 'Holocaust History' (see ch. 1 note 43).

[2] Yehuda Bauer, *The Holocaust in Historical Perspective* (London, 1978), p. 31. The chapter from which the quotation is taken is an attack on the 'mystification' (as Bauer put it) of the Holocaust. Bauer himself distinguished (pp. 31–5) between genocide – 'forcible, even murderous denationalization' – and the 'uniquely unique' Holocaust – 'total murder of every one of the members of the community'. I have to confess that I do not find the definitions or distinction very convincing or analytically helpful.

[3] Dawidowicz, *War*, p. 17.

Jewish relations, to the pathology of the modern state and an attempt to understand the thin veneer of 'civilization' in advanced industrial societies. Specifically applied to the Nazi Dictatorship, this demands an examination of complex processes of rule, and a readiness to locate the persecution of the Jews in a broader context of escalating racial discrimination and genocidal tendencies directed against various minority groups. This is not to forget the very special place which the Jews occupied in the Nazi doctrine, but to argue that the problem of explaining the Holocaust is part of the wider problem of how the Nazi regime functioned, in particular of how decisions were arrived at and implemented in the Nazi State.

The central issue remains, therefore, how Nazi hatred of the Jews became translated into the practice of government, and what precise role Hitler played in this process. Deceptively simple as this question sounds, it is the focal point of current controversy on 'the Holocaust' and forms the basis of the following enquiry, which attempts to survey and then evaluate recent research and interpretation.

Interpretations

Historians in both parts of Germany after the war came only slowly to concern themselves with anti-semitism and the persecution of the Jews. It was only in the wake of the Eichmann trial in Israel and the revelations of concentration camp trials in the Federal Republic that serious historical work on the Holocaust advanced in West Germany. Even then, historical scholarship and public 'enlightenment' on the fate of the Jews found only a muted echo in the German population, and popular consciousness was reached only through the showing of the American filmed 'soap-opera' dramatization of the Holocaust on West German television in 1979.[4] In the GDR, too, scholarly work on the persecution of the Jews effectively dates from the 1960s, though the subsuming, in the marxist–leninist conception of history, of race hatred within the nature of the class struggle and imperialism meant that down to the upheavals of 1989 few important works specifically on the Holocaust appeared.[5] The publications of Kurt Pätzold, while remaining firmly anchored within the marxist–leninist framework, marked a significant advance in GDR scholarship in this field.[6]

The major impulses to research and to scholarly debate have, therefore, been initiated outside Germany – in the first instance by Jewish scholars in Israel and

---

[4] See the excellent historiographical survey by Konrad Kwiet, 'Zur historiographischen Behandlung der Judenverfolgung im Dritten Reich', *MGM* (1980), Heft 1, pp. 149–92, here esp. pp. 149–53; and the valuable study by Otto Dov Kulka, 'Major Trends and Tendencies of German Historiography on National Socialism and the "Jewish Question" (1924–1984)', *Yearbook of the Leo Baeck Institute* 30 (1985), pp. 215–42. For recent, thorough analyses of the, by now, massive extent of research on most aspects of the Holocaust, see the essays by: Saul Friedländer, 'From Anti-Semitism to Extermination. A Historiographical Study of Nazi Policies towards the Jews and an Essay in Interpretation', *Yad Vashem Studies* 16 (1984), pp. 1–50; and Michael Marrus, 'The History of the Holocaust. A Survey of Recent Literature', *JMH* 59 (1987), pp. 114–60. Most comprehensively, there is now Michael Marrus, *The Holocaust in History* (London, 1988).
[5] See Konrad Kwiet, 'Historians of the German Democratic Republic on Antisemitism and Persecution', *Yearbook of the Leo Baeck Institute* 21 (1976), pp. 173–98.
[6] See Kurt Pätzold, *Faschismus, Rassenwahn, Judenverfolgung* (East Berlin, 1975), and 'Vertreibung' (see ch. 3 note 56).

other countries, and secondarily from non-Jewish historians outside Germany. However, even where the initial stimulant to debate emanated from non-German writers – and the controversies stirred up by Hannah Arendt's publications on the Eichmann trial[7] and more recently David Irving's attempt to whitewash Hitler's knowledge of the 'Final Solution'[8] provide merely the most spectacular examples – ensuing discussion in the Federal Republic has been strongly influenced by the intellectual climate of German historical writing on Nazism which we have already examined. Hence, the contours of the debate about Hitler and the implementation of the 'Final Solution' – the subject of this chapter – are again peculiarly West German, even where valuable contributions have been made by foreign scholars.

The interpretational divide on this issue brings us back to the dichotomy of 'intention' and 'structure' which we have already encountered. The conventional and dominant 'Hitlerist' approach proceeds from the assumption that Hitler himself, from a very early date seriously contemplated, pursued as a main aim, and strived unshakeably to accomplish the physical annihilation of the Jews. According to such an interpretation, the various stages of the persecution of the Jews are to be directly derived from the inflexible continuity of Hitler's aims and intentions; and the 'Final Solution' is to be seen as the central goal of the Dictator from the very beginning of his political career, and the result of a more or less consistent policy (subject only to 'tactical' deviation), 'programmed' by Hitler and ultimately implemented according to the Führer's orders. In contrast, the 'structuralist' type of approach lays emphasis upon the unsystematic and improvised shaping of Nazi 'policies' towards the Jews, seeing them as a series of *ad hoc* responses of a splintered and disorderly government machinery. Although, it is argued, this produced an inevitable spiral of radicalization, the actual physical extermination of the Jews was not planned in advance, could at no time before 1941 be in any realistic sense envisaged or predicted, and emerged itself as an *ad hoc* 'solution' to massive, and self-induced, administrative problems of the regime.

The interpretation of the destruction of European Jewry as the 'programmatic' execution of Hitler's unchangeable will has an immediate (though actually superficial) attractiveness and plausibility. It marries well with the views of those historians who incline to explanations of the Third Reich through the development of a specifically German ideology, where a great deal of weight is attached, as a causal factor in Nazism's success, to the spread of anti-semitic ideas and an ideological climate in which Hitler's own radical anti-semitism could find appeal.[9] There is, of course, no difficulty in demonstrating the basic continuity and inner consistency of Hitler's violent hatred of the Jews – ranging from his entry into politics in 1919 to the composition of his Political Testament in the bunker at the end of April 1945 – voiced throughout in the most extreme language conceivable. The interpretation corresponds, too, to the

[7] See Hannah Arendt, *Eichmann in Jerusalem. A Report on the Banality of Evil* (London, 1963).
[8] David Irving, *Hitler's War* (London, 1977). See the devastating critique by Martin Broszat, 'Hitler und die Genesis der "Endlösung". Aus Anlaß der Thesen von David Irving', *VfZ* 25 (1977), pp. 737–75, esp. pp. 759 ff. Engl. trans., 'Hitler and the Genesis of the "Final Solution": An Assessment of David Irving's Theses' in H W. Koch, ed., *Aspects of the Third Reich* (London, 1985), pp. 390–429.
[9] E.g. George L. Mosse, *The Crisis of German Ideology* (London, 1964).

'totalitarianism' model where state and society were 'co-ordinated' to the level of executors of the wishes of Hitler, the unchallenged 'master in the Third Reich', who determined policy from above, at least in those spheres – like the 'Jewish Question' – where he had a paramount interest. Seen in this light, the logic of the course of anti-Jewish policy from the boycott and legislation of spring 1933 down to the gas chambers of Treblinka and Auschwitz seems clear. In crude terms, the reason why the Jews of Europe were murdered in their millions was because Hitler, the dictator of Germany, wanted it – and had done since he entered politics over two decades earlier.[10] It is in short an explanation of the Holocaust which rests heavily upon an acceptance of the motive force and autonomy of individual will as the determinant of the course of history.

Numerous influential works on the destruction of the Jews have advanced this or similar types of 'Hitlerist' approach. Lucy Dawidowicz, in her widely acclaimed *The War against the Jews*, for instance, declares that Hitler's idea for the 'Final Solution' went back to his experience in the Pasewalk hospital in 1918, and that by the time he wrote the second volume of *Mein Kampf* in 1925 he 'openly espoused his programme of annihilation' in words which 'were to become the blueprint for his policies when he came to power'. She writes of 'the grand design' in Hitler's head, the 'long-range plans to realize his ideological goals' with the destruction of the Jews at their centre, and that the implementation of his plan was subject only to opportunism and expediency. She concludes: 'Through a maze of time, Hitler's decision of November 1918 led to Operation Barbarossa. There never had been any ideological deviation or wavering determination. In the end only the question of opportunity mattered'.[11]

A similar inclination to a personalized explanation of 'the Holocaust' can be found, not unnaturally, in leading biographies of Hitler. Toland has Hitler advocating, as early as 1919, the physical liquidation of Jewry and transforming his hatred of the Jews into a 'positive political programme'.[12] Haffner, too, speaks of a 'cherished wish to exterminate the Jews of the whole of Europe' as being Hitler's aim 'from the beginning on'.[13] Fest relates the first gassing of Jews near Chelmno in Poland in 1941 to Hitler's own experience in the First World War and the notorious lesson he drew from it, as recorded in *Mein Kampf*, that perhaps a million German lives would have been saved if 12,000–15,000 Jews had been put under poison gas at the start of or during the war.[14] And Binion's 'psycho-historical' study argues that Hitler's mission 'to remove Germany's Jewish cancer and to poison out Germany's Jewish poison' emanated from his hallucination while recovering from mustard-gas poisoning at Pasewalk, when he allegedly traumatized his mother's death while under treatment from a Jewish doctor and brought this in hysterical association with

---

[10] Mason, 'Intention and Explanation' (see ch. 1 note 25), p. 32. See also the 'explanation' of the Holocaust given by Sarah Gordon, *Hitler, Germans, and the 'Jewish Question'* (Princeton, 1984), p. 316: the reasons Jews were killed in their millions was 'that power was totally concentrated in one man, and that man happened to hate their "race" '.

[11] Dawidowicz, *War*, pp. 193–208.

[12] John Toland, *Adolf Hitler* (New York, 1976), pp. 88–9.

[13] Haffner (see ch. 4 note 10), pp. 178–9.

[14] Fest (see ch. 4 note 8), vol. 2, p. 930 (Ullstein edn., Frankfurt am Main/Berlin, Vienna, 1976); Adolf Hitler, *Mein Kampf* (Munich, 1943 edn.), p. 772.

his trauma at Germany's defeat in 1918. Hitler 'emerged from his trance resolved on entering politics in order to kill the Jews by way of discharging his mission to undo, and reverse, Germany's defeat'. This was his 'main line political track' which ran from Pasewalk to Auschwitz.[15]

The same basic premise of the early formulation and unshakeable retention of Hitler's will to exterminate the Jews as sufficient explanation of 'the Holocaust' underlies Gerald Fleming's study, which seeks to document as fully as possible Hitler's personal responsibility for the 'Final Solution'. Though concentrating almost exclusively on the period of extermination itself, the introductory chapters deal with the growth of Hitler's anti-semitism. There, the claim is repeatedly made that 'a straight path' led from Hitler's personal anti-semitism and the development of his original hatred of the Jews to his personal liquidation orders during the war – 'a straight path from Hitler's anti-semitism as shaped in Linz in the period 1904–1907 to the first mass shootings of German Jews in Fort IX in Kowno on 25 and 29 November 1941'. Physical extermination, in Fleming's view, was the aim maintained continually by Hitler from his experience of the November Revolution in 1918 down to his end in the bunker, and at the beginning of the 1920s 'Hitler developed . . . a strategic plan for the realization of his political aim'.[16]

Unwavering continuity of aim, a dominance in shaping anti-Jewish policy from first to last, and the decisive role in the initiation and implementation of the 'Final Solution' are also attributed to Hitler in the most influential works of leading West German experts on the Third Reich. Though prepared to accord 'the historical situation a comparatively high rank in the implementation of National Socialist "Jewish Policy" ',[17] the 'programmatist' line (as it has been styled) sees Nazi anti-Jewish aims and measures as integrally linked to foreign policy, framed along with foreign policy in terms of long-range 'final goals', and advancing 'with inner logic, consistency, and in stages'.[18] Klaus Hildebrand summarizes the position clearly and concisely: 'Fundamental to National Socialist genocide was Hitler's race dogma. . . . Hitler's programmatic ideas about the destruction of the Jews and racial domination have still to be rated as primary and causative, as motive and aim, as intention and goal (*Vorsatz and Fluchtpunkt*) of the "Jewish Policy" of the Third Reich'.[19] For the Swiss historian Walter Hofer, 'it is simply incomprehensible how the claim can be made that the National Socialist race policy was not the realization of Hitler's *Weltanschauung*'.[20]

Hofer's remarks were part of a particularly aggressive critique of the 'structuralist' approach of 'revisionist' historians. The particular target of attack in this instance was Hans Mommsen, who is accused of not seeing because he does not want to see the obvious connection between the announcement of Hitler's

[15] Binion (see ch. 4 note 11), p. 85 and chs. 1, 4; Toland, p. 934.
[16] Gerald Fleming, *Hitler und die Endlösung. 'Es ist des Führers Wunsch'* (Wiesbaden/Munich, 1982), pp. 13–27 (where Hitler's 'straight path' is mentioned at least four times). An English translation is available: *Hitler and the Final Solution* (Oxford, 1986).
[17] Hildebrand, *Das Dritte Reich*, p. 178.
[18] Hillgruber, *Endlich genug?*, pp. 64–6 and p. 52 note 88.
[19] Hildebrand, *Das Dritte Reich*, p. 178.
[20] Hofer (see ch. 1 note 2), p. 14.

programme (in *Mein Kampf* and elsewhere) and its later realization.[21] Mommsen himself has argued forcefully in a number of essays that the implementation of the 'Final Solution' can by no means be attributed to Hitler alone, nor to purely ideological factors in the German political culture.[22] Rather, the explanation has to be sought in the peculiarly fragmented decision-making processes in the Third Reich, which made for improvised bureaucratic initiatives with their own inbuilt momentum, promoting a dynamic process of cumulative radicalization. In his view, the assumption that the 'Final Solution' had to stem from a 'Führer Order' is mistaken. Though unquestionably Hitler knew of and approved of what was taking place, such an assumption, argues Mommsen, flies in the face of his known tendency to let things take their own course and to put off decisions wherever possible. Moreover, it is not compatible with his conscious attempts to conceal his own personal responsibility, with his more subconscious suppression of actual reality even to himself – for all the violence of his propagandistic statements, he never spoke in concrete terms about the 'Final Solution' even in his intimate circle – nor with maintaining the fiction of 'labour deployment' and 'natural wastage' through work. Accordingly, concludes Mommsen, there could have been no formal 'Führer Order' – written or verbal – for the 'Final Solution' of the 'European Jewish Question'. References in the sources to an 'order' or 'commission' as opposed to a vague 'wish of the Führer' relate invariably to the '*Kommissarbefehl*' complex of orders of spring 1941. Though the mass shootings of Russian Jews derived from the '*Kommissarbefehl*' group of directives, they must be distinguished from the 'Final Solution' proper – the systematic extermination of European Jewry. And that the latter was based on a Hitler order is, in Mommsen's view, neither supported by the evidence, nor inherently likely. Rather, although Hitler was the 'ideological and political originator' of the 'Final Solution', a 'utopian objective' could be translated into hard reality 'only in the uncertain light of the Dictator's fanatical propaganda utterances, eagerly seized upon as orders for action by men wishing to prove their diligence, the efficiency of their machinery, and their political indispensability'.

An essentially similar interpretation was advanced by Martin Broszat in his penetrating analysis of the genesis of the 'Final Solution'.[23] Broszat argued that 'there had been no comprehensive general extermination order at all', but that 'the "programme" of extermination of the Jews gradually developed institutionally and in practice out of individual actions down to early 1942 and gained determinative character after the erection of the extermination camps in Poland (between December 1941 and July 1942)'. In Broszat's view, deportation of the Jews was still the aim until autumn 1941, and it was only in the light of the unexpected failure of the Blitzkrieg invasion of the Soviet Union that

[21] Hofer, p. 14.
[22] See Hans Mommsen, 'Nationalsozialismus oder Hitlerismus?', pp. 66–70; 'National Socialism: Continuity and Change', p. 179; 'Hitlers Stellung', pp. 61 ff. (full references above ch. 4 note 28), and esp. his outstanding essay 'Die Realisierung des Utopischen: Die "Endlösung der Judenfrage" im "Dritten Reich" ', *GG* 9 (1983), pp. 381–420, here esp. pp. 394–5 and notes 48–9, 399, 416–18. An extended version of this last essay is published in English translation, 'The Realization of the Unthinkable: the "Final Solution of the Jewish Question" in the Third Reich', in Gerhard Hirschfeld, ed., *The Policies of Genocide* (London, 1986), pp. 97–144.
[23] Broszat, 'Genesis' (see note 8, this chapter), pp. 753–7.

problems in the deportation plans and the inability of Gauleiter, police chiefs, SS bosses, and other Nazi leaders in the Occupied Territories to cope with the vast numbers of Jews transported to and concentrated in their domains that led to a growing number of 'local initiatives' being taken to liquidate Jews, which then gained retrospective sanction 'from above'. Following this interpretation, therefore, 'the destruction of the Jews arose, so it seems, not only out of a previously existent will to exterminate, but also as the "way out" of a cul-de-sac into which [the regime] had manoeuvred itself. Once begun and institutionalized, the practice of liquidation nevertheless gained dominant weight and led finally *de facto* to a comprehensive "programme" '.

Broszat went out of his way in this essay (as had Mommsen in his writings) to emphasize that his interpretation could in no sense be seen in moral terms as removing the responsibility and guilt for the 'Final Solution' from Hitler, who approved, sanctioned, and empowered the liquidation actions 'whoever suggested them'. However, it does mean that in terms of actual practice of the implementation of the 'Final Solution', Hitler's personal role can only be indirectly deduced.[24] And morally, this clearly extends the responsibility and culpability to groups and agencies in the Nazi State beyond the Führer himself.

The role of Hitler is reduced still further in the analysis of the GDR historian Kurt Pätzold, who also demonstrates clearly the gradual and late emergence of an extermination 'policy' arising from unco-ordinated but increasingly barbarous attempts to drive Jews out of Germany and German-ruled territory.[25] While his description of the process which led from the aim of expulsion to genocide matches 'structuralist' explanations of western historians, Pätzold relates this to a sense of dynamic 'purpose' and direction of the Nazi regime which sometimes appears to be missing from 'structuralist' accounts. Despite a ritualistic overemphasis upon the functional purpose of anti-Jewish measures in serving the interests of monopoly capital, Pätzold's treatment has the merit, it seems to me, of locating the destruction of the Jews as an element within the overall context of the ruthless and dehumanizing expansionist drive of the Nazi State. This is to turn round the 'Hitlerist' interpretation, where the purposeful direction of Nazism is attributed as good as exclusively to the ideology of the Führer, and where Nazi *Lebensraum* ambitions are regarded as subsumed within and ultimately subordinate to Hitler's manic determination to destroy the Jews.

The lack of a long-range extermination programme has also come to be accepted by leading Israeli experts on 'the Holocaust'. Yehuda Bauer, for instance, writes that 'Nazi policy towards the Jews developed in stages, but that does not mean that at any given turning point there were not other options open to the Nazis that were considered seriously; there developed in Nazi Germany only one clear idea regarding Jews that was accepted by all policy-makers, namely the idea that ultimately the Jews had no place in Germany'.[26] Such a position is a recognition of the findings of detailed historical research on the course of anti-Jewish policy during the 1930s, where thorough analysis has suggested that the 'road to Auschwitz', was a 'twisted' one, and not at all the

[24] Broszat, 'Genesis', pp. 756–7.
[25] Pätzold, 'Vertreibung' (see ch. 3 note 56).
[26] Bauer, p. 11.

'straight path' which Fleming and others have seen.[27] Karl Schleunes's conclusion was, in fact, that 'the figure of Adolf Hitler during these years of search is a shadowy one. His hand appears only rarely in the actual making of Jewish policy between 1933 and 1938. One can only conclude from this that he occupied his time with more important concerns. In part the vagaries and inconsistencies of Jewish policy during the first five years of Nazi rule stem from his failure to offer guidance'.[28] Absence of clear objectives led to varying and rival 'policies', all of which ran into difficulties. But there was no turning back on the 'Jewish Question', and it was in this fashion that Hitler's known ideological obsession with the Jews had the objective function – without Hitler having to lift a finger – of pushing a failure in one direction (boycott, legislation, 'Aryanization', or emigration) into a renewed effort to 'solve the problem'.[29] Once again, there is no doubting Hitler's moral responsibility, nor the role his intentions – real or *presumed* – played. But of a consistent implementation of ideological prerogatives, there is little or nothing to be seen: 'The Final Solution as it emerged in 1941 and 1942 was not the product of a grand design'.[30]

The exploration of Uwe Dietrich Adam, which had the added advantage of continuing the investigation into the wartime period down to the implementation of the 'Final Solution' itself, arrived at similar conclusions: 'The empirical facts confirm first of all that there can be no talk of a planned and directed policy in this field, that a comprehensive plan for the method, content, and extent of the persecution of the Jews never existed, and that the mass killing

---

[27] See particularly the works of Schleunes and Adam (see ch. 3 note 54).

[28] Schleunes, p. 258. This interpretation has been directly called into question in a well-researched article by David Bankier, 'Hitler and the Policy-Making Process in the Jewish Question', *Holocaust and Genocide Studies* 3 (1988), pp. 1–20. Bankier succeeds in demonstrating that Hitler did intervene in the 'Jewish Question' more often than has been thought, and that he showed from time to time interest even in the minutiae of anti-Jewish policy. Even so, Bankier takes the thrust of his findings too far in claiming that Hitler 'conceived, initiated, and directed the entire process' (p. 17), and his argument appears to be based in part on a misunderstanding (or exaggeration) of the structuralist (or functionalist) case he is attacking. No one, for example, doubts Hitler's pragmatism and opportunism in the 'Jewish Question', which Bankier is rightly keen to emphasize (pp. 5–8). Bankier's attack on the view (attributed to me among others) that Hitler was 'a moderate' in anti-semitic policy rests on a misunderstanding. Even the most ardent 'structuralist' would regard Hitler as the most radical of the radicals in sentiment and any 'moderation' – a term, incidentally, which Bankier himself uses on one occasion (p. 16) – as merely deployed for tactical purposes, a point I myself sought to emphasize in *The 'Hitler Myth'*, (see ch. 4 note 36) e.g. pp. 236, 239, 250–1. Nor has it ever been in dispute that Hitler's 'profound interest in all matters concerning Jews served as a guideline for state policy in the Jewish Question' (p. 11), or that 'Hitler's ideology was an undeniably powerful factor in the shaping of Nazi antisemitic policy' (p. 16). Within this framework, on which there can be little disagreement, the evidence cited by Bankier interestingly reveals instances of contradictions (p. 13) in Hitler's stance, as well as 'nondecisions' (pp. 10–11). Cases which Bankier cites of Hitler's intervention more often than not arise from points of contention where he is asked to settle a problem, and the generalization that 'it was in fact Hitler and not others who initiated radical measures' (p. 7) is overdrawn. Hitler's own words on 25 October 1941, which Bankier cites (p. 7, from H R. Trevor-Roper, ed., *Hitler's Table Talk* (London, 1953), p. 90; see Werner Jochmann, *Adolf Hitler. Monologe in Führerhauptquartier 1941–1944* (Hamburg, 1980), p. 108), that 'even with regard to the Jews, I have found myself remaining inactive' – for tactical reasons, let it again be stressed – are themselves an indication that radicalization in the 'Jewish Question' could occur in the absence of his close involvement in the direction of policy.

[29] See Schleunes, p. 259.

[30] Schleunes, Introduction, p. 2.

and extermination, too, was most probably not striven after *a priori* by Hitler as a political aim'. Unlike Broszat, Adam attributes the commencement of the 'Final Solution' to a personal order of Hitler in autumn 1941. However, in his view this has to be placed in the context of 'an inner development, which bound Hitler too in no small part'.[31]

At the root of the divergence in historical explanations of 'the Holocaust' summarized here lies the basic dichotomy between 'intention' and 'structure'. Was the systematic extermination of European Jewry the direct realization of Hitler's ideologically motivated 'design for destruction', which, after various stages in an exorable process of development, he set into operation through a written or, more likely, verbal 'Führer Order' sometime in 1941? Or did the 'Final Solution' emerge piecemeal, and without any command of Hitler, as 'an imperative result of the system of cumulative radicalization'[32] in the Third Reich? We turn now to a brief evaluation of these positions and an appraisal of some of the available evidence on which an interpretation must be based.

Evaluation

It seems important to re-emphasize at the outset that, despite claims sometimes made by those adopting a 'Hitlerist' interpretation, Hitler's continuous personal hatred of the Jews, his unique and central importance to the Nazi system in general and to the unfolding of its anti-Jewish policy in particular, and his moral responsibility for what took place are not at stake in the debate.

Historians favouring a 'structuralist' approach readily accept the overwhelming evidence that Hitler maintained a personal, pathologically violent hatred of Jews (whatever its derivation) throughout his political 'career', and recognize, too, the importance of that paranoid obsession *in determining the climate* within which the escalating radicalization of anti-Jewish policies took place. To put the counter-factual point at its crudest: without Hitler as head of the German State between 1933 and 1945, and without his fanaticism on the 'Jewish Question' as impulse and sanction, touchstone and legitimation, of escalating discrimination and persecution, it seems hardly conceivable that the 'Final Solution' would have occurred. This thought itself is sufficient to posit a fundamental link between Hitler and genocide. Moreover, the moral allegation against 'structuralist' historians – that they are 'trivializing' the wickedness of Hitler – is also misplaced. The 'structuralist' approach in no sense denies Hitler's personal, political, and moral responsibility for 'the Holocaust'. But it does broaden that culpability to implicate directly and as active and willing agents large sections of the German non-Nazi élites in the army, industry,

---

[31] Adam, *Judenpolitik*, pp. 313, 357–60. See also Uwe Dietrich Adam, 'An Overall Plan for Anti-Jewish Legislation in the Third Reich?', *Yad Vashem Studies* 11 (1976), pp. 33–55, here pp. 34–5. The lack of a long-term 'extermination plan' is fully upheld in the two most recent analyses, by Arno Mayer and Philippe Burrin. Though their interpretations differ in important respects, both argue that physical extermination only arose as a comprehensive 'solution' during the course of the Russian campaign. See Arno J. Mayer, *Why did the Heavens not Darken. The 'Final Solution' in History* (New York, 1989) and Philippe Burrin, *Hitler et les Juifs. Genèse d'un génocide* (Paris, 1989). An English translation of Burrin's book is in preparation.

[32] Mommsen, 'Realisierung', p. 399 note 65.

and bureaucracy alongside the Nazi leadership and Party organizations. In fact, if anything it is the apparent need to find a supreme culprit which comes close to trivializing *in terms of historical explanation* by diverting attention from the active forces in German society which did not have to be given a 'Führer Order' to turn the screw of Jewish persecution one thread further until extermination became the logical (and only available) 'solution'. The question of allocating guilt thus distracts from the real question the *historian* has to answer: precisely *how* genocide could happen, how an unbalanced, paranoid hatred and chiliastic vision became reality and implemented as horrific government practice.

Rather, the central areas of debate among historians are: whether evidence of Hitler's continued and consistent personal hatred is sufficient explanation in itself of the Holocaust (given a background of widespread racial anti-semitism and ideological hatred of Jews, and a corresponding readiness to carry out 'Führer Orders'); whether physical extermination was Hitler's aim from a very early date or emerged as a realistic idea only as late as 1941 or so – the last remaining option in 'solving the Jewish Question'; and finally, whether it was necessary for Hitler to do more than establish the underlying objective of 'getting rid of Jews' from German territory, and then sanction the unco-ordinated but increasingly radical steps of the various groups in the State who were seeking, often for their own reasons and by no means primarily motivated by anti-semitic ideology, to turn this distant objective into practical reality. These are open questions, not foregone conclusions or matters for dogmatic assertion.

A problem with the 'intentionalist' position – in particular with its extreme 'grand design' variant – is an implicit teleology which takes Auschwitz as a starting-point and looks backwards to the violent expression of Hitler's early speeches and writing, treating these as a 'serious declaration of intent'.[33] Because Hitler frequently spoke about destroying the Jews, and the destruction of the Jews actually took place, the logically false conclusion is drawn that Hitler's expressed 'intention' must have *caused* the destruction. In the light of hindsight, it is easy to attribute a concrete and specific meaning to the barbarous, but vague and fairly commonplace, generalities about 'getting rid' (*Entfernung*) or even 'extermination' (*Vernichtung*) of Jews, which were part and parcel of Hitler's language (and that of others on the *völkisch* Right) from the early 1920s onwards. Coupled with this is the problem of establishing empirically Hitler's initiation or direct instigation of shifts in policy towards fulfilment of his aims – a problem accentuated by Hitler's obvious desire not to be publicly associated with inhumane and brutal measures, and the secrecy and euphemistic language which camouflaged the 'Final Solution' itself. If 'programme', 'plan', or 'design' in the context of Nazi anti-Jewish policy are to have real meaning, then they ought to imply something more than the mere conviction, however fanatically held, that somehow the Jews would be 'got rid of' from German territory and from Europe as a whole, and the 'Jewish Question' solved. Before 1941, the evidence that Hitler had more than such vague and imprecise convictions is slender. Finally, the moral 'lesson' to be drawn from the 'Hitlerist' position – apart from the 'alibi' it provides for non-Nazi institutions in the Third Reich – is by no means obvious. Fleming's rather

---

[33] Mommsen, 'Nationalsozialismus oder Hitlerismus?', p. 67.

jejune moral conclusion based upon his 'intentionalist' account of the 'Final Solution' is that hatred feeds the animal instinct for destruction of human life which resides in us all.[34]

More important than such bland moralization is the question posed by 'structuralist' approaches, of how and why a political system in all its complexity and sophistication can within the space of less than a decade become so corrupted that it regards the implementation of genocide as one of its supreme tasks. The central issue here revolves around the nature of 'charismatic' politics – how Hitler's vaguely expressed 'intent' was interpreted and turned into reality by government and bureaucratic agencies which developed their own momentum and impetus. The 'structuralist' type of interpretation also has some weaknesses. The empirical data are seldom good enough to allow detailed reconstruction of the processes of decision-making, on which much of the argument resides. And the emphasis upon contingency, lack of planning, absence of co-ordination, governmental chaos, and the *ad hoc* 'emergence' of policy out of administrative disorder seems at times potentially in danger of neglecting the motive force of intention (however vaguely expressed) and distorting the focus of the regime's ideologically rooted thrust and dynamic drive. However, the 'structuralist' approach does provide the opportunity of *locating* Hitler's 'intentions' within a governmental framework which allowed the bureaucratic implementation of a loose ideological imperative, turning a slogan of 'get rid of the Jews' into a programme of annihilation. And concentration on the historical question of how 'the Holocaust' happened rather than, implicitly or explicitly, seeking to allocate guilt makes the issue of whether Hitler took the initiative at every turn, or whether a particular decision was his alone, seem less relevant and important.

During the pre-war years, as the evidence assembled and analysed by Schleunes and Adam convincingly demonstrates, it seems clear that Hitler took no specific initiative in the 'Jewish Question' and *responded* to rather than instigated the confused and often conflicting lines of 'policy' which emerged. The main impulses derived from the pressure 'from below' of Party activists, the internal organizational and bureaucratic dynamism of the SS–Gestapo–SD apparatus, the personal and institutional rivalries which found an outlet in the 'Jewish Question', and, not least, from economic interest in eliminating Jewish competition and expropriating Jewish capital.

The national boycott of Jewish businesses which took place on 1 April 1933 was organized chiefly as a response to the pressure of Party radicals, especially within the SA, during the wave of violence and brutality unleashed by the 'seizure of power'. The only 'plans' of the NSDAP for tackling the 'Jewish Question' which had been formulated before Hitler became Chancellor related to measures for legal discrimination and deprivation of civil rights.[35] Such vague and undetailed administrative 'plans' hardly accorded with the wild and

---

[34] Fleming, p. 206. See also p. 204 for his conclusion that those implementing Hitler's orders acted out of opportunism, servility, lack of character, and 'the petty-bourgeois zeal of a following whose idealism was abused'.

[35] Schleunes, p. 70; Adam, *Judenpolitik*, pp. 28 ff.

dangerous mood of Party activists in the post- 'seizure of power' euphoria of spring 1933. In these weeks, in fact, no directives at all on 'the Jewish Question' came either from the Reich Chancellory or from the Nazi Party head-quarters.[36] Meanwhile, the SA, whose 'enthusiasm' could hardly now be checked, had started its own anti-Jewish campaign of boycotts and violence. When Gestapo chief Rudolf Diels complained about the excesses of the Berlin SA, he was informed that 'for very human reasons, certain activity must be found which will satisfy the feelings of our comrades'.[37] Under pressure, Hitler reacted towards the end of March with the call for a general boycott against Jewish businesses and professions, starting on 1 April and to be orga-nized by a 14-man steering committee under the direction of Julius Streicher. As is well known, the boycott was a notable failure, and in the light of the negative echo abroad, the lack of enthusiasm among important sectors of the conservative power-élite (including President Hindenburg), and the cool indif-ference of the German people, it was called off after a single day and a co-ordinated national boycott was never again attempted. The shameful dis-criminatory legislation of the first months of the Dictatorship, aimed at Jews in the civil service and the professions, arose in the same climate and under the same pressures. Hitler's own direct role was a limited one dictated by the need he felt, despite his obvious approval of the boycott, to avoid association with the worst 'excesses' of the Party radicals. But the pace was forced by the momentum of the violence and illegalities, which produced their own compul-sion to provide *post facto* legitimation and sanction – a process which was to repeat itself in later stages of the persecution of the Jews.[38]

Following a relatively quiet period between the summer of 1933 and the beginning of 1935, a new anti-semitic wave began and lasted until the autumn of 1935. Again, the agitation was set in motion and sustained 'from below' through the pressure at Gau level and from activists in the Party and in Hitler Youth and SA units in the localities. One Gauleiter noted in his report that stirr-ing up the 'Jewish Question' had been useful in revamping the sagging morale of the lower middle class.[39] The agitation was, of course, backed by pro-paganda from the Party and from the State. But other than that, there was remarkably little intervention from either the Party's headquarters or from the Reich government before mid August, when the boycotts and violence were becoming recognizably counter-productive, both in the repercussions for the German economy and on account of the unpopularity of the frequent breaches of the peace. Hitler himself was hardly involved in any direct sense. Despite his radical instincts, he was effectively compelled in this phase – in the interests of 'order', of the economy, and of diplomatic relations – to recognize the neces-sity of bringing the damaging campaign to a close.[40] This had to be balanced against the need not to lose face with Party activists and the pressure to comply with Party demands for 'action' – particularly for legislation in line with the demands of the Party programme – in the 'Jewish Question'. The resulting

[36] Schleunes, p. 71.
[37] Cited in Schleunes, p. 74.
[38] Schleunes, pp. 92–102; Adam, *Judenpolitik*, pp. 64 ff., esp. p. 68.
[39] Marlis G. Steinert, *Hitlers Krieg und die Deutschen* (Düsseldorf/Vienna, 1970), p. 57.
[40] Adam, *Judenpolitik*, p. 121.

'compromise' was effectively the promulgation of the notorious 'Nuremberg Laws' in September 1935 – at one and the same time according with demands for clear guidance and 'regulation' of the 'Jewish Question', and a further turn of the discriminatory screw.

The creation of the Nuremberg Laws demonstrates clearly how Hitler and the Nazi leadership responded to the considerable pressures from below in their formulation of anti-Jewish policy at this date.

The agitation and violence of the spring and summer 1935 rekindled expectations within the Party of incisive anti-Jewish legislation.[41] Hints and half-promises of measures were made by Reich Minister of the Interior Frick and others, bureaucrats hurried to regulate discrimination which was already taking place, and bans on various Jewish activities introduced independently by the Gestapo also forced retrospective sanctions by the administrators. One area of discontent among Party agitators was the failure to introduce the long-awaited exclusion of Jews from German citizenship. Despite indications from the Reich Ministry of the Interior, where preparations were underway, the summer brought nothing to satisfy the hotheads. The other major issue whipped up by propaganda and agitation was that of mixed marriages and sexual relations between 'Aryans' and Jews. Again, illegal but sanctioned terroristic actions in cases of 'racial defilement' forced the pace and shaped the atmosphere. The urgent need for legislation was accepted by the regime's leaders at an important ministerial meeting chaired by Schacht on 20 August. Only the timing remained undecided. There were in fact already rumours in the foreign press in late August that the official proclamation might come at the Nuremberg Party Rally in September. Though such rumours turned out to be accurate, it is possible that they were at the time no more than intelligent speculation since it still appears that the decision to promulgate the laws at a special meeting of the Reichstag summoned to Nuremberg was taken only after the Rally had actually started – probably under renewed pressure from 'Reich Doctors' Leader' Gerhard Wagner who, apparently after talks with Hitler, announced on 12 September the intention of promulgating a 'Law for the Protection of German Blood'. From this point, as is well known, things moved fast. 'Experts' on the 'Jewish Question' were suddenly summoned to Nuremberg on 13 September and told to prepare a law regulating marriage between 'Aryans' and Jews. The sudden decision to promulgate anti-Jewish laws during the Rally seems to have been predominantly determined by questions of propaganda, presentation, and image. The Reichstag had been summoned to Nuremberg, where Hitler originally intended, in the presence of the Diplomatic Corps, to make an important statement on foreign policy, exploiting the Abyssinian conflict to articulate German revisionist demands. On the advice of Foreign Minister von Neurath, this plan was dropped on 13 September. A suitable replacement programme for the Reichstag and for Party consumption had rapidly to be found.[42] The

[41] This account of the genesis of the Nuremberg Laws is primarily based upon Adam, *Judenpolitik*, pp. 118–22, 126; Schleunes, pp. 120–1; and, especially, upon the analyses of Lothar Gruchmann, ' "Blutschutzgesetz" und Justiz. Zur Entstehung und Auswirkung des Nürnberger Gesetzes vom 15. September 1935', *VfZ* 31 (1983), pp. 418–42, here esp. pp. 428–33, and Otto Dov Kulka, 'Die Nürnberger Rassengesetze und die deutsche Bevölkerung im Lichte geheimer NS-Lage- und Stimmungsberichte', *VfZ* 32 (1984), pp. 582–624, here esp. pp. 614–20.

[42] Mommsen, 'Realisierung', p. 387 and note 20. See also, for this section, Adam, *Judenpolitik*, pp. 125 ff., and Schleunes, pp. 121 ff.

rather undramatic 'Flag Law' hardly matched the demands of the occasion. Hence, the 'Blood Law', now being frantically drafted, and a Reich Citizenship Law, drafted in an hour on 14 September, were brought in as a substantial offering to the Reichstag and the assembled Party faithful. Hitler himself, who chose the mildest of the four drafts of the 'Blood Law' presented to him, apparently preferred to remain in the background during the drafting, pushing the Racial Political Office to the forefront. His role was a characteristically vague and elusive one in the question of how to define 'a Jew', when a conference for this purpose met at Munich at the end of the month. Hitler confined himself to a long monologue on the Jews, announced that the definitional problem would be sorted out between the Reich Ministry of the Interior and the Party, and adjourned the conference. It was mid November before State officials and representatives of the Party could iron out a compromise solution – after Hitler had cancelled a further planned meeting in early November at which he had been expected to resolve the matter.[43]

Hitler continued to take no initiative in the 'Jewish Question' during the relatively quiet years of 1936-7, in which the rivalries mounted between the various agencies with an interest in Jewish affairs – the Ministry of the Interior, the Economics Ministry, the Foreign Ministry, the Four Year Plan Administration, the Rosenberg Agency, and, not least, the SS and Gestapo apparatus. A clear line of policy was as distant as ever. To go from Goebbels's informative diary record of these years, Hitler appears to have spoken directly about the Jews only infrequently, and then in general terms, as in November 1937, when, in a long discussion with Goebbels about the 'Jewish Question', he allegedly said: 'The Jew must get out of Germany, yes out of the whole of Europe. That will take some time yet, but will and must happen'. According to Goebbels, the Führer was 'firmly decided' on it.[44]

These comments followed only a few weeks after Hitler had made his first public attack on the Jews for some time in a rhetorical propaganda tirade against 'the Jewish–Bolshevik World Enemy' during the Party Rally in September 1937.[45] This was enough to set the tone for a renewal of antisemitic activity on a large scale. However, Hitler himself needed to do no more in order to stimulate the process of 'aryanization' of Jewish concerns in the interests of 'big business', which set in at the end of 1937 and where Göring was the chief driving-force, nor to direct the escalating wave of violence which followed the *Anschluß* and became magnified during the Sudeten crisis of the summer. The agitation and terror of the Party rank-and-file in the summer and autumn of 1938, together with the expulsion in October of some 17,000 Polish Jews living in Germany – a move itself prompted by actions of the Polish government to deny them re-entry into Poland – shaped the ugly atmosphere which exploded in the so-called 'Crystal Night' pogrom of 9-10 November. And, as is generally known, the initiator here was Goebbels, who sought to exploit the situation in an attempt to re-establish his waned favour and

---

43 Adam, *Judenpolitik*, pp. 135-40; Schleunes, p. 128. Bankier (p. 14) points out that the first implementation ordinances to the Nuremberg Laws, legally defining a Jew, were reshaped to conform with Hitler's view. But Hitler's uncertainty, then anxiety to reach a compromise solution, are confirmed by Goebbels's diary notes, – *Die Tagebücher von Joseph Goebbels* (see ch. 4 note 82), vol. 2, pp. 520-1, 536-7, 540-1, entries of 1 Oct., 7 and 15 Nov. 1935.
44 *Die Tagebücher von Joseph Goebbels*, vol. 3, p. 351, entry of 30 Nov. 1937.
45 Adam, *Judenpolitik*, p. 173.

influence with Hitler. Other than giving Goebbels the green light verbally, Hitler himself took care to remain in the background, and to accept no responsibility for actions which were both unpopular with the public and castigated (though of course not from humane motives) by Nazi leaders.[46]

Previously missing sections of Goebbels' diaries, recently discovered in archives in Moscow, cast new light on the instigation of the pogrom, and on the respective roles of Hitler and Goebbels. 'I put the matter before the Führer', Goebbels noted, in his description of the gathering of the Party faithful in the Old Town Hall in Munich on the evening of 9 November 1938. 'He decides: let the demonstrations carry on. Pull back the police. The Jews should for once be made to feel the full fury of the people.' 'That is right', continued the Propaganda Minister. 'Straightaway I give directions along those lines to police and Party.' Immediately afterwards, Goebbels gave his rabble-rousing speech to the Party leaders, who then raced to the telephones to set the 'action' in motion. 'Now the people will act', wrote Goebbels. Hitler, it is clear from the diaries, also gave the order that night for the immediate arrest of 20,000–30,000 Jews.[47] The following morning, 10 November, when Goebbels reported on the progress of the pogrom, Hitler showed full agreement: 'his views are very radical and aggressive', commented Goebbels. Hitler also approved 'with minor alterations' the decree which Goebbels prepared once it was felt the time had come to break off the 'action', and also indicated his wish for 'very sharp measures' against the Jews in the economic sphere – for the compulsory restoration of their businesses without any insurance contributions, and their subsequent gradual expropriation. Again, Goebbels then gave out the 'secret decrees' to put this into practice.[48]

'Crystal Night', concludes Schleunes, 'was a product of the lack of co-ordination which marked Nazi planning on Jewish policy and the result of a last-ditch effort by the radicals to wrest control over this policy'.[49] In propaganda terms, it was a failure. But, as usual, Nazi leaders, differing in their proposals for tackling the problem, concurred in the view that radical measures were needed. Jews were now excluded from the economy, and responsibility for 'the solution of the Jewish Question', though formally entrusted to Göring, was effectively placed in the hands of the SS. Emigration, which had significantly increased in the panic after the pogrom, remained the main aim, and was to be channelled through a central office set up in January 1939. The start of the war

---

46 See Adam, *Judenpolitik*, pp. 206–7; Schleunes, ch. 7 (esp. pp. 240 ff.). In general, for the pogrom and its aftermath, Rita Thalmann and Emmanuel Feinermann, *Crystal Night: 9–10 November 1938* (London, 1974). A more recent, well-researched if journalistic, account is Anthony Read and David Fisher, *Kristallnacht. Unleashing the Holocaust* (London, 1989). A good brief analysis, locating the pogrom in the historical context of anti-semitism and discrimination against Jews in Germany, is now provided by Hermann Graml, *Reichskristallnacht. Antisemitismus und Judenverfolgung im Dritten Reich* (Munich, 1988). Engl. trans., *Antisemitism and its Origins in the Third Reich* (Oxford, 1992). An excellent collection of essays, brought out in the 50th anniversary of the pogrom and summarising much recent research, is: Walter H. Pehle, ed., *Der Judenpogrom 1938. Von der 'Reichskristallnacht' zum Völkermord* (Frankfurt an Main, 1988). Engl. trans. *November 1938. From 'Kristallnacht' to Genocide* (New York/Oxford, 1991).
47 *Der Spiegel* 29 (1992), pp. 126, 128, entry of 10 Nov. 1938. The *Spiegel* extract for the date is fuller than the curiously abbreviated version in Ralf Georg Reuth, ed., *Joseph Goebbels. Tagebücher* (Munich, 1992) vol. 3, pp. 1281–2.
48 *Der Spiegel*, 29 (1992), p. 128, entry of 11 Nov. 1938.
49 Schleunes, p. 236.

did not alter this aim. But it did alter the possibilities of its implementation.

The war itself and the rapid conquest of Poland brought about a transformation in the 'Jewish Question'. Forced emigration was no longer an option, and plans, for instance, to try to 'sell' Jews for foreign currency were not now feasible. After working on the idea of making German territory 'free of Jews', the Nazis now of course had an additional three million Polish Jews to cope with. On the other hand, there was now little need for consideration of foreign reactions, so that treatment of Polish Jews – as 'eastern Jews' particularly despised and dehumanized, the lowest form of existence in a conquered enemy itself held in contempt – reached levels of barbarity far in excess of what had taken place in Germany or Austria. Moreover, the more or less free hand given to Party and police, untrammelled by legal restraints or worries about 'public opinion', provided wide scope for autonomous individual 'initiatives' in the 'Jewish Question'.

Before considering the debate about whether the 'Final Solution' was instigated by a single, comprehensive 'Führer Order', and when such an order might have been given, it seems important to glance briefly at the process of radicalization as it gathered momentum between 1939 and 1941.

An administrative decree of 21 September 1939, in which Heydrich laid down the general lines of Jewish persecution in Poland, distinguished between a long-term 'final aim' or 'planned overall measures' – not further elucidated and to remain strictly secret – and short-term 'preliminary measures' with the intention of concentrating the Jews in larger cities around railway junctions.[50] It would be mistaken to draw the conclusion that the vaguely indicated 'final aim' meant the programmed annihilation of the actual 'Final Solution' which later evolved. Clearly, however, the operative part of the decree related to the provisional concentration of Jews for further transportation. On Himmler's order a few weeks later, on 30 October, all Jews in the north-western part of Poland, now called the Warthegau and annexed to the Reich, were to be deported into the so-called *Generalgouvernement* – the rest of German-occupied Poland under the governorship of Hans Frank – in order to make housing and jobs available for the Germans to be settled there. Hans Frank had accordingly to be prepared to receive several hundred thousand deported Jews and Poles from the Warthegau.[51] The policy of forced expulsion led unavoidably to the establishment of ghettos – the first of which was erected at Łódź (Litzmannstadt) in December 1939. Almost at the same time, compulsory labour was introduced for all Jews in the *Generalgouvernement*. The twin steps of ghettoization and forced labour provided part of the momentum which was later to culminate in the 'Final Solution'.[52] For the present, it was presumed that the deportations from the annexed areas would bring about the rapid end of the 'Jewish Question' there, and that in the *Generalgouvernement* those Jews (including women and children) incapable of work should be confined to ghettos, and Jews available for hard labour should be assigned to forced labour camps. This decision, taken at a meeting of top SS leaders in January 1940, and

[50] Peter Longerich, ed., *Die Ermordung der europäischen Juden. Eine umfassende Dokumentation des Holocaust 1941–1945* (Munich, 1989), pp. 47–8.
[51] Kurt Pätzold, ed., *Verfolgong, Vertreibung Vernichtung. Dokumente des faschistischen Antisemitismus 1933 bis 1942* (Leipzig, 1983), pp. 239–40.
[52] Pätzold, 'Vertreibung', pp. 196–7; Mommsen, 'Realisierung', p. 406.

accepting the inevitable deaths of thousands through exhaustion, hunger, and disease, marks a point at which 'the murderous anti-semitic idea, previously existing in a general, abstract form, began to take the shape of a concrete project. The decision to murder millions had at this point still not been taken. But in thought and practice a step in that direction had been taken'.[53]

In early 1940 there were still substantial differences of opinion on finding a 'solution to the Jewish Question', and there was no sign of any clear or comprehensive programme. Obviously not anticipating an early 'solution', Hans Frank indicated in a speech in March that the Reich could not be rendered 'free of Jews' during the war.[54] A few months later, Frank was faced with a demand to receive quarter of a million inhabitants of the Łódź ghetto, whom Gauleiter Greiser of the Warthegau wanted to be rid of from his domains. Frank refused, at which one of Greiser's team declared ominously that the 'Jewish Question would have to be solved in some sort of way'.[55]

'Jewish policy' in mid 1940 – by which time West European Jews had also fallen into German hands and the real possibility of an overall European 'solution' had arisen – was still in a state of confusion. Eichmann still nurtured ideas of a comprehensive programme of emigration to Palestine.[56] Attempts to further the emigration of Jews from Germany itself (mainly via Spain and Portugal) continued to be promoted well into 1941.[57] However, arbitrary deportation of Jews from eastern areas of the Reich into the *Generalgouvernement* was banned by Göring in March 1940, after Hans Frank had refused to accept any further deportees.[58] And for the 'eastern Jews' – by far the majority under German rule – emigration was in any case not an option. In June 1940 Heydrich informed Foreign Minister Ribbentrop that the 'overall problem' of the approximately three and a quarter million Jews in German-ruled territory could 'no longer be solved through emigration' and that 'a territorial solution' was therefore necessary.[59] Jewish representatives were told that a reservation in an as yet undefined colonial territory was what the government had in mind.[60] A few days earlier Franz Rademacher, head of the Jewish desk of the Foreign Office, had presented plans to create the reservation in Madagascar – a suggestion apparently approved by Himmler, mentioned by Hitler in talks with Mussolini and Ciano that same month, and finally laid to rest only at the start of 1942.[61] The reservation plans were certainly taken seriously for a while, and in the light of recent research cannot be regarded as simply a camouflage for the early stages of the 'Final Solution' itself – though undoubtedly any reservation plan would have led to physical extermination.[62]

[53] Pätzold, 'Vertreibung', p. 196.
[54] Werner Präg and Wolfgang Jacobmeyer, eds., *Das Diensttagebuch des deutschen Generalgouverneurs in Polen 1939-1945* (Stuttgart, 1975), p. 147 (entry for 4 March 1940).
[55] *Das Dienstagebuch des deutschen Generalgouverneurs*, p. 264 (entry for 31 July 1940).
[56] Mommsen, 'Realisierung', p. 407.
[57] Pätzold, 'Vertreibung', pp. 199–200; Christopher Browning, *The Final Solution and the German Foreign Office* (New York, 1978), pp. 44; Helmut Krausnick *et al.*, *The Anatomy of the SS State* (London, 1968), p. 67.
[58] Browning, *Final Solution*, p. 46; Mommsen, 'Realisierung', p. 407; Pätzold, *Verfolgung*, p. 262.
[59] Pätzold, 'Vertreibung', p. 201.
[60] Mommsen, 'Realisierung', p. 407.
[61] Browning, *Final Solution*, pp. 38, 79.
[62] Mommsen, 'Realisierung', pp. 395 note 52, 408; Pätzold, 'Vertreibung', p. 206.

Towards the end of 1940 there was no end of the Jewish ghettos in Poland apparent in the foreseeable future. At the same time, the condition of the inhabitants was worsening daily, and coming to resemble the appalling carica-ture of Jewish existence portrayed in the nauseating propaganda film of 1940, *The Eternal Jew*.[63] From the point of view of the Nazi overlords, the acute problems of hygiene, food provisioning, accommodation, and administration attached to the ghettos called out for 'a relief from the burden and a solution'. Possible ways out were already being mooted: in March 1941 Victor Brack, a leading official in the Führer Chancellory who had been in charge of the so-called 'Euthanasia Action' which had liquidated over 70,000 mental patients and others in Germany between 1939 and 1941, proposed methods for steriliz-ing between 3,000 and 4,000 Jews a day.[64]

By this time, spring 1941, the Nazi and military leadership were fully engaged in the preparations for the invasion of (and expected rapid Blitzkrieg victory over) the Soviet Union. In the war against the Bolshevik arch-enemy, the 'Jewish problem' was to enter a new dimension – the last phase before the actual 'Final Solution'. The mass shootings of Russian Jews by the SS-*Einsatzgruppen* marked a radicalization of anti-Jewish policy, which Christopher Browning justifiably labelled 'a quantum jump'.[65] This brings us back to our central concern of Hitler's personal role in the genesis of the 'Final Solution'.

The inadequacy of the sources, reflecting in good measure the secrecy of the killing operations and the deliberate unclarity of the language employed to refer to them, has led to historians drawing widely varying conclusions from the same evidence about the timing and the nature of the decision or decisions to exter-minate the Jews. Eberhard Jäckel hints that a Hitler order for the extermination of the European Jews might have been given as early as summer 1940 – on the basis of a source, which he himself admits is not a good one (the memoirs of Himmler's masseur and *confidant* Felix Kersten). However, he adjudges spring 1941 to be the period when the key decisions were taken, in the context of preparations for the Russian campaign.[66] Krausnick writes of a 'secret decree . . . that the Jews should be eliminated' being issued by Hitler not later than March 1941, in the context of the directives to shoot the political com-missars of the Red Army.[67] Hillgruber points to a verbal order of Hitler to either Himmler or Heydrich by at latest May 1941 for the systematic liquidation of Russian Jews, and implies the issuing of an order extending this to all Euro-pean Jews before the end of July 1941, when Heydrich received from Göring the commission to undertake preparations for 'a total solution of the Jewish Question' in the German sphere of influence and to submit an overall plan of measures necessary 'for the accomplishment of the final solution of the Jewish

[63] See David Welch, *Propaganda and the German Cinema 1933–1945* (Oxford, 1983), pp. 292 ff.
[64] Pätzold, 'Vertreibung', p. 204.
[65] Browning, *Final Solution*, p. 8.
[66] Eberhard Jäckel, 'Hitler und der Mord an den europäischen Juden', in Peter Märthesheimer and Ivo Frenzel, eds., *Im Kreuzfeuer: Der Fernsehfilm 'Holocaust'. Eine Nation ist betroffen* (Frankfurt am Main, 1979), pp. 151–62, here p. 156; Jäckel, *Hitler in History*, pp. 51 ff; Eberhard Jäckel, *Hitlers Herrschaft* (Stuttgart, 1986), pp. 99 ff; and Eberhard Jäckel and Jürgen Rohwer, *Der Mord an den Juden im Zweiten Weltkrieg* (Stuttgart, 1985), pp. 9–17, 190–1.
[67] Krausnick, *Anatomy*, p. 60 (and see also p. 68).

question which we desire'.[68] Most leading accounts (for instance of Reitlinger, Hilberg, Dawidowicz, and Fleming) concur in indicating a decision by Hitler to implement the 'Final Solution' during the spring or more likely the summer of 1941, and seeing this incorporated in the Göring mandate of 31 July.[69] Christopher Browning, too, emphasizes the centrality of Göring's order as reflecting a decision which Hitler had taken in the summer to extend the killing to all European Jews. However, he relativizes Hitler's decision by seeing it more in the shape of a prompting initiative rather than a clear directive, which the Führer approved and sanctioned in October or November.[70] Adam argues for a decision by Hitler in the autumn rather than the summer, at a time when the German advance in Russia had halted and vague ideas of a 'territorial solution' east of the Urals had obviously become totally illusory.[71] A more radical position is adopted by Broszat, Mommsen, and Streit, who reject altogether the existence of a single, specific, and comprehensive 'Führer Order' – written or verbal – and place the emphasis upon the cumulative 'sanctioning' of 'de facto' exterminations, initiated by other agencies and wildly escalating, between the summer of 1941 and early 1942, out of which the 'Final Solution' proper – the systematic gassing in the extermination camps – 'evolved'.[72] A similar interpretation seems implicitly offered by Hans-Heinrich Wilhelm at the end of an exhaustive study of the *Einsatzgruppen*, when he writes of a Hitler decision in the summer of 1941, but only relating to 'eastern Jews', with gradual later extension and radicalization, though not without Hitler's express agreement.[73]

Two recent studies support the case for a later date – at the earliest by the late summer or autumn of 1941 – for the shift into all-out genocide, while reaching quite different conclusions about Hitler's role. Arno Mayer sees the

[68] Andreas Hillgruber, 'Die ideologisch-dogmatische Grundlage der nationalsozialistischen Politik der Ausrottung der Juden in den besetzten Gebieten der Sowjetunion und ihre Durchführung 1941-44', *German Studies Review* 2 (1979), pp. 264-96, here p. 273, and also pp. 277-8; Andreas Hillgruber, 'Die "Endlösung" und das deutsche Ostimperium als Kernstück des rassenideologischen Programms des Nationalsozialismus', in Funke (see ch. 3 note 27), pp. 94-114, here pp. 103-5. The text of Göring's order is in Hans Buchheim *et al., Anatomie des SS-Staates* (Olten/Freiburg, 1965), vol. 2, pp. 372-3.
[69] Reitlinger, pp. 82-6; Raul Hilberg, *The Destruction of the European Jews* (New Viewpoints edn, New York, 1973), pp. 177, 257, 262; Dawidowicz, *War*, p. 169; Fleming, p. 59. Hilberg has more recently been inclined to date the Hitler order to the two weeks or so immediately following the Göring mandate. See Jäckel and Rohwer (see ref. in note 66 this chapter), pp. 125-6, 137-8.
[70] Browning, *Final Solution*, p. 8, and Christopher Browning, 'Zur Genesis der "Endlösung". Eine Antwort an Martin Broszat', *VfZ* 29 (1981), pp. 97-109, here pp. 98, 108 (also now in Engl. trans.: 'A Reply to Martin Broszat regarding the Origins of the Final Solution', *Simon Wiesenthal Center Annual* 1 (1984), pp. 113-32). For Browning's position, see above all his *Fateful Months* (New York, 1985), ch. 1, 'The Decision Concerning the Final Solution'.
[71] Adam, *Judenpolitik*, pp. 312-13. A similar date is favoured in a recent article by Shlomo Aronson, 'Die dreifache Falle. Hitlers Judenpolitik, die Alliierten und die Juden', *VfZ* 32 (1984), pp. 51-2.
[72] Broszat, 'Genesis', pp. 753 note 26, 763 ff.; Mommsen, 'Realisierung', pp. 416 and note 148, 417; Christian Streit, review of Helmut Krausnick and Hans-Heinrich Wilhelm, *Die Truppe des Weltanschauungskrieges. Die Einsatzgruppen der Sicherheitspolizei und des SD 1938-1942* (Stuttgart, 1981) in *Bulletin of the German Historical Institute, London* 10 (1982), p. 17. In his earlier book, *Keine Kameraden. Die Wehrmacht und die sowjetischen Kriegsgefangenen 1941-1945* (Stuttgart, 1978), p. 126 and p. 355 note 274, Streit appears to favour Adam's argument, though he found Broszat's then recent 'Genesis' article also 'convincing'.
[73] Krausnick and Wilhelm, pp. 634-5. The decision-making process in the 'Final Solution' was the subject of a major international conference at Stuttgart in 1984, at which all interpretations were discussed. See Jäckel and Rohwer (ref. in note 66 this chapter).

threshold to systematic mass murder crossed only once the Nazi 'crusade' against Bolshevism ran into difficulties, broadly beginning around September 1941. Even at the Wannsee Conference of 20 January 1942, the Nazis were, in Mayer's view, still only feeling their way towards the 'Final Solution'.[74] Hitler plays no specific role in Mayer's treatment, in contrast to that of the Swiss historian Philippe Burrin, who places Hitler at the centre of his interpretation while according full weight to the circumstances in which the push for a territorial solution was transformed into systematic genocide. In Burrin's analysis, the increasing difficulties of 'Operation Barbarossa' are again seen as the spur to the lurch into genocide – a move he dates to around mid August in the Soviet Union, extended to the whole of European Jewry about a month later by Hitler's reversal of his earlier position that Jews could only be deported to the east following the defeat of the Soviet Union.[75]

As these varied interpretations of leading experts demonstrate, the evidence for the precise nature of a decision to implement the 'Final Solution', for its timing, and even for the very existence of such a decision is circumstantial. Though second-rank SS leaders repeatedly referred in post-war trials to a 'Führer Order' or 'Commission', no direct witness of such an order survived the war. And for all the brutality of his own statements, there is no record of Hitler speaking categorically even in his close circle of a decision he had taken to kill the Jews – though his remarks leave not the slightest doubt of his approval, broad knowledge, and acceptance of the 'glory' for what was being done in his name.[76] Interpretation rests, therefore, on the 'balance of probabilities.'[77] We need briefly to consider the evidence in this light.

Hitler did not need to issue directives or take clear initiatives in order to promote the process of radicalization in the 'Jewish Question' between 1939 and 1941. Rather, as we have seen, the momentum was largely stimulated by a combination of bureaucratic measures emanating from the Reich Security Head Office (whose administrative consequences were not clearly envisaged), and *ad hoc* initiatives taken 'on the ground' by individuals and agencies responsible for coping with an increasingly unmanageable task. Typical of Hitler's stance was his wish, expressed towards the end of 1940, that his Gauleiter in the East should be accorded the 'necessary freedom of movement' to accomplish their difficult task, that he would demand from his Gauleiter *after 10 years* only the single announcement that their territories were purely German, and would not enquire about the methods used to bring this about.[78] His own direct role was largely confined to the propaganda arena – to public tirades of hatred and dire but vague prognostications about the fate of the Jews. The most notorious of these is his Reichstag speech of 30 January 1939, when he 'prophesied' that the war would bring about the 'annihilation (*Vernichtung*) of the Jewish race in Europe' – a prophecy to which he made frequent reference in the years to

---

[74] Mayer (see ref. in note 31, this chapter), chs. 8–9.
[75] Burrin (see ref. in note 31, this chapter), chs. 4–5.
[76] See Mommsen, 'Realisierung', pp. 391 ff. It is uncertain whether and how far Hitler was directly informed about the actual details of the killings in the East (see p. 409 and note 117), even though directives had been given to keep him in the picture regarding the 'progress' of the *Einsatzgruppen* (see Fleming, p. 123; Krausnick and Wilhelm, p. 335). For Hitler's public references to the 'Final Solution', see Kershaw, The 'Hitler Myth', pp. 243–4.
[77] Broszat, 'Genesis', p. 753; Browning, 'Zur Genesis', pp. 98, 105, 109.
[78] Cited in Krausnick and Wilhelm, pp. 626–7.

come, and which he significantly post-dated to 1 September 1939, the day of the outbreak of war.[79] This itself reflected Hitler's mental merger of the war and his 'mission' to destroy the Jews, which reached its fateful point of convergence in the conception of the 'war of annihilation' against the Soviet Union.[80]

The barbarous preparations for the attack on the Soviet Union, which implicated the *Wehrmacht*, too, in the series of criminal directives associated with the *Kommissarbefehl* – the ordered shooting of political commissars in the Soviet army – included briefings of the leaders of the *Einsatzgruppen*, and their subunits the *Einsatzkommandos*, by Heydrich on the role they were to play in the wake of the advancing army. A number of *Einsatzkommando* leaders claimed after the war that it was during these briefings that they heard of the Führer order to exterminate the Russian Jews.[81] Most historians have accepted that some blanket empowering directive from Hitler to kill the Russian Jews lay behind Heydrich's verbal instructions, and that Heydrich's more limited written order to the Higher SS and Police Leaders in the Soviet Union of 2 July 1941 targeting the liquidation of 'radical elements' in the conquered population, among them 'Jews in party and state positions' was aimed at giving some sort of justification to the *Wehrmacht* or other authorities for the mass shootings.[82] Certainly the *Einsatzgruppen* killings were from the beginning far from confined to those in party and state offices. Already on 3 July, for example, the head of the *Einsatzkommando* in Luzk had around 1,160 Jewish men shot in order, as he said, to put his stamp on the town.[83] The death-squads of *Einsatzgruppe A* in the Baltic placed a particularly liberal interpretation on their mandate.[84] The *Einsatzgruppen* ultimately came to make a major contribution to the murder of in all over two million Russian Jews; *Einsatzgruppe A* alone reported the 'execution' of 229,052 Jews by the beginning of January 1942.[85] Their detailed monthly 'reports of events' belong to the most horrific surviving relics of the Third Reich.

The vast numbers of Russian Jews massacred speaks plainly in favour of a general commission from above, rather than simply local initiatives on the part of trigger-happy units of the *Einsatzgruppen*.[86] At the same time, there was in the early stages of the invasion evidently a lack of clarity among the heads of

---

[79] Hillgruber, 'Die ideologisch-dogmatische Grundlage', pp. 271, 285 ff.; Jäckel, 'Hitler und der Mord', pp. 160–2.

[80] See esp. Hillgruber's essays on this point, references note 66 this chapter.

[81] Alfred Streim, *Die Behandlung sowjetischer Kriegsgefangener im 'Fall Barbarossa'* (Heidelberg/Karlsruhe, 1981), pp. 74–80.

[82] Krausnick, *Anatomy*, pp. 60–4; Krausnick and Wilhelm, pp. 150 ff., 634; Hillgruber, 'Die ideologisch-dogmatische Grundlage', p. 243; Heinz Höhne, *The Order of the Death's Head* (Pan Books edn, London, 1972), pp. 329–30. For controversy about the nature of the orders given to the *Einsatzgruppen*, see Browning, *Fateful Months*, pp. 17–20. The text of the order of 2 July 1941 can be found in Peter Longerich, ed., *Die Ermordung der europäischen Juden* (Munich, 1989), pp. 116–18.

[83] Streim, p. 89 note 333.

[84] Burrin, pp. 122–3.

[85] Krausnick, *Anatomy*, p. 64; Krausnick and Wilhelm, p. 619. Wilhelm's conservative estimate of the total number of murdered Russian Jews, on the basis of the most exhaustive analysis possible of incomplete evidence, is 2.2 million (Krausnick and Wilhelm, pp. 618–22). The large proportion of these killed specifically by the *Einsatzgruppen* cannot be precisely determined.

[86] Krausnick and Wilhelm, p. 634.

the *Einsatzgruppen* and other leaders of SS, Party, and police in the eastern occupied territory both about the precise scope of their task and about the nature of any long-term solution to the 'Jewish problem'. It seems likely that during the various pre-invasion briefings of the *Einsatzgruppen* there had been talk of exterminating the Jews in the Russian territories to which they were about to be sent, but that such talk was couched in ambiguous terms capable of being understood in different ways.[87] At any rate, the evidence assembled by Alfred Steim and extended in Philippe Burrin's analysis is hard to reconcile with the transmission of a specific Führer order for the extermination of Russian Jewry *before* the beginning of 'Operation Barbarossa' and suggests that the killing instructions to the *Einsatzgruppen* were initially limited, probably indeed along the lines of Heydrich's directive of 2 July 1941.

The early post-war court testimony of *Einsatzkommando* leaders about the prior existence of a Führer order has been shown to be demonstrably false, concocted to provide a unified defence of the leader of *Einsatzgruppe D*, Otto Ohlendorf, at his trial in 1947.[88] More reliable subsequent testimony by those directly involved has indicated with a high degree of plausibility that there was no knowledge of a general liquidation order before the march into the Soviet Union, and that such a mandate was provided only several weeks after the beginning of the Russian campaign.[89] There was little logic, as Streim has pointed out, in trying to stir up the local population to unleash pogroms against the Jews (which had been part of Heydrich's verbal briefings) had a general extermination order already been in existence. Moreover, at the beginning of 'Barbarossa' the guidelines of Heydrich's written order of 2 July were for the most part *broadly* adhered to.[90] Compared with the scale of the killing from around mid August onwards, the numbers shot by units of the *Einsatzgruppen* in the first weeks after the invasion were *relatively* small and overwhelmingly confined to male Jews. For example, the exceptionally brutal *Einsatzkommando 3* operating in Lithuania killed 4,239 Jews, of which 135 were women, during the month of July 1941. In August, this rose to 37,186 killed, as many as 32,430 of them after the middle of the month, while in September the victims totalled 56,459, including 26,243 women and 15,112 children.[91] The actual practice of the *Einsatzgruppen* corresponds, therefore, to the significant indicators of post-war testimony and to a number of pieces of documentary evidence that the 'Führer order' was transmitted to the *Einsatzkommandos* sometime during the month of August.[92] However, the mandate to extend the killing now to all Jews irrespective of gender and age – with its notorious culmination in the mass shooting of 33,771 Jewish men, women, and children at Babi-Yar near Kiev on 29–30 September 1941 – was not, it seems, given at a specific time in a single centralized meeting addressed by Heydrich or Himmler. Rather, it seems to have been conveyed by Himmler in discussions with the Higher Police and SS-Leaders in the eastern territories, passed by them to the leaders of the *Einsatzgruppen*, and further transmitted in individual

[87] Krausnick and Wilhelm, p. 627; Streim, pp. 88–9.
[88] Streim, p. 80.
[89] Streim, p. 83.
[90] Streim, p. 84.
[91] Burrin, pp. 124–5; see also p. 128.
[92] Streim, pp. 85–6.

briefings to the heads of the *Einsatzkommandos*.[93] That the extension of the killing in August 1941 had Hitler's approval seems unquestionable. The nature and form of the 'Führer order', and whether it amounted to an initiative by Hitler himself or was any more than the granting of approval to a suggestion by Heydrich or Himmler, is impossible to establish.

A hint that the possibility was being mooted, even before the *Einsatzgruppen* had begun their massacres of Russian Jews, of a 'solution' involving all European Jews is given in Eichmann's circular of 20 May 1941, advising of Göring's ban on Jewish emigration from France and Belgium (in order not to block any further possible emigration of German Jews) and mentioning the imminent proximity of the 'final solution of the Jewish problem' which was 'doubtless to come'.[94] It was, however, over two months later, after the death-squads had been rampaging in the Soviet Union for almost six weeks, that Heydrich received an order from Göring to prepare for 'a total solution of the Jewish question'.[95] As we noted earlier, this authorization, initiated by Heydrich and drafted for him by Eichmann for Göring's signature in the context of the expected imminent victory over the Soviet Union,[96] has frequently been interpreted as giving voice to a Hitler directive marking *the* order for the 'Final Solution'. This interpretation seems unconvincing.

Whether Hitler was directly consulted about the Göring order to Heydrich is itself doubtful. Since the order technically amounted to no more than an extension of the authority which Heydrich had been granted by Göring in 1939, Hitler's further approval was not strictly necessary.[97] In any case, as Burrin has convincingly argued, it seems almost certain that this order did *not* mark the shift to all-out genocide, but still formed part of the intention to bring about a comprehensive territorial 'solution' once the war in the east was over.[98] At the end of July 1941, victory over the USSR seemed a matter of weeks rather than months away, and Heydrich was doubtless keen to establish beyond question his authority in the administration of the 'Jewish Question', which had initially derived from the mandate Göring had given him on 24 January 1939. For his part, Hitler still adhered throughout August 1941 to the view that Jews would be deported to the east only after the end of the Russian campaign.[99] In mid September, Hitler then changed his mind and ordered the earliest possible deportation of Jews from Germany, Austria, and Czechoslovakia. The reasons for the *volte fâce* are unclear. Demands were certainly being made by Rosenberg among others to deport Jews to the east. And Hitler seems to have been gloomy around this time about the slowing advance in the east, with the mounting possibility of a prolonged struggle. He reverted in his inner circle in precisely these weeks to the lessons to be drawn from Germany's defeat in 1918 and the need to destroy the 'elements' which had undermined Germany's chance of victory in the First World War.[100] And by September of course, as

93 Streim, pp. 89–93.
94 Pätzold, *Verfolgung*, pp. 288–9; Krausnick, *Anatomy*, p. 67; Reitlinger, p. 84; Fleming, p. 57.
95 Longerich, *Ermordung*, p. 78; Reitlinger, p. 85.
96 See Jäckel and Ruhwer, p. 15.
97 See Mommsen's comments on the Göring order, 'Realisierung', pp. 409 and 417 note 149; and Browning, 'Zur Genesis', p. 105 and *Fateful Months*, pp. 22.
98 Burrin, pp. 129–31.
99 Burrin, pp. 137–8.
100 Burrin, pp. 138–9, 164–5, 168–9, 173–4.

we have noted, full-scale genocide had already been embarked upon by the *Einsatzgruppen* in the Soviet Union. The case, then, for a linkage between the physical extermination which was already comprehensively taking place in the east, the inability to bring about a territorial solution in the foreseeable future, and the mandate which Heydrich had already obtained to organize an overall solution to the 'Jewish problem' in all areas under German occupation was by September 1941 becoming a compelling one. Much speaks, therefore, in favour of Burrin's conclusion that, in the circumstance, Hitler's approval of deportation to the east was tantamount to an order to kill the Jews of Europe.[101] Perhaps, indeed, no further Führer order was given, or was needed.

The summer and autumn of 1941 were characterized by a high degree of confusion and contradictory interpretations of the aims of anti-Jewish policy by the Nazi authorities. It was a period of experimentation and resort to 'self-help' and 'local initiatives' in liquidating Jews, particularly once the transportations from the Reich and from the west of Europe had (in this case clearly on Hitler's orders) started rolling eastwards in autumn 1941, persuading Nazi bosses in Poland and Russia to adopt radical *ad hoc* measures - liquidation - to cope with the countless numbers of Jews from the west pouring into their domains and randomly deposited on their doorsteps.[102] Meanwhile the killing process was escalating rapidly - and not just in the 'Jewish Question'. Christian Streit has demonstrated how the *Wehrmacht* willingly collaborated in the multiplying barbarity of the 'war of annihilation' through its close co-operation with the *Einsatzgruppen* and by its direct involvement in the liquidation of almost two-thirds of the Soviet prisoners-of-war to fall into German hands.[103] It was initially to house Soviet captives that the then small concentration camp at Auschwitz was expanded, and the first experiments with the gas chambers there had as their victims not Jews but Soviet war prisoners.

The confusion, contradictions, and improvisations of the summer and autumn 1941 are, however, compatible with a date of September 1941 for the decision to commence the 'Final Solution' - the physical extermination of the Jews in the whole of German-occupied Europe. Rudolf Höss (the Commandant of Auschwitz), it is true, recalled after the war receiving the extermination order from Himmler in the summer of 1941. But Höss's testimony can not be relied upon, and in this case much points to the conclusion that he has erroneously pre-dated events by a year and was really referring to the summer of 1942.[104] Eichmann's testimony in Israel in 1960 was also at times inaccurate, but he claimed to remember vividly Heydrich communicating to him two or three months after the invasion of the Soviet Union that 'the Führer has ordered the physical extermination of the Jews'.[105] Though not conclusive evidence in itself, this testimony is supported by the circumstantial evidence of

---

[101] Burrin, p. 141.

[102] Brozat, 'Genesis', pp. 750 ff.; see also Mommsen, 'Realisierung', pp. 410–12.

[103] Streit, *Keine Kameraden* (see note 70 this chapter); see the review of Streit's book by Hans Mommsen, *Bulletin of the German Historical Institute, London* 1 (1979), pp. 17–23. On the behaviour of the German troops on the eastern fronts, see esp. Omer Bartov, *The Eastern Front, 1941–45, German Troops and the Barbarisation of Warfare* (London, 1985); and Omer Bartov, *Hitler's Army* (Oxford, 1991).

[104] Burrin, p. 193 note 15.

[105] Jochen von Lang, *Das Eichmann-Protokoll, Tonbandaufzeichnungen der israelischen Verhöre* (Berlin, 1982), p. 69.

the timing of the extermination developments in the autumn.[106]

On 23 October 1941 the Gestapo circulated Himmler's order banning all further Jewish emigration.[107] In the same month permission was granted to the Reich Commissar for the *Ostland* (Baltic), Hinrich Lohse, to liquidate Jews incapable of work – also those deported from Germany itself – by carbon-monoxide gassing in extermination vans devised by Viktor Brack of the Chancellory of the Führer, who had developed the gassing techniques while head of the 'euthanasia action'.[108] By this date, it is clear that not only the SS leadership, but also the Foreign Office, the Ministry for the Occupied Eastern Territories, and the Chancellory of the Führer were in the picture.[109] Construction of the extermination camp at Belzec and the extermination complex at Auschwitz-Birkenau began in all probability around November or December 1941. At the end of November the first German Jews were shot at Riga, and at the beginning of the following month the first gassings in the extermination vehicles attached to the camp at Chelmno in Poland took place. Himmler reportedly told his masseur, Kersten, in mid November that the extermination of the Jews was imminent, and at the end of the month Heydrich sent out the invitations for the Wannsee Conference, initially planned for 9 December, then postponed until 20 January 1942, whose purpose was to regulate and co-ordinate an extermination policy already underway.[110] Though some questions of method, technique, and organization had to be clarified at the conference, there could by December 1941 be little lingering doubt about the aim of anti-Jewish policy. Hans Frank told Nazi leaders in the *Generalgouvernement* that month that since Jews could not be deported from their area they had better see to their liquidation themselves.[111] And in reply to a request as to whether all Jews in the east irrespective of age, sex, and economic requirements should be liquidated, Lohse, head of the Baltic region, was told: 'The Jewish question has probably been clarified by now through verbal discussions. Economic considerations are to be regarded as fundamentally irrelevant in the settlement of the problem'.[112]

The conclusion which Browning draws from this confused evidence is that Hitler approved in late October or November 'the extermination plan he had solicited the previous summer'.[113] The evidence corresponds better, however, to the interpretation advanced by Phillippe Burrin that the Führer order to kill the Jews of Europe was given about September 1941, and was probably synonymous with the order to deport the Jews to the east. Whether such an order arose directly from Hitler's own initiative or whether – which on balance

---

[106] Browning, 'Zur Genesis', pp. 100–1.
[107] Pätzold, *Verfolgung*, p. 306; Krausnick, pp. 68–9. Himmler had noted the ending of emigration following a telephone conversation with Heydrich five days earlier, see Burrin, p. 146.
[108] Krausnick, *Anatomy*, pp. 96–8; Browning; 'Zur Genesis', pp. 101–2; Fleming, pp. 81–4.
[109] Browning, 'Zur Genesis', p. 102. The Foreign Ministry's 'Jewish experts' apparently still imagined that 'a basic solution of the Jewish Question' would be brought about after the war (whose end was at that date presumed to be imminent): see Browning, *Final Solution*, p. 66.
[110] This chronology follows Browning, 'Zur Genesis', pp. 106–7.
[111] Browning, 'Zur Genesis', p. 107.
[112] *IMT*, 32, pp. 436–7, Docs. 3663-PS, 3666-PS.
[113] Browning, 'Zur Genesis', p. 107 ('A Reply', p. 126). See also the balanced assessment of Wolfgang Scheffler, 'Zur Entstehungsgeschichte der "Endlösung" ', *APZ* (30 Oct. 1982), pp. 3–10.

seems more likely – it was prompted by the Reich Security Head Office, cannot be ascertained and is in itself not important. But if distinguishable at all from the deportation order, the order setting the 'Final Solution' in train almost certainly took the form of a verbal blanket empowering. This was all that was needed. The rest could be left to Himmler, Heydrich, and their minions.

Though Hitler's precise role remains hidden in the shadows, and – given the nature of the cloudy evidence – some points of interpretation are inevitably still debatable, the findings of recent research have allowed a relatively high level of consensus to emerge from the earlier scholarly disagreement on the complex process of decision-making which led to the full emergence of the 'Final Solution' by the spring of 1942. Summarizing the above evaluation, it is now widely accepted that the orders given to the *Einsatzgruppen* prior to the invasion of the Soviet Union marked a vital step in the direction of genocide. Disagreement remains on the content of the instructions, though the evidence assembled by Streim and Burrin carries conviction that an initially somewhat limited mandate for liquidation became widened once the *Einsatzgruppen* were actually at work in the Soviet Union and was expanded into a full genocidal remit around August 1941. It is also now generally accepted that central direction was plainly visible from the start, though much scope was left to local initiatives to force the pace; and that the whole development was unthinkable without some sort of order by Hitler, first in the case of the Soviet Union then to instigate the full-scale 'Final Solution', though such orders may well have amounted to no more than a signal to Himmler and Heydrich – perhaps, as Browning suggests, amounting to no more than a 'nod of the head' – rather than a specific and unequivocal command.[114] Finally, it is generally agreed that, despite overlapping and confused developments between mid 1941 and spring 1942, the basic contours of the annihilation programme were already taking clear shape by October 1941. There is still disagreement about when precisely the decision to extend the extermination programme to the whole of European, not just Russian, Jewry was taken, whether such a decision was taken with Germany seemingly predestined to glorious triumph in the eastern campaign or as vengeful retaliation for a faltering 'crusade' against the Jewish–Bolshevic enemy. The balance of the evidence points, as the above remarks have indicated, in the direction of Burrin's dating of late summer 1941.

Relating this discussion of the genesis of the 'Final Solution' to the polarized 'Hitlerist' and 'structuralist' interpretations – the one emphasizing a Hitler order as the culmination of a planned long-term programme directed towards extermination, the other stressing a process of permanent improvisation as a way out of self-made administrative difficulties – one would have to conclude that neither model offers a wholly satisfactory explanation, and that some room for compromise is obvious.

For all the unparalleled barbarity of his language, Hitler's direct actions are difficult to locate. Though his hatred of the Jews was undoubtedly a constant, the relationship of this hatred to actual policy changed considerably over time as the policy options themselves narrowed. Hitler himself took relatively little part in the overt formulation of that policy, either during the 1930s or even the genesis of the 'Final Solution' itself. His major role consisted of setting the

---

[114] Browning, 'Zur Genesis', p. 105; *Fateful Months*, p. 22.

vicious tone within which the persecution took place and providing the sanction and legitimation of initiatives which came mainly from others. More was not for the most part necessary. The vagaries of anti-Jewish policy both before the war and in the period 1939–41, out of which the 'Final Solution' evolved, belie any notion of 'plan' or 'programme'. The radicalization could occur without any decisive steerage by Hitler. His influence was, however, all-pervasive, and his direct intervention in anti-Jewish policy was on occasion crucial. Above all, his dogmatic, unwavering assertion of the ideological imperative – 'getting rid of the Jews' from Germany, then finding a 'final solution to the Jewish question' – which had to be translated into bureaucratic and executive action, was the indispensable prerequisite for the escalating barbarity and the gradual transition into full-scale genocide.

Without Hitler's fanatical will to destroy Jewry, which crystallized only by 1941 into a realizable aim to exterminate physically the Jews of Europe, the Holocaust would almost certainly not have come about. But it would also not have become reality, as Streit has emphasized,[115] without the active collaboration of the *Wehrmacht* – the one force still capable of checking the Nazi regime; or, for that matter, without the consent ranging to active complicity of the civil service bureaucracy, which strived to meet the requirements of spiralling discrimination, or the leaders of Germany's industries, who manufactured the death machinery and set up their factories at the concentration camps.[116] And within the SS–SD–Gestapo organizational complex, it was less the outright racial fanatics so much as the ambitious organizers and competent administrators like Eichmann and ice-cold executioners like Höss who turned the hellish vision into hell on earth.[117]

The lengthy but gradual process of depersonalization and dehumanization of Jews, together with the organizational chaos in the eastern territories arising from the lack of clear central direction and concept, the hording together in the most inhumane circumstances of increasing masses of 'non-persons', provided the context in which mass killing, once it had been instigated in the Russian campaign, was applied *ad hoc* and extended until it developed into full-scale annihilation. At the same time, the 'Final Solution' did not simply emerge from a myriad of 'local initiatives': however falteringly at first, decisive steps were taken at the centre to co-ordinate measures for total extermination. Such central direction appears for the most part to have come from the Reich Security Head Office, though undoubtedly the most important steps had Hitler's approval and sanction.

---

[115] Streit, *Keine Kameraden*, esp. chs. 3, 6, 13.

[116] See ch. 3 for a brief discussion of the economic context in which the 'Final Solution' came about.

[117] Hannah Arendt's controversial report of the Eichmann trial ended: 'The trouble with Eichmann was precisely that so many were like him, and that the many were neither perverted nor sadistic, that they were, and still are, terribly and terrifyingly normal' (Arendt, *Eichmann* (see this chapter note 7), p. 253; see also pp. 18–31). According to their editor, Höss's autobiographical recollections reveal him as a 'petty-bourgeois, normal person' rather than a sadistic brute: Martin Broszat, ed., *Kommandant in Auschwitz. Autobiographische Aufzeichungen des Rudolf Höß* (dtv-edn., Munich, 1978), p. 15. Ideological anti-semitism seems at best to have provided a secondary motive in these cases, as it does in the career of Franz Stangl, Commandant at Treblinka death-camp: see Gitta Sereny, *Into that Darkness* (London, 1974). However, it has to be added that there is no intrinsic contradiction between ideological conviction and managerial talent.

Hitler's 'intention' was certainly a fundamental factor in the process of radicalization in anti-Jewish policy which culminated in extermination. But even more important to an explanation of the Holocaust is the nature of 'charismatic' rule in the Third Reich and the way it functioned in sustaining the momentum of escalating radicalization around 'heroic', chimeric goals while corroding and fragmenting the structure of government. This was the essential framework within which Hitler's racial lunacy could be turned into practical politics.

This examination of the complex development of racial policy, lying at the very heart of Hitler's *Weltanschauung*, has shown that, while it would be meaningless to speak to him as a 'weak dictator', it is also misleading to regard the Third Reich as a dictatorship with a coherent, unitary command structure providing for the regulated and centrally directed consistent implementation of Hitler's will. It remains to turn our attention to the area where Hitler's directing hand seems most evident: foreign policy.

6

Nazi Foreign Policy: Hitler's 'Programme' or 'Expansion without Object'?

Several important aspects of German foreign policy in the Third Reich are still unresolved issues of scholarly debate. In this sphere too, however, interpretations – especially among West German scholars – have come to be divided in recent years around the polarized concepts of 'intention' and 'structure', which we have encountered in other contexts. Research in the GDR before the revolution of 1989–90 showed no interest in this division of interpretation, and proceeded on the basis of predictably different premises, concentrating on documenting and analysing the expansionist aims of Germany's industrial giants – a task which was accomplished with no small degree of success. Nevertheless, with all recognition of the imperialist aspirations of German capitalism, explanations which limit the role of Hitler and other leading Nazis to little more than that of executants of big business aims have never carried much conviction among western scholars. Conventional orthodoxy in the West, resting in good measure upon West German scholarship, has in fact, as we saw in an earlier chapter, tended to turn such explanations on their heads in advocating an uncompromising 'primacy of politics' in the Third Reich. And whatever the nuances of interpretation, Hitler's own steerage of the course of German aggression in accordance with the 'programme' he had outlined (for those with eyes to see) in *Mein Kampf* and the *Second Book* is generally and strongly emphasized. Parallel to explanations of the Holocaust, outright primacy is accorded to Hitler's ideological goals in shaping a consistent foreign policy whose broad outlines and objectives were 'programmed' long in advance.

Such an interpretation has in recent years been subjected to challenge by historians seeking to apply a 'structuralist' approach to foreign policy as to other aspects of Nazi rule – even if the 'structuralist' argument appears in this area to be on its least firm ground. Exponents of a 'structuralist' approach reject the notion of a foreign policy which has clear contours unfolding in line with a Hitlerian ideological 'programme' in favour of an emphasis upon expansion whose format and aims were unclear and unspecific, and which took shape in no small measure as a result of the uncontrollable dynamism and radicalizing momentum of the Nazi movement and governmental system. In this gradual and somewhat confused process of development – as in the 'Jewish Question' – terms such as '*Lebensraum*' served for long as propaganda slogans and 'ideological metaphors' before appearing as attainable and concrete goals. Again, the *function* of Hitler's foreign-policy image and ideological fixations rather than his direct personal intervention and initiative is stressed. And rather than picturing Hitler as a man of unshakeable will and crystal-clear vision, moulding events to his liking in accordance with his ideological aims, he is portrayed as 'a man of improvization, of experiment, and the spur-of-the-moment bright idea'.[1] Any 'logic' or inner 'rationality' of the course of German foreign policy gains its appearance, it is argued, only teleologically – by looking at the end results and interpreting these in the light of Hitler's apparently prophetic statements of the 1920s.

Before attempting a brief evaluation of Hitler's role in the making of foreign policy decisions, the part played by his ideological fixations in determining the development of foreign policy, and the extent of Nazi expansionist ambitions, we need to examine in rather greater detail the main trends in historiography and the arguments of leading exponents of the interpretations just indicated.

Interpretations

Exactly what objectives Hitler was pursuing has long been a matter of debate among experts on German foreign policy. Two long-standing areas of controversy – whether Hitler was an ideological visionary with a 'programme' for aggression or merely a supremely 'unprincipled opportunist', and whether his foreign policy aims were novel and revolutionary or in essence a continuation of traditional German expansionism – can be seen in embryonic fashion in the antagonistic positions taken up long ago by the British historians Trevor-Roper and Taylor. While Taylor argued (somewhat capriciously as usual) that 'in international affairs there was nothing wrong with Hitler except that he was a German',[2] Trevor-Roper was among the first historians to deduce – what now seems fairly commonplace – a fundamental and unmoveable consistency in Hitler's ideas and in fact to take Hitler seriously as a genuine man of ideas which, however repulsive, were novel and broke through traditional boundaries of political thinking.[3] In a way, both views were traceable to different readings of (among other texts) the sometimes ambivalent comments of Hermann Rauschning, the former President of the Danzig Senate.[4] It was, of course, soon pointed out that there was no necessary contradiction between the interpretations as they stood: Hitler could be seen both as a fixated ideologue, and as a man with a particular talent for exploiting the needs of opportunities which were presented to him in foreign affairs.[5]

Once advanced, however, the conception of Hitler as a fanatical visionary pursuing his defined objectives with relentless consistency rapidly established

---

[1] Hans Mommsen, review of Jacobsen (see ch. 4 note 31), p. 183.

[2] A.J.P. Taylor, *The Origins of the Second World War* (Harmondsworth, 1971), p. 27.

[3] H.R. Trevor-Roper, 'Hitlers Kriegsziele', *VfZ* 8 (1960), pp. 121–33.

[4] See Hermann Rauschning, *Hitler Speaks* (London, 1939) and *The Revolution of Nihilism* (New York, 1939). Indispensable to an evaluation of Rauschning's evidence is Theodor Schieder, *Hermann Rauschnings 'Gespräche mit Hitler' als Geschichtsquelle* (Opladen, 1972). Wolfgang Hänel, *Hermann Rauschnings 'Gespräche mit Hitler' – Eine Geschichtsfälschung* (Ingolstadt, 1984), casts grave doubt on the authenticity of Rauschning's evidence. But see on this the comment of Martin Broszat, 'Enthüllung? Die Rauschning-Kontroverse' in his *Nach Hitler. Der Schwierige Umgang mit unserer Geschichte* (Munich, 1986), pp. 249–51.

[5] See Alan Bullock, 'Hitler and the Origins of the Second World War', in Esmonde M. Robertson, ed., *The Origins of the Second World War* (London, 1971), pp. 189–224, here esp., pp. 192–3.

itself. Major studies, especially those exploring German foreign policy, were now erected on the premise that Hitler's expansionist ideology had to be regarded with deadly seriousness, and that the underestimation of Hitler within and outside Germany had been one fatal key to his success. The emphasis which Trevor-Roper had laid upon the seriousness of Hitler's *Lebensraum* plans for eastern Europe was now extended by Günter Moltmann who, for the first time, advanced the argument that Hitler's aims were not confined to Europe but were quite literally directed at world mastery for Germany.[6] This claim was soon more systematically worked out in Hillgruber's analysis of Hitler's war strategy, published in 1963, in which the concept of a three-stage plan (*Stufenplan*) for establishing German hegemony first over the whole of Europe, then over the Middle East and other British colonial territory, and finally – at a distant future date – over the USA and with that the entire world, was advanced as the basis of Nazi foreign policy.[7] The heuristic device of the 'stage by stage plan' set the tone for most later influential work on foreign policy, prominent among which was Klaus Hildebrand's massive study of German colonial policy.[8] More recently, the 'world domination' thesis has been further supported in analyses of German naval plans, grandiose architectural projects, and policies towards Britain's Middle-Eastern possessions.[9]

A 'sub-debate' rumbled on between the 'continentalists' (such as Trevor-Roper, Jäckel, and Kuhn), who saw Hitler's 'final aims' as comprising the conquest of *Lebensraum* in eastern Europe, and the 'globalists' (Moltmann, Hillgruber, Hildebrand, Dülffer, Thies, Hauner, and others), whose interpretation – the dominant one – accepted nothing short of total world mastery as the extent of Hitler's foreign ambitions. Common to both positions, however, was the emphasis upon the intrinsically related components of conquest of *Lebensraum* and racial domination as programmatic elements of Hitler's own *Weltanschauung* and as the essence of his politics. Concepts such as that of the 'stage by stage plan' (*Stufenplan*) or 'programme' are, it is emphasized, not intended to denote a 'timetable' for world domination, but rather to encapsulate 'the essential driving forces and central aims of Hitler's unshakeable foreign policy (conquest of *Lebensraum*, racial domination, world power status), without mistaking the "improvisation" of the Dictator and the high measure of his tactical flexibility'.[10] Whether 'continentalist' or 'globalist', German foreign policy, in the interpretations summarized so far, was Hitler's foreign policy. One historian, for instance, advancing a representative view of Hitler's personal role in determining Nazi foreign policy, sees him 'within the framework of the totalitarian state' as 'not only the final arbiter but also its

[6] Günter Moltmann, 'Weltherrschaftsideen Hitlers', in O. Bruner and D. Gerhard, eds., *Europa und Übersee. Festschrift für Egmont Zechlin* (Hamburg, 1961), pp. 197–240.
[7] Hillgruber, *Hitlers Strategie* (see ch. 1 note 17).
[8] Hildebrand, *Vom Reich zum Weltreich* (see ch. 1 note 17).
[9] Jost Dülffer, *Weimar, Hitler und die Marine. Reichspolitik und Flottenbau 1920–1939* (Düsseldorf, 1973); Jochen Thies, *Architekt der Weltherrschaft. Die 'Endziele' Hitlers* (Düsseldorf, 1976); Milan Hauner, *India in Axis Strategy: Germany, Japan, and Indian Nationalists in the Second World War* (Publications of the German Historical Institute, London, Stuttgart, 1981).
[10] Klaus Hildebrand, 'Die Geschichte der deutschen Außenpolitik (1933–1945) im Urteil der neueren Forschung: Ergebnisse, Kontroversen, Perspektiven', ('Nachwort' to the fourth edition of his *Deutsche Außenpolitik 1933–1945. Kalkül oder Dogma?* (Stuttgart etc., 1980), pp. 188–9. Hildebrand has consistently advanced this view in many publications.

chief animator'.[11] So important was the Führer to the development of German foreign policy that the same historian, Milan Hauner, in another essay expounding the aim of world dominion, felt it necessary to 'warn the reader that in this survey the name "Hitler" will be frequently used in place of "Germany" ' – the apogee of the 'Hitlerist' interpretation; for such, in his view, 'was the charismatic appeal of this man and the totalitarian character of his power, that Hitler can justifiably be seen as the personification of Germany's will-power from the moment he assumed full control over her foreign and military affairs'.[12] Hauner ends by repeating Norman Rich's epithet of Hitler as 'master in the Third Reich'. Equally uncompromising is the statement of Gerhard Weinberg, one of the foremost authorities of Nazi foreign policy, at the end of his exhaustive diplomatic history of the pre-war years: 'The power of Germany was directed by Adolf Hitler. Careful analyses by scholars have revealed internal divisions, organizational confusions, jurisdictional battles, institutional rivalries, and local deviations behind the façade of monolithic unity that the Third Reich liked to present to its citizens and to the world in word and picture. The fact remains, however, that the broad lines of policy were determined in all cases by Hitler himself. Where others agreed, or at least did not object strenuously, they were allowed the choice of going along or retreating into silence, but on major issues of policy the Führer went his own way'.[13]

Serious attempts to challenge this dominant orthodoxy which emphasizes the autonomy of Hitler's programmatic aims in determining foreign policy have come from a number of different directions. They might conveniently be fitted into three interlocking categories:

(i) Rejection of any notion of a 'programme' or 'plan in stages', denial of concrete and specific long-range foreign policy aims, and portrayal of Hitler as a man of spontaneous response to circumstances – not far removed from the image of the 'unprincipled opportunist' – with a central concern in propaganda exploitation and the protection of his own prestige.

(ii) The claim that Hitler was not a 'free agent' in determining foreign policy, but was subjected to pressures from significant élite groups (*Wehrmacht* leadership, industry etc.), from a variety of agencies involved in making foreign policy, from the demands of the Party faithful for action consonant with his wild promises and propaganda statements (with the corresponding need to act to maintain his Führer image), from the international constellation of forces, and from mounting economic crisis.

(iii) The view that foreign policy has to be seen as a form of 'social imperialism' an outward conveyance of domestic problems, a release from or compensation for internal discontent with the function of preserving the domestic order.

The most radical 'structuralist' approach, that of Hans Mommsen, returns in part, in its emphasis on Hitler's improvised, spontaneous responses to

---

[11] Milan Hauner, 'The Professionals and the Amateurs in National Socialist Foreign Policy: Revolution and Subversion in the Islamic and Indian World', in Hirschfeld and Kettenacker (see ch. 1 note 23), pp. 305–28, here p. 325.
[12] Milan Hauner, 'Did Hitler want a World Dominion?', *JCH* 13 (1978), p. 15.
[13] Gerhard Weinberg, *The Foreign Policy of Hitler's Germany. Starting World War II* (Chicago/London, 1980), p. 657.

developments which he did little directly to shape, to the early view of the German Dictator as little more than a gifted opportunist. In Mommsen's view, 'it is questionable, too, whether National Socialist foreign policy can be considered as an unchanging pursuit of established priorities. Hitler's foreign policy aims, purely dynamic in nature, knew no bounds; Joseph Schumpeter's reference to "expansion without object" is entirely justified. For this very reason, to interpret their implementation as in any way consistent or logical is highly problematic. . . In reality, the regime's foreign policy ambitions were many and varied, without clear aims, and only linked by the ultimate goal: hindsight alone gives them some air of consistency' – a danger implicit in such concepts as 'programme' or 'stage-by-stage plan'.[14] According to Mommsen, Hitler's behaviour in foreign as in domestic and anti-Jewish policy was shaped largely – apart, that is, from the demands of the international situation – by considerations of prestige and propaganda. Seen in this light, then, Nazi foreign policy was 'in its form domestic policy projected outwards, which was able to conceal (*überspielen*) the increasing loss of reality only by maintaining political dynamism through incessant action. As such it became ever more distant from the chance of political stabilization'.[15]

A not dissimilar interpretation was advanced by Martin Broszat, who also saw little evidence of a design or plan behind Hitler's foreign policy.[16] Rather, the pursuit of *Lebensraum* in the East – parallel to the case of anti-semitism – has, he argued, to be regarded as reflecting Hitler's fanatical adherence to the need to sustain the dynamic momentum he had helped unleash. In foreign policy this meant above all breaking all shackles of restraint, formal bonds, pacts or alliances, and the attainment of complete freedom of action, unrestricted by international law or treaty, in German power-political considerations. The image of unlimited land in the East, according with traditional mythology of German colonization, with utopian ideals of economic autarky, re-agrarianization, and the creation of a master-race, meant that *Lebensraum* (matching as it did also expansionist aims of the First World War) was perfectly placed to serve as a metaphor and touchstone for German power-politics in which, as in the 'Jewish Question' and by equally circuitous route, the distant symbolic vision gradually emerged as imminent and attainable reality. The absence of any clear thinking by Hitler before 1939 on the position of Poland, despite the fact that its geographical situation ought to have made it a central component of any concrete notions of an attack on the Soviet Union, is seen by Broszat as one example of the nebulous, unspecific, and essentially 'utopian' nature of Hitler's foreign policy goals. He reached the conclusion, therefore, that 'the aim of winning *Lebensraum* in the east had until 1939 largely the function of an ideological metaphor, a symbol to account for ever new foreign political activity'. Ultimately, for Broszat, the plebiscitary social dynamic of the 'Movement', which in the sphere of foreign policy pushed Hitler and the regime inexorably in the direction of turning the *Lebensraum* metaphor into

---

[14] Mommsen, 'National Socialism: Continuity and Change', p. 177; see also his 'Ausnahmezustand', p. 45 and *Adolf Hitler*, pp. 97, 102 (full references, ch. 4 note 28).

[15] Mommsen, 'Ausnahmezustand', pp. 43–5.

[16] See Broszat, 'Soziale Motivation' (see ch. 4 note 27), esp. pp. 407–9.

reality, was, in its demand for ceaseless action, the only guarantee of any form of integration and diversion of 'the antagonistic forces' in the Third Reich. As a consequence, it was bound to veer further and further from rational control, and to end in 'self-destructive madness'. And though Hitler remains indispensable to the explanation of developments, he ought not to be envisaged as an autonomous personality, whose arbitrary whim and ideological fixations operated independently of the social motivation and political pressures of his mass following.

Tim Mason's interpretation, which we already encountered in chapter 4, can be regarded as a third variant of 'structural' approaches to Nazi foreign policy. In Mason's view, the domestic–economic crisis of the later 1930s greatly restricted Hitler's room for manoeuvre in foreign affairs and war preparation, and an inability to come to terms with the growing economic crisis forced him back on the one area where he could take 'clear, world-historical decisions': foreign policy.[17] More recently, Mason again argued that the later 1930s bore more the hallmarks of confusion than of a programmatic line of development in Hitler's foreign policy.[18] Mason's own emphasis on the 'legacy of 1918' and the compulsion this brought to bear on German foreign as well as domestic policy meant that for him – as in somewhat different ways for Mommsen and Broszat – Nazi foreign policy and the war itself could be seen under the rubric of the 'primacy of domestic politics', as a barbarous variant of social imperialism.[19]

Other historians have in recent years also attempted to diffuse what they regard as an unduly Hitler-centric treatment of German foreign policy by applying 'polycratic' or 'pluralist' models to the decision-making processes in foreign affairs. Wolfgang Schieder, for instance, took as a case-study the circumstances of Germany's decision in July 1936 to intervene in the Spanish Civil War, arguing that the crucial factor in determining intervention was Göring's interest in acquiring Spanish raw materials. The initial pressure for participation – against German foreign ministry advice – came from representatives of the Party's *Auslandsorganisation*, who engineered an audience with Hitler between opera performances at the Bayreuth Festival. Hitler himself took no initiative before deciding to intervene after deliberations (which excluded the foreign ministry) with Göring, Blomberg, and Canaris. Schieder's conclusion was that Nazi policy on the Spanish Civil War, 'while not an arbitrary product of chance decisions', was 'also not the calculated result of long-term planning', but rather a combination of both, as, he suspected, was Nazi foreign policy in general. In his opinion, any notion of a 'programmatic' Hitlerist foreign policy had to see it on two levels: ideologized global aims, in which Hitler showed 'unusually fanatical consistency'; and relatively definable objectives, where Hitler was extremely flexible and where concrete decisions followed. In this sense, Hitler's foreign policy could be interpreted neither as the putting into operation of a long-term programme, nor simply as the product of an 'objectless nihilism'. Rather, it consisted of 'a frequently contradictory mixture of

---

[17] Mason, *Sozialpolitik* (see ch. 4 note 62), p. 40.
[18] Mason, 'Intention and Explanation' (see ch. 1 note 25), pp. 32–3.
[19] Mason, *Sozialpolitik*, p. 30, and 'The Legacy of 1918' (see ch. 4 note 27), p. 218.

dogmatic rigidity in fundamentals and extreme flexibility in concrete matters', between which, however, there was no necessary connection.[20] The trouble with Schieder's case-study, as he himself realized, was that since Spain did not play a primary role in Hitler's ideological constructs and whatever long-term strategic thinking he might have had, a convincing *general* case could hardly be drawn from this example. Furthermore, Hitler's own considerations in this issue, as opposed to those of Göring, do appear to have been primarily ideological – the 'fight against Bolshevism' – which on the whole tends to confirm rather than contradict any argument about consistency in his thought, motivation, and policy-making. And whatever the influence of Göring (and War Minister Blomberg), the decision to involve Germany in the Spanish arena appears to have been taken by Hitler alone.

Other approaches to what has been somewhat misleadingly dubbed 'pluralistic' foreign policy formulation also seem compatible with the 'intentionalist' interpretation. Hans-Adolf Jacobsen, for example, and more recently Milan Hauner, have analysed the many agencies involved in foreign policy, with their different functions and policy emphases. Jacobsen was prepared to accept that centrifugal forces influenced 'the structure of the totalitarian system' far more than pure will and directives to ideological unity, and saw the presence of 'lack of system' and 'administrative chaos' also in the sphere of foreign policy. Nevertheless, it is mistaken in his view to attribute the development of foreign policy to absence of planning or pure opportunism. Rather, there was a consistent basic line in foreign policy common to all individuals or groups involved in the formulation of foreign policy, where here – as in other branches of policy – they were striving to put into concrete form what they presumed to be Hitler's intentions (which Jacobsen interprets as the striving for a racially new formation of Europe, a revolutionary goal consistently held by Hitler since the 1920s).[21] Milan Hauner reached similar conclusions. Conflict between the Foreign Office professionals and other agencies with a finger in the foreign policy pie was not about different conceptions of foreign policy, but was merely a part of the tug-of-war for power and influence which was endemic to the Nazi system. Once more, there was no contradiction between such institutional or personal rivalries together with the conflicting interests and influences which ensued, and the developments of a central line of policy-making in which Hitler's personal role was the decisive element.[22]

[20] Wolfgang Schieder, 'Spanischer Bürgerkrieg und Vierjahresplan. Zur Struktur nationalsozialistischer Außenpolitik', in Wolfgang Michalka, ed., *Nationalsozialistische Außenpolitik* (Darmstadt, 1978), pp. 325–59; see also, William Carr, *Hitler. A Study in Personality and Politics* (London, 1978), p. 52; Gerhard Weinberg, *The Foreign Policy of Hitler's Germany. Diplomatic Revolution in Europe 1933–36* (Chicago/London, 1970), pp. 288–9; and Hans-Henning Abendroth, 'Deutschlands Rolle im Spanischen Bürgerkrieg', in Funke (see ch. 3 note 27), pp. 471–88, here pp. 473–7, where Hitler's ideological interest is advanced as the main cause of Germany's entry, with Göring initially opposed. Unduly sharp in his criticism of Schieder is Hofer (see ch. 1 note 2), pp. 12–13. Conflicting interpretations of Göring's role and motivation in the decision to support Franco can be found in the recent biographies of Göring: Stefan Martens, *Hermann Göring* (Paderborn, 1985), pp. 65–7; and Alfred Kube, *Pour le mérite und Hakenkreuz* (Munich, 1986), pp. 163–6.
[21] Hans-Adolf Jacobsen, 'Zur Struktur der NS-Außenpolitik 1933–1945', in Funke (ch. 3 note 27), pp. 137–85, here esp, pp. 169–75. Hitler's consistent 'striving towards a goal' (*Zielstrebigkeit*) is emphasized even more sharply in Jacobsen's massive monograph on Nazi foreign policy – a point strongly criticized by Hans Mommsen in his review of this work (see ch. 4 note 31 for references).
[22] Hauner, 'Professionals', p. 325.

The notion of 'concept pluralism' – a rather grandiose term to imply that there were a number of different views among the leaders of the Third Reich about the foreign policy Germany should pursue – has recently been taken a step further by Wolfgang Michalka in his analysis of Ribbentrop's own foreign-policy ideas and influence upon Hitler. Michalka argues that from the mid 1930s onwards an anti-English rather than essentially anti-Russian policy provided the main thrust of Ribbentrop's own conception of foreign policy – one which was more pragmatically power-political than directly aligned to Hitler's fixation in race ideology. He demonstrates how, in the later 1930s, Hitler's increasing recognition of the failure to win over England allowed Ribbentrop a considerable scope for exerting influence, culminating in the signing of the Non-Aggression Pact with the Soviet Union in 1939. This temporary and opportunistic use of Ribbentrop's 'conception' between 1939 and 1941 was in Michalka's view, however, bound to founder ultimately on the primacy of Hitler's racial 'programme' directed at the Soviet Union. Ultimately, therefore, Michalka comes down on the side of a very 'intentionalist' position, if one moderated by looking to important influences upon the Dictator.[23]

None of the 'structural–functionalist', 'concept pluralist', or 'polycratic' approaches to foreign policy which we have rapidly summarized here has shaken the conviction of the 'intentionalists' (or 'programmatists') that the character and consistency of Hitler's ideology was the crucial and determining element in the equation. Indeed, as we have just seen, the leading studies of the varying centres of influence in the formation of foreign policy all come down ultimately to similar or compatible conclusions. Klaus Hildebrand, articulating as ever the 'programmatist' line in its clearest and most forthright form, rejects 'revisionist' interpretations on four grounds: *(1)* They ignore the relatively high degree of autonomy of Hitler's programme, whose aims were formulated by the Dictator himself as intentions which were then put into effect. *(2)* Anti-semitism and anti-Bolshevism were not in the first instance functional in character, but ought to be regarded as primary and autonomous, 'real' political aims. *(3)* The 'revisionists' stand in danger in this respect of mistaking the consequences of Hitler's policies for their motives. *(4)* The dynamic of the system, which, Hildebrand accepts, Hitler could control only with increasing difficulty, never posed the Dictator with unacceptable fundamental alternatives, but rather pushed him 'programmatically' in the direction of the 'final aims' which he had set, even if affecting the realization of these goals.[24]

Though each of these assertions is, of course, open to debate, the important fourth point suggests that – as in the case of domestic and race policy – interpretations are less far apart than they appear to be at first sight, and that therefore some degree of synthesis seems possible. An evaluation of the debate

[23] See Wolfgang Michalka, 'Die nationalsozialistische Außenpolitik im Zeichen eines "Konzeptionen-Pluralismus" – Fragestellungen und Forschungsaufgaben', in Funke, pp. 46–62; 'Vom Antikominternpakt zum Euro-Asiatischen Kontinentalblock. Ribbentrops Alternativkonzeptionen zu Hitlers außenpolitischem "Programm" ', in Michalka, ed., *Nationalsozialistische Außenpolitik*, pp. 471–92; and his major work, *Ribbentrop und die deutsche Weltpolitik 1933–1940. Au-enpolitische Konzeptionen und Entscheidungsprozesse im Dritten Reich* (Munich, 1980). See also summaries of his position in English: 'Conflicts within the German Leadership on the Objectives and Tactics of German Foreign Policy, 1933–9', in Wolfgang J. Mommsen and Lothar Kettenacker, eds., *The Fascist Challenge and the Policy of Appeasement* (London, 1983), pp. 48–60; and 'From the Anti-Comintern Pact to the Euro-Asiatic Bloc: Ribbentrop's Alternative Concept of Hitler's Foreign Policy Programme', in Koch, *Aspects of the Third Reich*, pp. 267–84.
[24] Hildebrand, 'Nachwort' (see note 10, this chapter), p. 191.

on the aims and execution of German foreign policy in the Third Reich might focus on three central issues: *(1)* Were the key decisions in the sphere of foreign policy taken by Hitler himself? Did they simply voice a consensus which had already been reached, or were they taken in the face of weighty advice offering alternative policy? And to what extent was Hitler curtailed in his freedom of action in taking foreign policy decisions? *(2)* How far is it possible to see in the course of German foreign policy an inner consistency (subject to tactical 'deflections') determined by Hitler's ideological obsessions, without imposing this consistency in teleological fashion? *(3)* Was the extent of Hitler's foreign policy ambition European or literally world domination? The following pages provide an attempt to assess the arguments and evidence for answering these questions.

Evaluation

I

These seems little disagreement among historians that Hitler did personally take the 'big' decisions in foreign policy after 1933. Even the most forceful 'structuralist' analyses accept that Hitler's 'leadership monopoly' was far more in evidence in the foreign-policy decision-making process than in the realm of domestic policy.[25] There is less agreement, however, about the extent to which Hitler stamped a peculiarly personal mark on the development of German foreign affairs and whether 1933 can be seen to indicate a break in German foreign policy deriving from Hitler's own ideological pre-possessions and 'programme'.[26] The question of the continuity or discontinuity of German foreign policy after 1933 lies, therefore, at the centre of the first part of our enquiry.

Whatever the differences in interpretation, there has been a general readiness since the publication of Fritz Fischer's work in the early 1960s to accept that Germany's expansionist aims form one of the continuous threads linking the Bismarckian and especially the Wilhelmine era with the Third Reich. The clamour for massive expansion and subjection of much of central and eastern Europe, as well as overseas territories, to German dominance was by the early years of the twentieth century not confined to a few extremists, but featured in the aspirations and propaganda of heavily supported and influential pressure groups.[27] It was reflected during the war itself in the aims of the German High Command – aims which can certainly be seen as a bridge to Nazi *Lebensraum*

---

[25] Mommsen 'Ausnahmezustand', p. 43. See also the comments of Mason, *Sozialpolitik*, p. 40. Broszat's work leaves no doubt that he also sees Hitler as the actual executant of Nazi foreign policy.
[26] In addition to the works referred to in ch. 2 note 60, see on the 'continuity question' in German foreign policy, Jacobsen, *Nationalsozialistische Außenpolitik* (ch. 4 note 31) and Konrad H. Jarausch, 'From Second to Third Reich: The Problem of Continuity in German Foreign Policy', *CEH* 12 (1979), pp. 68–82. Of direct importance and relevance is Hans-Jürgen Döscher, *Das Auswärtige Amt im Dritten Reich* (Berlin, 1987).
[27] See esp. Geoff Eley, *Reshaping the German Right. Radical Nationalism and Political Change after Bismarck* (New Haven/London, 1980), and Roger Chickering, *We Men Who Feel Most German: a Cultural Study of the Pan-German League 1886–1914* (London, 1984). The imperialist tradition in Germany is thoroughly explored by Woodruff D. Smith, *The Ideological Origins of Nazi Imperialism* (Oxford, 1986).

policy. Defeat and the loss of territory in the Versailles settlement kept alive expansionist demands on the Right, and encouraged revisionist intentions and claims, which seemed legitimate to the majority of Germans. The popular success of Hitler in the foreign policy arena after 1933 was based squarely upon this continuity of a consensus about the need for German expansion which extended from the power élite to extensive sections of society (with the general exception of the bulk of the now outcast and outlawed adherents of the left-wing parties). This is the context in which the role of Hitler in the formulation of German foreign policy after 1933 has to be assessed.

The most significant steps in German foreign policy during the first year of Nazi rule were the withdrawal from the League of Nations in October 1933, and the reversals in relations with Russia and Poland which had taken place by the beginning of 1934. Obviously, these developments were not unconnected with each other. Together they represented a break with past policy which conceivably could have taken place under a different Reich Chancellor – say Papen or Schleicher – but which, at the same time, in the manner, timing, and speed it came about owed not a little to Hitler's own direction and initiatives.

In the decision to leave the Geneva disarmament conference and the League of Nations, not much more than the timing was Hitler's. The withdrawal was inevitable given the generally accepted commitment to rearmament (which would have been high on the agenda of any nationalist–revisionist government in Germany at that time), and Hitler acted in almost total concert with leading diplomats, the army leadership, and the other dominant revisionist forces in the country.[28]

In the case of Poland, Hitler played a greater role personally – initially in the teeth of the traditional foreign ministry line, against revisionist instincts, and against the wishes of Party activists in Danzig – in steering a new course of *rapprochement*. While Foreign Minister von Neurath, representing the traditional approach, argued at a Cabinet meeting in April 1933 that 'an understanding with Poland is neither possible nor desirable',[29] Hitler was prepared to explore the possibilities of a new relationship with Poland, especially following initial feelers put out by the Polish government in April. The withdrawal from the League of Nations made a *rapprochement* more urgently desirable from the point of view of both sides. Again it was a Polish initiative, in November 1933, which accelerated negotiations. Agreement to end the long-standing trade war with Poland – a move which satisfied many leading German industrialists – was followed by a decision, taking up an original suggestion of Hitler himself, to embody the new relationship in a non-aggression treaty, which came to be signed on 26 January 1934. The Polish minister in Berlin wrote to his superiors in December that 'as if by orders from the top, a change of front toward us is taking place all along the line'.[30] While Hitler was by no means isolated in his new policy on Poland, and while he was able to exploit an obvious desire on Poland's part for a *rapprochement*, the indications are that he personally played a dominant role in developments and that he was not thinking *purely* opportunistically but had long-term possibilities in mind. In a mixture of admiration and scepticism, the German ambassador in Bern, von Weizsäcker,

---

[28] See Weinberg, *Diplomatic Revolution* (see note 20, this chapter), pp. 159–67.
[29] Cited in Weinberg, *Diplomatic Revolution*, p. 62.
[30] Cited in Weinberg, *Diplomatic Revolution*, p. 73.

wrote shortly afterwards that 'no parliamentary minister between 1920 and 1933 could have gone so far'.[31]

The mirror image of the changing relations with Poland in 1933 were those with the Soviet Union. After the maintenance during the first few months of Nazi rule of the mutually advantageous reasonably good relations which had existed since the treaties of Rapallo (1922) and Berlin (1926) – despite some deterioration even before 1933 and the anti-communist propaganda barrage which followed the Nazi takeover – Hitler did nothing to discourage a new basis of 'natural antagonism' towards the Soviet Union from the summer of 1933 onwards.[32] This development, naturally conducive ideologically to Hitler and matching the expectations of his mass following, took place against the wishes both of the German foreign ministry and – despite growing fears and suspicions – of Soviet diplomats, too. When, however, suggestions came from the German foreign ministry in September 1933 for a renewed *rapprochement* with the Soviet Union, Hitler himself rejected it out of hand, stating categorically that 'a restoration of the German–Russian relationship would be impossible'.[33] In like fashion, and now supported by the opportunistic foreign minister von Neurath, he personally rejected new overtures by the Soviet Union in March 1934 – a move which prompted the resignation of the German ambassador to the Soviet Union.[34] In this case, too, Hitler had not acted autonomously, in isolation from the pressures within the Nazi Party and the ranks of its Nationalist partners for a strong anti-Russian line. But he had certainly been more than a cypher or a pure opportunist in shaping the major shift in German alignment, here as in relations with Poland.

More than in any other sphere of foreign policy, Hitler's hand was visible in shaping the new approach towards Britain. As is well known, this was also the area of the most unmitigated failure of German foreign policy during the 1930s. The first major (and successful) initiative led to the bilateral naval treaty with Britain concluded in 1935. Hitler's personal role was decisive both in the formation of the idea for the treaty, and in its execution. Von Neurath thought the idea 'dilettante' and correspondingly found himself excluded from all negotiations and not even in receipt of the minutes. Hitler's insistence also carried the day on the nature of German demands, which were lower than those desired by the German navy. In the light of criticism to be heard in the foreign ministry and in the navy, signs of growing coolness towards the idea in Britain, and the absence of any notable influence from economic interest groups, an armaments lobby, or the *Wehrmacht*, Hitler's own part – and to a lesser extent that of Ribbentrop – was the critical factor.[35] Hitler himself, of course, attached great importance to the treaty as a step on the way towards the British alliance he was so keen to establish.

[31] Cited in Jost Dülffer, 'Zum "decision-making process" in der deutschen Außenpolitik 1933–1939', in Funke, pp. 186–204, here p. 190 note 12. See also Carr, *Hitler*, pp. 48–9; Weinberg, *Diplomatic Revolution*, pp. 57–74.
[32] See Carr, *Hitler*, p. 50.
[33] Cited in Weinberg, *Diplomatic Revolution*, p. 81. See also William Carr, *Der Weg zum Krieg* (Nationalsozialismus im Unterricht, Studieneinheit 9, Deutsches Institut für Fernstudien an der Universität Tübingen, Tübingen, 1983), pp. 17–18.
[34] Weinberg, *Diplomatic Revolution*, pp. 180–3; Carr, *Der Weg zum Krieg*, pp. 18–19.
[35] This section is based largely on Dülffer's analysis, 'Zum "decision-making process" ' (see note 31), pp. 191–3.

The remilitarization of the Rhineland – and with it the breaking of the provisions of Versailles and Locarno – was again an issue which would have been on the agenda of any revisionist German government. The question was already under abstract discussion between the army and foreign ministry by late 1934, and before that Hitler had played with the idea of introducing a demand for the abolition of the demilitarized zone into the disarmament negotiations that year. The issue was revived by the foreign ministry following the ratification of the French–Soviet pact in May 1935, and Hitler mentioned it as a future German demand to the English and French ambassadors towards the end of the year. A solution through negotiation was by no means without prospect of success, and corresponded to the traditional revisionist expectations of Germany's conservative élites. Hitler's main contribution in this case was timing – he claimed he had been originally thinking in terms of a reoccupation in early 1937 – and a decision for the theatrical coup of immediate military reoccupation rather than a lengthier and less dramatic process of negotiation. The opportunistic exploitation of the diplomatic upheaval – which Hitler feared would be shortlived – arising from Mussolini's Abyssinian adventure was coupled with internal considerations: the need to lift popular morale, revitalize the sinking élan of the Party, and to reconsolidate the support for the regime which various indicators suggested had seriously waned by early 1936.[36] Though a surprisingly large body of diplomatic and military 'advisers', along with leading Nazis, shared the secret planning for the reoccupation, the decision was Hitler's alone, and was taken after much worried deliberation and again in the face of coolness from the foreign ministry and nervousness on the part of the military. Jost Dülffer's conclusion, that 'Hitler was the actual driving force' in the affair, seems undeniable.[37]

In the case of Austria, which along with Czechoslovakia had an intrinsic economic and military–strategic significance according with Nazi ideological expansionist ideas, early Nazi policy of supporting the undermining of the State from within was shown to be a disastrous failure, and was promptly ended, following the assassination of the Austrian Chancellor Dollfuss in July 1934. The Austrian question thereafter took a subordinate place to the improvement of relations with Italy in foreign-policy thinking until the latter part of 1937. In the actual *Anschluß* crisis which unfolded in March 1938, it was Göring rather than Hitler who pushed the pace along – probably because of his interest in seizing Austrian economic assets and avoiding the flight of capital which a prolonged crisis would have provoked.[38] Before the events of February and March 1938, the indications are that Hitler was thinking in terms of subordination rather than the outright annexation of Austria. In fact, he appears to have taken the decision for annexation only *after* the military invasion had

---

[36] See Dülffer, 'Zum "decision-making process" ', p. 196; Manfred Funke, '7. März 1936. Fallstudie zum außenpolitischen Führungsstil Hitlers', in Michalka, *Nationalsozialistische Außenpolitik*, pp. 277–324, here pp. 278–9; Orlow, *Nazi Party*, vol. 2 (ch. 4 note 57), pp. 174–6. I try to indicate some of the internal problems facing the regime around this time and the possible links with foreign policy in my contribution, 'Social Unrest and the Response of the Nazi Regime 1934–1936', to Francis R. Nicosia and Lawrence D. Stokes, *Germans against Nazism* (Oxford, 1991), pp. 157–74.
[37] See Dülffer, 'Zum "decision-making process" ', pp. 194–7, and in general Weinberg, *Diplomatic Revolution*, pp. 239–63.
[38] Weinberg, *Starting World War II* (see note 13, this chapter), p. 299 note 170.

occurred – characteristically, under the impact of the delirious reception he had encountered in his home town of Linz.[39] While this points to Hitler's spontaneous, *reactive* decisions even in vitally important matters, and though the chain of developments in the crisis weeks again shows his opportunistic and *ad hoc* exploitation of favourable circumstances, it would be insufficient to leave it at that. The evidence suggests that Göring and Wilhelm Keppler, whom Hitler had placed in charge of Party affairs in Austria in 1937, both believed that Hitler was determined to move on the Austrian question in spring or summer 1938.[40] Goebbels' diary entries also record Hitler speaking about imposing a solution by force 'sometime' on a number of occasions in August and September 1937,[41] and of course Austria formed an important part of Hitler's thinking in November 1937, according to the notes which Colonel Hossbach made of the meeting with top military leaders.[42] In this case too, therefore, Hitler had played a prominent personal role in determining the contours for action, even if his part in the actual events – which could not have been exactly planned or foreseen – was opportunistic, even impulsive.

The remaining events of 1938 and 1939 are sufficiently well known to be summarized briefly. The Sudeten crisis of summer 1938 again illustrates Hitler's direct influence on the course of events. Although traditional power politics and military–strategic considerations would have made the neutralization of Czechoslovakia a high priority for any revisionist government of Germany, it was Hitler's personal determination that he would 'smash Czechoslovakia by military action'[43] – thereby embarking on a high-risk policy in which everything indicates he was not bluffing – that, because of the speed and danger rather than the intrinsic nature of the enterprise, seriously alienated sections of the regime's conservative support, not least in the army. Only the concessions made to Hitler at the Munich Conference deflected him from what can justifiably be regarded as *his* policy to wage war *then* against Czechoslovakia. As is well known, it was Hitler – learning the lessons of Munich – who rejected any alternative to war in 1939, whereas Göring, the second man in the Reich, attempted belatedly to defer any outbreak of hostilities.

Our first set of questions about Hitler's influence on the making of decisions in foreign policy has met with a fairly clear response – and one which would be further bolstered if we were to continue the survey to embrace foreign, strategic, and military affairs during the war years. Whereas in domestic matters Hitler only sporadically intervened in decision-making, and in anti-Jewish policy, which was ideologically highly conducive to him, felt unwilling for prestige reasons to become openly involved, he showed no reluctance to unfold new initiatives or to take vital decisions in the field of foreign policy. In some important areas, as we have seen, he not only set the tone for policy, but pushed through a new or an unorthodox line despite suspicion and objections, par-

[39] Carr, *Hitler*, p. 55.
[40] Weinberg, *Starting World War II*, pp. 287–9.
[41] *Die Tagebücher von Joseph Goebbels*, vol. 3, pp. 223, 263, 266, entries of 3 Aug., 12 Sept., 14 Sept. 1937. The 'overrunning' of Czechoslovakia was also mentioned in the entry of 3 Aug. 1937 and the forceful solution of the Czech question on a number of occasions in these months before the Hossbach meeting.
[42] *IMT*, 25, pp. 402 ff.
[43] *IMT*, 25, p. 434.

ticularly of the foreign ministry. There is no sign of any foreign-policy initiative from any of the numerous agencies with an interest in foreign affairs which could not be reconciled with – let alone flatly opposed – Hitler's own thinking and intentions. Evidence of a 'weak dictator' is, therefore, difficult to come by in Hitler's actions in the foreign-policy arena.

Any 'weakness' would have to be located in the presumption that Hitler was the captive of forces limiting his ability to take decisions. Certainly there were forces at work, both within and outside Germany, conditioning the framework of Hitler's actions, which, naturally, did not take place in a vacuum as a free expression of autonomous will. The pressures of foreign-policy revisionism and rearmament, for instance, which would have preoccupied any German government in the 1930s and demanded adjustments to the international order, developed in the years after 1933 a momentum which substantially restricted Germany's options and ran increasingly out of control. The arms race and diplomatic upheaval which Germany had instigated, gradually imposed, therefore, their own laws on the situation, reflected in Hitler's growing feeling and expression that time was running against Germany. Built into Germany's accelerated armaments production were additional economic pressures for German action, confirming the prognosis that war would have to come about sooner rather than later. The nature of his 'charismatic' authority and the need not to disappoint the expectations aroused in his mass following also constrained Hitler's potential scope for action. Finally, of course, and most self-evidently of all, the relative strength and actions of other powers, and strategic–diplomatic considerations imposed their own restrictions on Hitler's manoeuvrability – though these restrictions diminished sharply in the immediate pre-war years.

Hitler's foreign policy was, therefore, in no way independent of 'structural determinants' of different kinds. These, however, pushed him if anything still faster on the path he was in any case determined to tread. When all due consideration is given to the actions – and grave mistakes – of other governments in the diplomatic turmoil of the 1930s, the crucial and pivotal role of Germany as the active catalyst in the upheaval is undeniable. Many of the developments which took place were in certain respects likely if not inevitable as the unfinished business of the First World War and the post-war settlement. The continuities in German foreign policy after 1933 are manifest, and formed part of the basis of the far-reaching identity of interest – certainly until 1937–8 – of the conservative élites with the Nazi leadership, rooted in the pursuit of a traditional German power policy aimed at attaining hegemony in central Europe. At the same time, important strands of discontinuity and an unquestionable new dynamism were also unmistakable hallmarks of German foreign policy after 1933 – such that one can speak with justification of a 'diplomatic revolution'[44] in Europe by 1936. Hitler's own decisions and actions, as we have seen, were central to this development.

In the framework of foreign policy decision-making, Jost Dülffer's conclusions seem apposite:[45] *(1)* The influence of the old leadership élites waned in correspondence with the growing influence of the 'new' Nazi forces. *(2)* Though

---

[44] The sub-title of the first of Weinberg's two-volume study of Nazi foreign policy (see note 20, this chapter).
[45] Dülffer, 'Zum "decision-making process" ', pp. 200–3.

not undertaken autonomously and in a social vacuum, the major initiatives in German foreign policy in the 1930s can be traced to Hitler himself. *(3)* Economic factors contributed to the framework within which decisions had to be made, but did not play a *dominant* role in Hitler's decisions. *(4)* Hitler cannot be seen as simply a machiavellian opportunist, but rather advanced a consistent anti-Soviet policy (until 1939), necessitating a realignment of Germany's relations with Poland and Britain.

This suggestion of an inner consistency directed at war against the Soviet Union brings us to the second question of our enquiry.

II

We have established that Hitler actively intervened and personally played a central role in shaping German foreign policy during the 1930s. The interpretation that the course of German foreign policy had an inner consistency determined more than any other factor by Hitler's ideology remains, however, open to dispute. Historians have put forward three (in some ways interlinked) alternative explanations.

The first is that Hitler's ideological motivation, while basically unchanging, was not the decisive factor. Rather, Hitler articulated and represented the expansionist–imperialist demands of the German ruling class and made possible the imperialist war sought after by monopoly capital. Hitler had a certain functional role, therefore, but a similar course of action would have unfolded even without him. There can be no doubting, of course, the expansionist aims of influential sectors of the German military, economic, and bureaucratic élites. However, as we saw in considering foreign-policy decision-making earlier in this chapter, it would be short-circuiting the evidence to give the impression that the course of foreign policy was a foregone conclusion after 1933, that it followed closely and at all points the perceived wishes and interests of the traditional élites, that genuine policy options even with the context of revisionism were not available at crucial junctures, and that Hitler himself did not take a prominent part in deciding policy options. Certainly Hitler was never out of step with the *dominant* sectors within the élites. But that does not mean he was their captive. The dominance of particular factions within the élites was itself related to the speed with which they could attune to policy initiatives and make them their own, as well as to their ability to influence the formulation of policy in the first place. The evidence suggests, therefore, that German expansionism in the 1930s was an inevitability, but that its precise direction and dynamic was not independent of Hitler's personal role.

A second approach lays the weight of explanation on the 'primacy of domestic politics', accepting an underlying consistency in foreign affairs, but seeing this less in the implementation of Hitler's ideology than in the need to preserve and uphold the domestic social order. This, too, seems inadequate as a general interpretation. Again, we have accepted in earlier chapters that domestic pressures undoubtedly contributed to the character and the timing of some foreign-policy initiatives, especially in the earlier years of the regime. Domestic, as well as diplomatic, considerations seem to have played a part, for instance, in the decision to reoccupy the Rhineland in March 1936. But there was no such pressure dictating other major developments or shifts in policy, such as the Non-Aggression Treaty with Poland in 1934 or the Naval Treaty with Britain the following year. And by the later 1930s the mounting economic problems appear to have corroborated, not caused, the direction of foreign policy, and, indeed, to have been in no small part a product of it. The evidence is suggestive, therefore, of a total interdependence of domestic and foreign policy, in which domestic considerations helped shape the parameters of foreign-policy action – though to a diminishing extent; and, vice versa, in which foreign policy objectives heavily determined the nature and aims of domestic policy.[46] Ideologically, and practically, foreign and domestic policy were so fused that it seems quite misplaced to speak of a primacy of one over the other: there was no contradiction between the imperialist and social imperialist aims of the regime, and there is no means analytically of separating them. Nor does it appear satisfactory to perceive Nazi aims as lying in the preservation of the *existing* social order, however unclear and nebulous the social ambitions of any 'new order' might have been.

A final alternative explanation argues that German foreign policy had no single, clear direction, that it simultaneously pursued a variety of basically unconnected objectives, and that it was characterized by Hitler's own dilettante opportunism which, in the context of a fragmented political system, produced a diminishing sense of reality and an accelerating nihilistic momentum. Even among historians favouring a 'structuralist' interpretation of foreign policy, Hans Mommsen, it has to be said, seems alone in advancing such an argument so emphatically.[47] Martin Broszat, the other foremost exponent of the 'structuralist' approach, appears, as we saw earlier, to accept the existence of a more or less consistent 'directional force' aimed at expansion in the east, though in his view this served only the function of an 'ideological metaphor'.[48] This raises the question of whether, in fact, the debate about the existence and consistency of foreign-policy objectives has not been falsely polarized by the vagueness of some of the key terms employed by historians. While, for example, 'intentionalists' naturally reject categorically the view that Hitler was simply an opportunist and improviser without basic orientation or goal, their own frequent usage of concepts such as 'programme' (sometimes begun with a capital letter and with the inverted commas omitted), 'basic plan' (*Grund-Plan*), or 'stage by stage plan' (*Stufenplan*), is not without problems.[49] These terms, it is often emphasized, do not imply detailed blueprints for action. Rather, they are, it seems, meant to suggest only that Hitler had fixed ideas in the sphere of foreign policy (especially *Lebensraum*), to which he clung obsessively from the 1920s; that as Führer he directed foreign policy in accordance with these ideas; and that, although having a clear target in mind (above all conquest of the Soviet Union) and a basic strategy for reaching that target (the alliance with Britain), he had no concrete design worked out. The gap between this view and Broszat's suggestion that *Lebensraum* in the East was so

[46] See esp. Erhard Fordran, 'Zur Theorie der internationalen Beziehungen – Das Verhältnis von Innen-, Außen- und internationaler Politik und die historischen Beispiele der 30 er Jahre', in Erhard Fordran *et al.*, eds., *Innen- und Außenpolitik unter nationalsozialistischer Bedrohung* (Opladen, 1977), pp. 315–61, here esp. pp. 353–4.

[47] See Mommsen, 'National Socialism: Continuity and Change', p. 177 and *Adolf Hitler*, esp. p. 93.

[48] Broszat, 'Soziale Motivation', pp. 406–9.

[49] The rather contorted passage of Klaus Hildebrand, 'Hitlers "Programm" und seine Realisierung 1939–1942', in Funke, pp. 63–93, here p. 65, suggests some of the difficulties of formulating a clear definition of Hitler's 'programme'.

vacuous a notion that it served merely as a directional guide to action (*Aktions-richtung*)[50] certainly exists, but is perhaps less wide than at first sight. The gap seems unbridgeable only if *exclusive* weight is attached to *either* intention *or* function as a factor determining the course of foreign policy. While it could indeed be argued that *Lebensraum* served the function of an ideological metaphor in providing the Movement with a directional focus for action, it seems inadequate to view this function as the sole or even main *raison d'être* of foreign policy, to deny that there was indeed a genuine reality to Nazi foreign policy aims, a reality which was at least in part shaped by Hitler's ideological aims and intentions.[51] However vague the notion, *Lebensraum* did mean something concrete – even if the way there was uncharted: war against the Soviet Union. Hitler's words and actions in the period 1933–41 are consistent with the interpretation that he was convinced that such a war would come about, that although he did not know how or when, it would be sooner rather than later, that he was steering German foreign policy towards that goal, and that he was attempting to shape German society for participation in that war.

As we saw earlier, the basic orientation of German foreign policy was shifted as early as 1933, when Hitler determined that 'natural antagonism' should shape relations with the Soviet Union. In autumn 1935, according to Alfred Sohn-Rethel's account, talk at the 'fireside discussions' with leaders of the army and economy of stifling rearmament expenditure was invariably countered by Göring's reminder to Hitler about his coming war against the Soviet Union.[52] The beginning of the Spanish Civil War must have contributed to Hitler's growing preoccupation with this idea in 1936. His secret memorandum on the Four Year Plan, compiled in the summer, rested on the basic premise that 'the showdown with Russia is inevitable',[53] and the recently published Goebbels diaries reveal how much the coming clash with Russia was on Hitler's mind in the years 1936 and 1937. In June, according to Goebbel's diary notes, Hitler spoke of a coming conflict between Japan and Russia, after which 'this colossus will start to totter [*ins Wanken kommen*]. And then our great hour will have arrived. Then we must supply ourselves with land for 100 years'. 'Let's hope we're ready then' (added Goebbels) 'and that the Führer is still alive. So that action will be taken'.[54] In November the same year, Goebbels recorded: 'After dinner I talked thoroughly with the Führer alone. He is very content with the situation. Rearmament is proceeding. We're sticking in fabulous sums. In 1938 we'll be completely ready. The show-down with Bolshevism is coming. Then we want to be prepared'.[55] Less than a month later, set in the context of the Spanish Civil War, Hitler portrayed the danger of Bolshevism to his cabinet in a three-hour meeting, arguing (according to Goebbels' account): 'Europe is already divided into two camps. We can't go back any longer. . . . Germany can only wish that the danger be deferred till we're ready. When it comes, seize the opportunity [*zugreifen*]. Get into the

---

[50] Broszat, 'Soziale Motivation', p. 403.
[51] This is clearly accepted by Broszat, 'Soziale Motivation', p. 403.
[52] Sohn-Rethel (see ch. 3 note 19), pp. 139–41.
[53] 'Denkschrift Hitlers über die Aufgaben eines Vierjahrplans', *VfZ* 3 (1965), pp. 204–10, here p. 205.
[54] *Die Tagebücher von Joseph Goebbels*, vol. 2, p. 622, entry of 9 June 1936.
[55] *Die Tagebücher von Joseph Goebbels*, vol. 2, p. 726, entry of 15 Nov. 1936.

paternoster lift at the right time. But also get out again at the right time. Re-arm, money can play no role'.[56] According to his reported comments in February 1937, Hitler expected 'a great world showdown' in five or six years' time.[57] In July Goebbels reported Hitler's puzzlement over the purges in the Soviet Union and his view that Stalin must be mad. Hitler's alleged comments ended: 'But Russia knows nothing other than Bolshevism. That is the danger which we will have to knock down sometime'.[58] In December Hitler repeated the same sentiments about Stalin and his supporters, concluding: 'Must be exterminated [*Muß ausgerottet werden*]'.[59] Finally, there is the well-known comment by Hitler to the Swiss Commissioner to the League of Nations, Carl Burckhardt, in 1939: 'Everything that I undertake is directed against Russia. If those in the West are too stupid and too blind to understand this, then I shall be forced to come to an understanding with the Russians to beat the West, and then, after its defeat, turn with all my concerted force against the Soviet Union'.[60] That Hitler was saying this in the knowledge that the message would be relayed to the West does not detract from its basic reality.

The cosmic struggle with Bolshevism gradually became imminent reality, just as the vision of destroying the Jews had emerged as a realizable goal. In neither case do Hitler's 'intentions' come near providing a full or satisfactory explanation. But the chances of either coming about without those 'intentions' would have been diminished – greatly so in the case of the extermination of the Jews, to a much lesser extent in the case of the war against the Soviet Union. The 'twisted road' to this ideological 'war of annihilation' needs no emphasis. The only strategy was the alliance with Britain. By the mid 1930s that had failed irretrievably, and any 'policy', 'programme', or 'basic plan' worth the name was in tatters – resulting in fact by 1939 in forced, if temporary, alliance with the arch-enemy and a state of war with the would-be 'friend' which had spurned him. Only in these conditions, the reverse of what had been hoped for, could the war against the Soviet Union, from summer 1940 onwards, be planned, not merely 'targeted'. And despite German supremacy in western Europe, the unresolved problem of the United States was by then looming ever larger in the background.

III

The debate about the extent of Hitler's long-term ambitions whether he wanted world dominion or whether his final goal was 'merely' the conquest of *Lebensraum* in the East – has a rather artificial ring about it. As we noted earlier, the view has generally prevailed since the publications of Moltmann and especially Hillgruber in the 1960s that Hitler's intentions stopped at nothing short of German mastery of the entire globe, a goal to be achieved in stages and perhaps

[56] *Die Tagebücher von Joseph Goebbels*, vol. 2, p. 743, entry of 2 Dec. 1936.
[57] *Die Tagebücher von Joseph Goebbels*, vol. 3, p. 55, entry of 23 Feb. 1937. See also the entries of 28 Jan. 1937, where Hitler reportedly hoped to have six years, but would act before then if an advantageous situation arose, and 16 Feb. 1937, where he expected 'the great world struggle' in 'several years' ' time (pp. 26, 45).
[58] *Die Tagebücher von Joseph Goebbels*, vol. 3, p. 198, entry of 10 July 1937.
[59] *Die Tagebücher von Joseph Goebbels*, vol. 3, p. 378, entry of 22 Dec. 1937.
[60] Cited in Hildebrand, *Foreign Policy* (see ch. 1 note 17), p. 88; Carl J. Burckhardt, *Meine Danziger Mission 1937–1939* (dtv-edition, Munich, 1962), p. 272.

not accomplished until long after his death. Some leading historians have, however, doggedly held to the view that Hitler's final aim was that which he had expressed consistently throughout practically his whole career: the attainment of *Lebensraum* at the expense of Russia. One might question at the outset whether this difference of interpretation reflects much more than the weighting historians have attached to the *relative* clarity and consistency of the focus on the East in Hitler's thinking as compared with his more nebulous and sporadic musings on the long-term possibilities (and inevitability) of further expansion following the expected German victory over Bolshevism. There are indeed few grounds for doubting that Hitler did at times entertain 'world domination' thoughts. It is less clear, however, what significance such notions had for formulating practical policy. We suggested earlier that, while the term *Lebensraum* indeed possessed a metaphorical quality, and that neither Hitler nor anyone else had a clearly worked-out conception of what precisely it would amount to, it also did have a concrete meaning in denoting war against the Soviet Union and the need to prepare as much as possible for such a struggle. Thoughts of this war, however unclear the path to it might have seemed, were never far from the minds of Hitler and the top Nazi and army leadership, and practical military, strategic, and diplomatic consequences ensued. Whether vague megalomaniac meanderings about future global domination can be seen in the same light might be intrinsically doubted; even more so, whether such notions ought to be elevated to the status of a 'programme', let alone 'grand strategy'.[61]

In its most forthright formulation, the 'world mastery' thesis claims that 'at no time between 1920 and 1945 did [Hitler], as his statements prove, lose sight of the aim of world domination'[62] – an aim which, another historian adds, he wanted to achieve 'in a series of blitz campaigns, extending stage by stage over the entire globe'.[63] The main supporting evidence comprises Hitler's early writings (especially his *Second Book* of 1928), Rauschning's version of Hitler's monologues in 1932–4, the *Table Talk*, audiences with foreign diplomats, aspects of military planning during the years 1940–1, and – as has been more recently emphasized – the deductions to be drawn from Hitler's monumental architectural plans, and long-term naval planning. We need briefly to consider the strength of this evidence.

Hitler's *Second Book* raises the spectre of a contest for hegemony at some point in the distant future between the United States of America and Europe. His view was that the USA could only be defeated by a racially pure European state, and that it was the task of the Nazi movement to prepare 'its own fatherland' for the task.[64] Before this time, the United States had attracted little of Hitler's attention. His early speeches and writings (including *Mein Kampf*) contain few references to America going beyond conventional and

[61] The latter term is used by Hauner, 'World Dominion' (see note 12, this chapter), p. 23.
[62] Thies, *Architekt* (see note 9, this chapter), p. 189. See also his essays: 'Hitler's European Building Programme', *JCH* 13 (1978), pp. 413–31; 'Hitlers "Endziele": Zielloser Aktionismus, Kontinentalimperium oder Weltherrschaft?', in Michalka, *Nationalsozialistische Außenpolitik*, pp. 70–91; and 'Nazi Architecture – A Blueprint for World Domination: The Last Aims of Adolf Hitler', in David Welch, ed., *Nazi Propaganda. The Power and the Limitations* (London, 1983), pp. 45–64.
[63] Hauner, 'World Dominion', p. 23.
[64] Telford Taylor, ed., *Hitler's Secret Book* (New York, 1961), p. 106.

general denunciation for its part in the First World War and the peace settlement.[65] By the late 1920s views of a long-term threat from America to Germany were fairly commonplace, and it was in this climate that Hitler expressed his vague notions about the great conflict between the German-dominated Eurasian empire and the USA in the distant future.[66] Hitler's image of America, vague as it was, did not in fact remain constant. By the early 1930s, under the impact of the Depression, America was taken to be a weak, racially mongrel state which would be incapable of engaging again in a European war, and whose only hope of salvation lay in German-Americans rejuvenated by Nazism.[67] By the later 1930s American distaste for Nazi racial and religious policy had confirmed Hitler's assessment of the USA's debility. He did not at this stage regard the United States as an actual or potentially strong military power to be feared by Germany; his vision remained primarily continental, and he paid little attention in concrete terms to areas outside Europe.[68] If the vague idea of a future conflict with the USA remained, it had no practical importance in policy formulation.

Evidence of Hitler's 'programme' for global mastery for the period between the *Second Book* and the later 1930s is dependent on references to 'world domination', or to Germany being the 'greatest power in the world', in a few public speeches – in which presumably the propaganda effect was the greatest consideration – and in private conversations subsequently recapitulated by participants (and which cannot, in their printed form, be regarded as accurate *verbatim* records of what transpired).[69] Of the latter category, Hermann Rauschning's *Hitler Speaks*, published in 1939 (at a timely date for western propaganda purposes), is the most important. Though it cannot be taken as accurate to the last word as a record of what Hitler actually said there is nothing in it which is not consonant with what is otherwise known of Hitler's character and opinions.[70] There are, indeed, passages in Rauschning in which Hitler pontificates, for instance, about the future German domination of Latin America and the exploitation of the treasures of Mexican soil by Germany. As Rauschning himself pointed out, however, Hitler was on such occasions invariably repeating, on the basis of no detailed information, banal popular images of these countries. He added that Hitler had always been a *poseur*, so that it was difficult to know how serious he was about any comments he made.[71] German relations with Latin America in the 1930s turned out, not surprisingly, to have nothing to do with Hitler's wild visions and megalomaniac mouthings.[72] Again these cannot be seen as falling within the framework of any 'plan' or 'strategy'.

Jochen Thies has recently argued that evidence for the consistency of Hitler's 'world domination' aim between 1920 and 1945 can best be found in his plans

---

[65] Weinberg, *Diplomatic Revolution*, p. 21.
[66] Dietrich Aigner, 'Hitler und die Weltherrschaft', in Michalka, *Nationalsozialistische Außenpolitik*, pp. 49–69, here p. 62.
[67] Weinberg, *Diplomatic Revolution*, pp. 21–2; Rauschning, *Hitler Speaks*, pp. 75–7.
[68] Weinberg, *Starting World War II*, pp. 252–3; *Diplomatic Revolution*, p. 20.
[69] Thies, 'Hitlers "Endziele" ', p. 78 note 45 and see also pp. 72–3; and Aigner, pp. 53–4.
[70] This is the general tenor of Theodor Schieder's conclusion: see note 4, this chapter.
[71] Rauschning, *Hitler Speaks*, pp. 69–75, 138.
[72] See Weinberg, *Starting World War II*, pp. 255–60.

for the erection of representative buildings on a monumental scale, as images of German strength which would last for up to 10,000 years.[73] Clearly, they were intended as symbols of Germany's lasting world-power status and are testimony to Hitler's grandiose vision of German potential. But it seems to be stretching the argument to see the building plans themselves as an unambiguous reflection of a consistent 'programme' leading to 'world domination'.

Rather more convincing is the view that the growing proximity of war and the inability to cement the intended alliance with Britain, at the same time, however, the growing confidence derived from a series of diplomatic coups, led in the later 1930s to Hitler giving greater strategic consideration to a range of possibilities which could emerge from armed conflict, in which Germany's struggle might take on a global character. He hinted at this on a number of occasions to his generals from 1937 onwards.[74] From this time, too, he began to show more interest in naval strategy, culminating in the Z-Plan of January 1939, in which Hitler's insistence on the building of a huge battle-fleet by 1944 (as opposed to the navy's preference for U-Boats, which made a better offensive weapon against Britain, and in detriment to steel allocations for the army and *Luftwaffe*) has been taken to point beyond a war with Britain to a future German mastery of the oceans and the inevitability of global conflict.[75] At the same time, the inconsistency and ambiguity of Hitler's 'global' thinking is shown by his lack of interest in inciting revolution in the Islamic world and actively supporting nationalist undermining of British rule in India.[76]

More specific evidence of Hitler's strategic global thinking is largely confined to the war period, especially to the years 1940–1. By this time, however, Hitler was largely *reacting* (not wholly consistently) to circumstances which he had indeed done much to bring about, but which were now rapidly going beyond any measure of his control. It is difficult, therefore, to relate strategic considerations at this date directly to the earlier vague utterances about 'world domination'.[77] As Hillgruber argued, planning for the war against the Soviet Union (much though Hitler wanted the war ideologically), and the urgent need of a speedy victory, was conditioned strategically by the necessity of bringing Britain to the peace table, keeping America out of the war, and ending the war in the only way possible to Germany's advantage.[78] Convinced that America (whose image in Hitler's eyes had again shifted from one of weakness back to one of strength) would enter the war by 1942 at the latest, the overriding need was to have done with the eastern war in order to be in a position to fend off the United States. At the height of his powers, Hitler thought for a short while of 'destroying' America in tandem with Japan, and of stationing long-range bombers in the Azores in autumn 1941 in order to attack the USA. But with the imminent entry of America into the war, and the German offensive stuck

[73] Thies, *Architekt*, and 'Hitlers "Endziele" ', esp. pp. 83–4.
[74] Thies, 'Hitlers "Endziele" ', pp. 86–8.
[75] Jost Dülffer, 'Der Einfluß des Auslandes auf die nationalsozialistische Politik', in Fordran *et al.* (note 46, this chapter), pp. 295–313, here p. 302; Hauner, 'World Dominion', p. 27; Carr, *Hitler*, p. 131. For a sceptical view of the weight attached to the Z-Plan, see Aigner, pp. 60–1.
[76] Hauner, 'Professionals' (note 11, this chapter) and his *India in Axis Strategy* (note 9, this chapter).
[77] Andreas Hillgruber, 'Der Faktor Amerika in Hitlers Strategie 1938–1941', *APZ* (11 May 1966), p. 4.
[78] Hillgruber, 'Amerika', p. 13.

in the Russian mud, he reverted to the vague notion of a showdown with the USA 'in the next generation', declared war on the USA in a futile gesture, and told the Japanese ambassador two months later that he still did not know how to conquer the United States.[79] Further musings during the remainder of the war about 'world domination' after a hundred years of struggle, of a later ruler of Germany being 'master of the world', and of an 'unshakeable conviction' that German world mastery would ultimately be attained,[80] were pipe-dreams not evidence of a *Stufenplan*. As the Third Reich was collapsing in ruins and the Red Army stood at the gates of Berlin, Hitler returned to more modest targets: the destruction of Bolshevism, the conquest of 'wide spaces in the East', and a continental *Lebensraum* policy as opposed to the acquisition of overseas colonies. His last message to the army, a day before his suicide, was equally utopian: it should fight to 'win territory for the German people in the East'.[81]

It seems necessary to draw a distinction between strategic aims and vague and visionary orientations for action. The evidence for Hitler's strategic global thinking is concentrated in the years immediately prior to the war, when his underlying concept of the alliance with Britain had collapsed, and in the first years of the war, when faced with the increasingly likely entry of the United States into the conflict. Before those years, there are only grey visions of a cosmic struggle at some dim and distant time in the future. After those years, there are again glimmers of a far-off utopia, now presumably compensating for the reality of inevitable and crushing defeat. To label this a 'programme' for world mastery seems inappropriate. As Rauschning saw, however, Nazism could not have ceased its 'perpetual motion';[82] its internal and external dynamism could never have brought stability or subsided into stagnation; not least, Hitler's own social darwinist interpretation of existence itself as struggle, transmuted into the titanic struggles of nations in which there was no half-way between total victory and complete destruction, added a decisive component which was wholly compatible with short-term opportunistic exploitation but quite irreconcilable with long-term rational calculation and planning. In this respect, perhaps, 'expansion without object' (following the presumed victory over the Soviet Union) fits the ethos of Nazism and corresponds to Hitler's utopian dreams far better than does the concept of a 'programme' for world domination.

Our survey of differing interpretations of Hitler's contribution to shaping domestic, anti-Jewish, and foreign policy in the Third Reich is now completed. In each case, we have argued, Hitler's 'intentions' *and* impersonal 'structures' are both indispensable components of any interpretation of the course of German politics in the Nazi State. And there is no mathematical formula for deciding what weighting to attach to each factor. We have seen that Hitler shaped initiatives and personally took the major decisions in foreign policy, though this was less frequently the case in domestic affairs or even in anti-Jewish policy. In domestic matters his uneven intervention was usually

[79] Hillgruber, 'Amerika', pp. 14–21. See also Jäckel, *Hitler in History*, ch 4, and William Carr, *Poland to Pearl Harbor. The Making of the Second World War* (London, 1985), esp. pp. 167–9.
[80] See Meir Michaelis, 'World Power Status or World Dominion?', *The Historical Journal* 15 (1972), pp. 331–60, here p. 351.
[81] Cited in Michaelis, pp. 351, 357.
[82] See Michaelis, p. 359.

prompted by varied and often conflicting requests for his authorisation for legislative or executive action; in the 'Jewish Question' his main contribution consisted of setting the distant target, shaping the climate, and sanctioning the actions of others; in foreign policy he *both* symbolized the 'great cause' which motivated others *and* played a central role personally in the course of aggression. Hitler's ideological aims were one important factor in deciding the contours of German foreign policy. But they fused for the most part in the formulation of policy so inseparably with strategic power-political considerations, and frequently, too, with economic interest that it is usually impossible to distinguish them analytically. And alongside Hitler's personality, the *function* of his Führer role was also vital to the framing of foreign policy and determining the road to war in its legitimation of the struggle towards the ends it was presumed he wanted. It legitimized the self-interest of an army leadership only too willing to profit from unlimited rearmament, over-ready to engage in expansionist plans, and hopeful of a central role for itself in the State. It legitimized the ambitions of a foreign office only too anxious to prepare the ground diplomatically for upturning the European order, and the various 'amateur' agencies dabbling in foreign affairs with even more aggressive intentions.[83] It also legitimized the greed and ruthlessness of industrialists only too eager to offer plans for the economic plunder of much of Europe. Finally, it provided the touchstone for the wildest chauvinist and imperialist clamour from the mass of the Party faithful for the restoration of Germany's might and glory. Each of these elements – from the élites and from the masses – bound in turn Hitler and the Nazi leadership to the course of action, gathering in pace and escalating in danger, which they had been partly instrumental in creating. The complex radicalization, also in the sphere of foreign policy, which turned Hitler's ideological dreams into living nightmares for millions can, thus, only inadequately be explained by heavy concentration on Hitler's intentions divorced from the conditions and forces – inside and outside Germany – which structured the implementation of those intentions.[84]

[83] See Hauner, 'Professionals'. For an example of 'local initiatives' of 'amateurs' making the running in the Balkans, see Weinberg, *Diplomatic Revolution*, p. 23 note 81.
[84] I have sought to address these problems rather more fully in my *Hitler. A Profile in Power* (London, 1991).

7

The Third Reich: 'Social Reaction' or 'Social Revolution'?

Assessing the nature and extent of Nazism's impact on German society is one of the most complex – and most important – tasks facing historians of the Third Reich. And, clearly, the social impact of an ideologically doctrinaire and ruthlessly repressive authoritarian state has potential implications extending far beyond the geographical and chronological confines of Germany under Nazism.

A differentiated understanding of German society in the Third Reich has become possible since the 1960s, when serious scholarly research in this field was first carried out. The major advances, however, came as late as the 1970s, when the source base was massively extended, and continue to the present time. The huge expansion and attractiveness of *Alltagsgeschichte* ('history of everyday life') or *Geschichte von unten* ('history from below') in West Germany during the past two decades has provided a plethora of detailed empirical studies – of greatly varying quality – of the experience of different social groups, frequently in a local or regional context, during the Nazi Dictatorship. A good deal of material is, therefore, now available for examining the social impact of Nazism. That there are often major difficulties of interpretation built into the sources emanating from such a political system goes without saying. As in other issues we have considered, however, the problems and perspectives of interpretation are far more closely related to different theoretical starting-points and unbridgeable ideological divisions among historians. The debate is characterized by fundamental disagreements about the very nature of Nazism, its social aims and intentions; about the criteria and methods needed to evaluate change under Nazism; and about the terms used to define that social change.

Part of the problem rests in the eclectic nature and internal contradictions of the Nazi Party itself, its ideology and its social composition. There are considerable difficulties involved even in attempting to define clearly what its social goals and objectives were, and in distinguishing these ends from the means necessary to attain them, which in practice often seem to have produced the diametrically opposite result. Hence, Nazism has been interpreted by some leading historians as genuinely revolutionary in content, and branded by others as quintessentially counter-revolutionary; some have regarded it as a modernizing force despite archaic, reactionary aspects of its ideology; others as violently anti-modernist, or – paradoxically – 'revolutionary reaction'; while still others have found no cause to see in Nazism any other than plain social reaction.[1] In

[1] A number of the contradictory positions are summarized in Francis L. Carsten, 'Interpretations of Fascism', in Laqueur (ch. 2 note 3), pp. 457–87, here esp. pp. 474 ff.

any event there is a genuine question mark over the extent to which Nazi 'social ideology' ought to be regarded as a serious declaration of intent as opposed to mere manipulative propaganda.

A second part of the problem derives from the complexity of attempting to construct some type of 'balance-sheet' of social change in Germany under Nazism. While some aspects of 'social change', such as rate of social mobility, can be measured with difficulty, changes in attitudes, mentality, and value systems can only be qualitatively assessed on the basis of evidence which is far from ideal for the purposes. Moreover, the time-scale is extremely short. The Third Reich lasted only 12 out of its scheduled 1,000 years, and six of those were war years. Since war, especially on the scale of the Second World War, contains its own momentum for rapid social change promoted by massive destruction, displaced populations, mobilization and demobilization, and post-war expectations, there is an obvious problem involved in extrapolating such change from that which was intended by the Nazi system (even accepting that the war itself was a product of Nazism). It is necessary, therefore, to try to distinguish between change which the Nazi regime brought about directly, and that which indirectly and even unintentionally stemmed from Nazism. A further difficulty is how to relate such change to long-term secular changes in society which were taking place in Germany as elsewhere in the industrial era. It has even been suggested that in order to assess social change under Nazism it would be necessary to build a counter-factual model to estimate what change would have come about by 1945 had Nazism never existed.[2] This raises the further question: are we trying to assess whatever social change did take place under Nazism against our understanding of what we presume Nazism was setting out to accomplish, against what might have happened without Nazism, against the rate and nature of change in other industrial societies at the same time, or against some notional 'ideal type' of development?

The third part of the problem is definitional. As is frequently the case in the social, political, and historical sciences, the terms and concepts used are often imprecise, capable of more than a single interpretation, or ideologically 'loaded'. To use the term 'revolution', it has been said, 'is to enter a semantic minefield',[3] and one furthermore in which personal predilections for what might be taken to constitute a 'revolution' – in particular a 'social revolution' – evidently play a determining role. While it may be reasonable to object that 'revolution' need not be something 'positive', 'progressive', or 'morally commendable', nor confined to marxist terms of fundamental alteration to the economic substance of a society,[4] this negative point brings us little closer to defining precisely what *would* constitute a 'social revolution'. It goes almost without saying that 'reaction' and 'counter-revolution' are hardly 'cleaner' as intellectual concepts.

Certainly, terms like 'social change' or 'social development' are more neutral, though they are so vague in themselves that they only become operable when attached to some theory or concept of change over time. Only marxist

[2] Matzerath and Volkmann (see ch. 2 note 53), p. 109 (comment of T. Sarrazin).
[3] Jeremy Noakes, 'Nazism and Revolution', in Noel O'Sullivan, ed., *Revolutionary Theory and Political Reality* (London, 1983), pp. 73–100.
[4] See Karl Dietrich Bracher, 'Tradition und Revolution im Nationalsozialismus', in his *Zeitgeschichtliche Kontroversen* (see ch. 1 note 40), pp. 62–78, here esp., pp. 66–70.

theories and modernization theories suggest themselves as possible explanatory models.

Marxist theorists tend to restrict their analyses of 'social change' primarily to alterations in the structure of the mode of production – that is, in modern times, in the structure of capitalism – and to the state of the 'class struggle', with a corresponding tendency to underplay change in social forms or culture unless the economic substance of society has also been transformed. Notions of 'social change' locked into a marxist approach quickly, therefore, lose their tones of vagueness, but also their tones of intellectual neutrality.

Alternative explanations of 'social change', which have commended themselves in varying degrees to non-marxist or 'liberal' historians, are linked to 'modernization' approaches. The concept of 'modernization' – a product of American social science – attempts to embrace the various elements of cultural, political, and socio-economic development which gained their major impulse with the industrial and French revolutions in western Europe, transforming the 'traditional' societies of the West and gradually of large sections of the globe into 'modern societies'. This transformation includes a huge growth in the quantity and availability of goods and services; increasing access to these goods and services; growth in social differentiation, more complex division of labour, and increased specialization in function; and a heightened capacity for the institutional regulation of social and political conflict.[5] Although modernization approaches have become greatly refined since their early rather unsophisticated usage, they remain eclectic, imprecise, and open to widely differing subjective weightings attached to some of the fundamental premises and concepts used. The implicit or explicit linkage of modernization approaches to 'ideal types' suggested by western liberal democracies, the relative neglect of class conflict, and the relegation of economic structures to only one – if very important – component of 'social change' add to the highly debatable nature of the 'modernization' concept in its conventional usages, and make it generally unacceptable to marxist scholars.

Any attempt to evaluate Nazism's impact on German society must grapple with the difficulties we have raised. Before attempting our own evaluation, we need to survey the main divisions in interpretation among historians who have tackled the problem.

Interpretations

Resting on the basic premise that Hitler-Fascism was the dictatorship of the most reactionary elements of the German ruling class, it is hardly surprising that pre-1989 GDR historiography accorded short shrift to notions that the Third Reich brought about change in German society amounting to a 'social

[5] See Werner Abelshauser and Anselm Faust, *Wirtschafts- und Sozialpolitik. Eine national-sozialistische Sozialrevolution?* (Nationalsozialismus im Unterricht, Studieneinheit 4, Deutsches Institut für Fernstudien an der Universität Tübingen, Tübingen, 1983), p. 4; Matzerath and Volkmann, p. 95. For an evaluation of modernization theories and applicability in historical writing, Hans-Ulrich Wehler, *Modernisierungstheorie und Geschichte* (Göttingen, 1975) is invaluable. Helmut Kaelble *et al.*, *Probleme der Modernisierung in Deutschland. Sozialhistorische Studien zum 19. und 20. Jahrhundert* (Opladen, 1978), apply modernization models explicitly to German social development.

revolution'. While a heavy concentration upon organized communist resistance groups imposed a strait-jacket on research into wider aspects of the social history of the Third Reich, the possibility of long-term, 'modernizing' consequences of Nazism for German society remained, of course, a 'non-question' for GDR historians. Modernization theories were regarded as merely a bourgeois pseudo-doctrine of industrial society, so lacking in definition as to be purely subjective in application, anti-marxist by intent and implication, euphemizing fascism in regarding it as (even unwittingly) a 'push into modernity', and, in so far as Nazism is taken to have been instrumental in promoting a 'social revolution', arbitrarily distorting and misusing the concept of revolution for a phenomenon which was blatantly counter-revolutionary.[6] Conceptions built into modernization theories of 'progress' within capitalist society – and not in the direction of marxist–leninist socialism – were clearly irreconcilable with the emphasis placed upon the continuities of imperialist monopoly capitalism outliving the Third Reich and ensuring the reactionary character of the Federal Republic. From this starting-point, it is obvious that questions of any lasting or long-term impact of the Third Reich on the development of German society were irrelevant for GDR historiography. Genuine social revolution could, they argued, come about only under the aegis of marxism–leninism. In the case of Germany, this had allegedly taken place through the agency of the Red Army and the Socialist Unity Party (SED), while the reaction continued in new guise under a different political system of bourgeois domination in the Federal Republic.

Without sharing this fundamentalist position, western marxist and marxist-influenced historical writing – which has contributed to some of the most important work on the social history of the Third Reich during the past two decades – is equally impatient with suggestions of a 'social revolution' under Nazism. The historical balance, it is argued, is clear: Nazism destroyed working-class organizations, reshaped class relations by greatly strengthening the position of employers, who were backed with all the weight of a repressive police state, and kept down living standards while providing for soaring profits.[7] Clear though this balance may be, however, it arguably marks the beginning, not the end, of the enquiry. The Nazi regime unquestionably enjoyed until well into the war a degree of popularity and active support which cannot adequately be explained by the manipulative power of propaganda or the heavy repression of a police state. It has to be accepted that Nazism made real – if partial – inroads into wide sectors of German society, not excluding the working class, and that a considerable degree of material as well as affective integration into the Nazi State was attained, even though Catholic, communist, and socialist subcultures proved relatively resistent and impenetrable barriers. Recognition of considerable and extensive Nazi penetration, which in itself is, of course, perfectly compatible with marxist approaches, requires explanation which does not block any notion of Nazism's impetus for social change (even if only of a negative kind through its massive destructive drive) on the grounds

[6] See Gerhard Lozek and Rolf Richter, 'Zur Auseinandersetzung mit vorherrschenden bürgerlichen Faschismustheorien', in Gossweiler and Eichholtz, *Faschismusforschung* (ch. 1 note 27), pp. 417–51, here pp. 427–9; and Gerhard Lozek *et al.*, eds., *Unbewältigte Vergangenheit. Kritik der bürgerlichen Geschichtsschreibung in der BRD* (East Berlin, 1977), pp. 340–1.
[7] See e.g. the comments of Ernest Mandel, in Trotsky, *Struggle* (ch. 2 note 16), p. 13.

that Nazism equalled social reaction. Recent research on the social bases of Nazi support before 1933 has, in fact, completely undermined earlier generalizations about the backward-looking, reactionary (in a literal sense) nature of Nazism's mass backing, and has emphasized the strong, dynamic motivation for radical social change and undeniable 'modern' tendencies and aspirations among the socially heterogeneous support for the NSDAP.[8] Nazi support was no mere search for a return to yesteryear, whatever restorative tendencies were undoubtedly *also* present. The pressures in the Movement for social change, even if inchoate and pressing in different directions, could not have been totally ignored or repressed after 1933, even if that had been the intention of the Nazi leadership. Moreover, even on the most cursory common-sense ground, the Germany – even taking the nascent Federal Republic alone – of the later 1940s or early 1950s was, with all recognition of the numerous and inevitable continuities, a very different place and a very different society from the Germany of 1933. Whatever the complexities of the enquiry, the question of whether Nazism marked a caesura in Germany's social development, or left a lasting legacy in its impact on social and political values and attitudes, is, therefore, a legitimate one to pose.

Two works by non-marxist 'liberal' scholars, the German sociologist Ralf Dahrendorf and the American historian David Schoenbaum, appeared at roughly the same time in the mid 1960s and attempted to answer the question, by quite different routes, by arguing that the Third Reich did indeed produce a 'social revolution', the main feature of which was 'the break with tradition and thus a strong push toward modernity'.[9]

For Dahrendorf, Nazism completed the social revolution in Germany that had been 'lost in the faultings of Imperial Germany and again held up by the contradictions of the Weimar Republic'.[10] The substance of the revolution was 'modernity', by which he understood in essence the structures and values of western liberal–democratic society. Such a revolution had, he argued, naturally not been intended by the Nazis, whose social ideology rested on a recovery of past values. But in practice, their *Gleichschaltung* ('co-ordination') of German society had destroyed German 'tribal loyalties', breaking traditional anti-liberal religious, regional, family, and corporative bonds, had reduced élites to a 'monopolistic clique', and had levelled down social strata to the equalizing status of the *Volksgenosse*, the 'people's comrade'. In order to retain power, in fact, Nazi 'totalitarianism' had been compelled to turn against all traces of the social order that provided the basis of conservative authoritarian rule. Through the destruction of traditional loyalties, norms, and values, concluded Dahrendorf, Nazism 'finally abolished the German past as it was embodied in Imperial Germany. What came after it was free of the mortgage that burdened the Weimar Republic at its beginning, thanks to the suspended revolution. There could be no return from the revolution of National Socialist times'.[11] Unwittingly, therefore, Nazism had paved the way for a liberal–democratic society in post-war West Germany.

Dahrendorf's highly influential interpretation was contained in a single

---

[8] See e.g. Broszat, 'Zur Struktur der NS-Massenbewegung' (ch. 2 note 53).
[9] See ch. 2 note 26 for references; quotation from Dahrendorf, p. 403.
[10] For this paragraph, see Dahrendorf, pp. 402–18 (quotation from p. 403).
[11] Dahrendorf, p. 418.

chapter of his sociological analysis of modern Germany. David Schoenbaum's stylishly written study, on the other hand, was entirely directed at an examination of what he called 'Hitler's social revolution'.[12] Confining his investigation to the years 1933–9, Schoenbaum omitted from consideration any changes deriving from the wartime period, but in a complex discussion, developed an argument which, though more thoroughly researched, came close to Dahrendorf's. Schoenbaum's main thesis, in his own words, was that 'the Third Reich was a double revolution . . . of means and ends. The revolution of ends was ideological – war against bourgeois and industrial society. The revolution of means was its reciprocal. It was bourgeois and industrial since, in an industrial age, even a war against industrial society must be fought with industrial means and bourgeois are necessary to fight the bourgeoisie'.[13] This paradox runs through Schoenbaum's entire analysis, a crucial element of which is the distinction between what he termed 'objective' and 'interpreted social reality'. While 'objective social reality', he argued, 'was the very opposite of what Hitler had presumably promised and what the majority of his followers had expected him to fulfil' – with greater urbanization, industrialization, concentration of capital, inequality of income distribution, and the preservation of social divides – 'interpreted social reality' reflected 'a society united like no other in recent German history, a society of opportunities for young and old, classes and masses, a society that was New Deal and good old days at the same time'.[14] On this premise, Schoenbaum argued that 'Hitler's social revolution' amounted to the destruction of the traditional relationship between class and status: 'In the Third Reich, relative approximation of class and status came to an end', since 'in the wonderland of Hitler Germany' nobody knew 'what was up and what was down'.[15] Thus, too, workers' 'loss of *liberté* . . . was practically linked with the promotion of *égalité*', so that, though we might regard their status as one of slavery, 'it was not necessarily slavery from the point of view of a contemporary'.[16] The collapse of the status–class underpinning was, in fact, sufficient for Schoenbaum to go still further, and to argue that 'in the resultant collision of ideological and industrial revolution, traditional class structure broke down', so that one could speak of a 'classless reality of the Third Reich'.[17] As these remarks demonstrate, Schoenbaum is arguing for 'a revolution of class and a revolution of status at the same time' – amounting in class terms to unprecedented social mobility, in status terms even to 'the triumph of egalitarianism'.[18] The contrast between such an interpretation and marxist approaches – as typified by Franz Neumann's view that 'the essence of National Socialist social policy consists in the acceptance and strengthening of

[12] Hans Mommsen provides a good critical commentary in a 'Nachwort' to the German edition of Schoenbaum's book, *Die braune Revolution. Eine Sozialgeschichte des Dritten Reiches* (Munich, 1980, edn.), pp. 352–68.
[13] Schoenbaum (ch. 2 note 26), pp. xxi–xxii.
[14] Schoenbaum, pp. 285–6.
[15] Schoenbaum, pp. 280–1.
[16] Schoenbaum, pp. 110–11.
[17] Schoenbaum, p. 283.
[18] Schoenbaum, pp. 272–3. This view of a 'socialist side' of Nazism, which fostered the progressive breakdown of status privilege and class barriers has been very influential, particularly when assisted by the multiplier-effect of mass-circulation works such as Haffner, *Anmerkungen* (ch. 4 note 10), pp. 48–53.

the prevailing class character of German society' – could hardly be starker.[19]

Nazi 'social ideology' has generally been regarded by historians either as nothing more than a propagandistic sham, or as serious in intent but impossible to implement because of its internal contradictions. Hence, marxist writers have usually stressed the distinction between social base and social function, of a heavily lower-middle-class mass movement but a regime which consistently 'betrayed' its mass support in the interests of big capital.[20] Alternatively, Schoenbaum's argument is followed, emphasizing the paradox that anti-industrial social ends needed industrial social means. In an influential essay, Henry Turner pushed this paradox further than Schoenbaum had been prepared to take it, in accepting Nazi ideology at its word as an absolutely serious intent to do away with modern society, in which modern means would be used to bring about anti-modern conclusions through a successful war.[21] As Turner saw it, 'by reducing Germany's need for industry and thus for industrial workers, and by providing fertile soil upon which these displaced workers and others could be resettled, the acquisition of *Lebensraum* was expected to open the way to a vast new wave of German eastward colonization comparable to that of the Middle Ages, making possible a significant degree of de-urbanization and de-industrialization'.[22] Of course, the conquest of *Lebensraum* could only come about through a vast industrial war, and the Nazis, therefore 'practised modernization out of necessity in order to pursue their fundamentally anti-modern aims'. Once realized, the goal of *Lebensraum* would have rendered them largely unnecessary.[23] The Nazi solution of escape from the modern world by a 'desperate backward leap' could therefore be characterized as 'a utopian form of anti-modernism – utopian in the double sense of being a visionary panacea and being unrealizable'.[24] The last point seems the most important one: the vision was wholly unrealizable. Turner seems, in fact, in danger of attributing a rationality and cohesion to Nazi 'anti-modern aims' which is scarcely warranted in the light of the gulf between the actual reality of the 'New Order' in eastern Europe and the visionary pipe-dreams of Himmler or Darré; the character of the development of German industry and technology during the war; and the fact that modern armaments were going to remain an absolute necessity for the perpetual struggle to defend conquered territory and continue expansion, ingrained in Hitler's philosophy. Naturally, of course, speculation about an illusory future can say little about Nazism's *actual* impact on German society.

A further discussion of this problem, by Werner Abelshauser and Anselm Faust, adopted a position not far from the interpretations of Dahrendorf and Schoenbaum.[25] Again, Abelshauser and Faust were prepared to consider

---

[19] Neumann, *Behemoth* (ch. 2 note 5), p. 298.
[20] E.g. Kühnl, *Formen bürgerlicher Herrschaft* (ch. 2 note 13), pp. 80 ff., 118 ff.; and, more crudely, Reinhard Opitz, 'Die faschistische Massenbewegung', in Kühnl, *Texte* (ch. 2 note 13), pp. 176–90. For a summary and assessment of this type of argument, see Saage, *Faschismustheorien* (ch. 2 note 33), pp. 131 ff., and Adelheid von Saldern, *Mittelstand im Dritten Reich. Handwerker – Einzelhändler – Bauern* (Frankfurt am Main/New York, 1979), pp. 9–15, 234 ff.
[21] Turner, 'Fascism and Modernization' (ch. 2 note 52), pp. 117–39.
[22] Turner, 'Fascism and Modernization', pp. 120–2.
[23] Turner, 'Fascism and Modernization', pp. 126–7.
[24] Turner, 'Fascism and Modernization', pp. 120–1.
[25] See note 5, this chapter, for full reference.

Nazism's effects as part of a 'social revolution' – a concept they use in the sense of long-term but incisive processes of change in social and economic life, as in 'industrial revolution', 'Keynesian revolution', and 'modernizing revolutions', and attributing to Nazism 'not more and not less than the role of a catalyst of modernization, in that it exploded with force the bonds of tradition, region, religion, and corporation which were so specially pronounced in Germany'.[26] In their interpretation, Nazi social and economic policy was in a two-fold sense a means of social revolutionary change: both in anticipating the 'Keynesian revolution' of post-war German capitalism by its policies of economic stimulus to master the slump; and in its imposed 'co-ordination', which destroyed the trade unions, subordinated the employers to the primacy of politics of the authoritarian state, and thus altered the life of the Germans in the shortest possible time more decisively than the Revolution of 1918–19 had been able to do.[27]

Still operating with the concept of 'modernization', but now within the framework of a consciously theoretical model, Horst Matzerath and Heinrich Volkmann arrived at conclusions differing from those of both Turner and Abelshauser and Faust, in a stimulating if contentious conference paper published in 1977.[28] They argued strongly for the value of applying the concept of modernization to Nazism by considering the degree of both quantitative and qualitative economic, social, and political change between 1933 and 1939, using indicators of modernization such as those we discussed earlier in this chapter.

Their findings suggested a picture of contradictions, featuring in all sectors of their modernization model the continuation or accentuation of earlier trends, but also anti-modern counter-developments, especially in the political sphere (such as anti-parliamentary, anti-emancipatory, and anti-participatory measures).[29] They rejected the notion of a 'social revolution' as proposed by Dahrendorf and Schoenbaum, and built instead on aspects of Talcott Parsons's hypothesis, formulated as long ago as 1942. Parsons had argued that Nazism arose out of a conflict between modern economic and social structures and traditional value systems and patterns of socialization, producing an 'anomie' which found effect not in adjustment to changing reality, but in irrational flight to a radical denial of the new and the modern through resort to an extreme version of traditional values.[30] Taking Parsons's hypothesis a stage further, Matzerath and Volkmann argued that Nazism was structurally determined by the conditions which produced the Movement: the aggressive reaction of traditional values against modernity in the shape of 'the accelerated change of the economic, social, and political system, sharpened through an acute crisis unleashed through war, defeat, inflation, depression, and the danger of an alternative system', all primarily manifesting themselves in the social anxieties and resentments featured in Nazi ideology.[31] Thus, Nazi ideology functioned as 'a suitable instrument for the mobilization of sensitive strata of the population affected by problems of modernization'. Since, however, Nazism in power

---

[26] Abelshauser and Faust, p. 116.
[27] Abelshauser and Faust, p. 118.
[28] Matzerath and Volkmann (see ch. 2 note 53).
[29] Matzerath and Volkmann, pp. 95–7.
[30] See ch. 2 note 18 for references.
[31] Matzerath and Volkmann, p. 98.

was unable to produce any positive or constructive social concept, but had destroyed all alternative concepts derived from the previous system, a new basis of legitimation was necessary. This was found in the diversion of inherited conflicts on to internal and external opponents, who were used in turn to justify the central aims of the system – establishment of a totalitarian apparatus of domination and preparation of a war of brutal conquest. This meant the destruction of traditional loyalties and the distortion to the point of destruction of traditional values. Nevertheless, the 'anti-modernity' of Nazism ought not to be misunderstood as the programmatic reconstruction of pre-modern conditions (as Turner, for instance, had seen it), or as a 'conservative revolution'. Rather, according to Matzerath and Volkmann, 'National Socialism is the attempt at a special path out of the problems of modernization into the utopia of a third way, beyond the internal social crises and conflicts of the parliamentary democratic capitalist society, and beyond the concept – releasing anxiety and aggression – of a communist total alteration [of society], but essentially without giving up the capitalist and industrial economic bases of this development'.[32] Such a definition accords, in the authors' view, with the partly modern, partly anti-modern ambivalent reality of Nazism. Even so, Matzerath and Volkmann reach the conclusion that the partially modernizing effects of Nazism cannot be seen as the result of conscious modernizing policies, and in fact ought best to be described as 'pseudo-modernization'. Moreover, and an important point in the overall argument, the Nazi regime was incapable of developing any lasting structures. Through its inability to recognize social conflict and cope with it, the system was incapable of producing stability with change. Even as an 'exceptional or transitional form of social organization in a stress phase of modernization', Nazism was 'dysfunctional': 'it was not a roundabout path to modernization, but the expression of its failure, the historical cul-de-sac of a process, whose steerage problems had overtaxed the social capacities'.[33]

In their emphasis on Nazism's inbuilt failure to produce lasting social structures, Matzerath and Volkmann were returning by a circuitous route to something approaching the position which Rauschning had impressionistically – and from an entirely different vantage point – reached in the later 1930s, when he claimed that Nazism could bring about only a 'revolution of nihilism'.[34] In essence, this corresponds, too, to Winkler's argument that 'the greatest social caesura which National Socialism brought about is its collapse', and that nothing of the social change which took place during the Dictatorship itself compared in its significance with the devastation of the last year of the war and total defeat, with their far-reaching consequences for the two German societies which replaced the Third Reich.[35] A similar conclusion was reached by Jeremy Noakes who, in a thorough examination of the whole problem, argued that whatever was revolutionary about Nazism lay in the destruction and self-destruction which were inevitable corollaries of its irrational goals: 'Arguably, therefore, the Nazi revolution was the war – not simply because the war

---

[32] Matzerath and Volkmann, p. 99.
[33] Matzerath and Volkmann, p. 100.
[34] The title of his book (see ch. 6 note 4), first published in German as *Die Revolution des Nihilismus* (Zurich, 1938).
[35] Winkler, 'Vom Mythos der Volksgemeinschaft' (ch. 3 note 47), p. 490.

accelerated political, economic, and social change to a degree which had not occurred in peacetime, but more profoundly because in war Nazism was in its element. In this sense, Nazism was truly 'a revolution of destruction' – of itself and of others on an unparalleled scale'.[36]

The approaches we have rapidly summarized here can be subsumed under three main categories of interpretation:

(i) One central interpretation, favoured especially but not only by marxist historians, is that, whatever superficial changes were made in social *forms* and institutional *appearances* in the Third Reich, the fundamental *substance* of society remained unchanged, since the position of capitalism was strengthened and the class structure enhanced, not broken down, by Nazism.

(ii) In contrast, an influential interpretation advanced by 'liberal' scholars suggests that the changes in the structures of society and in social values brought about directly or indirectly by Nazism were so profound that it is not going too far to regard them as a 'social revolution'.

(iii) A third position can be distinguished from both of these interpretations, though in practice it comes closer to the second than to the first. It is argued here, that whatever changes Nazism itself brought about did not in any sense amount to a 'social revolution'. Its social effects were, in fact, contradictory – some 'modernizing', others reactionary. Nevertheless, the Third Reich did have important consequences for post-war society, especially in the nature of its own total collapse and destruction, bringing down with it authoritarian structures which had dominated Germany since Bismarck's era, and wreaking such havoc, dislocation, and upheaval that in radically different ways new starts were necessary in the eastern and the western zones of defeated Germany.

We can now consider these interpretations in the light of recent research on the social history of the Third Reich.

Evaluation

An evaluation of the social impact of Nazism must begin with the nature and social dynamic of the Nazi movement.

As innumerable studies have shown, it is simplistic to regard the Nazi movement as no more than the direct product and instrument of reactionary capitalist forces. It was the outgrowth of extreme socio-political unrest and disaffection, with a most heterogeneous mass following ideologically integrated only through radical negative protest (anti-marxism, anti-Weimar, anti-Semitism) coupled with a chiliastic, pseudo-religious vision of a 'national awakening' – socially expressed through the vague (and ultimately also negative) 'idea' of the 'national community' (*Volksgemeinschaft*). The appeal before 1933 of the 'national community' slogan – symbolizing the transcending of class, denominational, and political divisions through a new ethnic unity based on 'true' German values – is undeniable. Socially, it reflected not only

---

[36] Noakes, 'Nazism and Revolution', p. 96. See also Peukert, *Volkgenossen* (ch. 2 note 45), p. 294 for emphasis on the socially 'destructive forces and effects' of Nazism, out of which a more 'modern' society arose following the end of the regime and the war.

the desire to banish the cancer of marxism, but also to overcome the rigid immobility and sterility of the old social order by offering mobility and advancement through merit and achievement, not through inherited social rank and birthright. The mood of social protest was at its most radical, as is well known, among young Germans, where the drive and élan of the Nazi movement held especial appeal.[37]

Before 1933 the one uniting aim of the dynamic but unstable and ramshackle Nazi movement was to gain power. The 'seizure of power', however, could only be attained through the collaboration of the ruling élites. The relative strength of these groups in the early period of Nazi rule, together with the regime's allocated priority to rearmament, ensured that sectional interests within the Party (such as those of small retailers or craftsmen) were inevitably sacrificed where they did not fit the needs of Germany's large-scale (especially armaments-geared) capitalist enterprises. The perceived lingering challenge to the 'social order' posed by the SA had its sting removed through the liquidation of Röhm and other SA leaders in the so-called 'Night of the Long Knives' in June 1934. But although pruned of its socially most 'dangerous' elements, the Nazi Party and ancillary organizations were hardly a source of stability. Deprived of any real governing function after 1933, the role of the amorphous Nazi movement was confined largely to providing action for activists through social control tasks, propaganda of the word and 'deed', and the whipping up of acclamation for the Führer's 'achievements'. The disappointment of many social aspirations in the Third Reich was compensated to some degree by the channelling of pent-up energies into activism levelled at helpless and disparaged minorities who formed the racial and social pariahs of the 'national community'. Alongside the escalating discrimination against Jews and other 'outcasts', the subordination of sectional material interests within the Party to the overriding 'national' goals of the Führer was equally inevitable. Everything had to be subsumed in the preparation for the inexorably coming struggle. But, obsessively single-minded in ends, Hitler was wholly eclectic in means. Thus, there could be no thought of destroying Germany's industry to satisfy the needs of archaic *Mittelstand* interests or romantic peasant idealists in the Party.[38] The Party's ideologues and representatives of Party sectional interests with their own ideas about what the 'national community' should look like were

[37] On the social 'drive' of Nazism before 1933, Broszat's articles, 'Soziale Motivation' (ch. 4 note 27) and 'Zur Struktur der NS-Massenbewegung' (ch. 2 note 53) provide stimulating general interpretations. The most valuable insights on the appeal to youth – apart from works on the youth movement and on Nazi youth organizations such as Peter D. Stachura, *Nazi Youth in the Weimar Republic* (Santa Barbara/Oxford, 1975) and *The German Youth Movement 1900–1945* (London, 1981) – have emerged from recent work, differing in orientation and interpretation, on the SA: Peter H. Merkl, *The Making of a Stormtrooper* (Princeton, 1980); Conan Fischer, *Stormtroopers. A Social, Economic, and Ideological Analysis 1929–1935* (London, 1983); Richard Bessel, *Political Violence and the Rise of Nazism. The Storm Troopers in Eastern Germany 1925–1934* (New Haven/London, 1984); and Jamin (ch. 2 note 29). I have attempted briefly to evaluate some recent interpretations of the appeal of Nazism in 'Ideology, Propaganda, and the Rise of the Nazi Party', in Peter D. Stachura, *The Nazi Machtergreifung* (London, 1983), pp. 162–81.

[38] For Hitler's social aims, see Noakes, 'Nazism and Revolution', pp. 76 ff. A thorough survey of Hitler's thinking on social developments is provided by Rainer Zitelmann, *Hitler. Selbstverständnis eines Revolutionärs* (Hamburg/Leamington Spa/New York, 1987), although Zitelmann is over-inclined to treat Hitler's utopian 'social ideas', predicated upon final victory and European hegemony, as firm plans for a revolutionary 'modernizing' transformation of German society.

invariably shunted on to sidelines sooner or later – the fate of Feder, Wagener, Darré, and Rosenberg. Unlike such Party 'theoreticians', Hitler had no real interest in social structures as long as they were not dangerous or obstructive. Long-term, it is true, his own views were dominated by vague notions of a racial élite, rule by those who had proved themselves fit to rule, and the passing of social groups for whom he had little but contempt (such as the aristocracy, and 'captains of industry'). But in the real world of the short-term, Hitler was uninterested in tampering with the social order. Just like industry and capitalism, social groups were there to serve in their different ways the political goal of the struggle for 'national survival'. In any case, quite apart from Hitler's own predilections, the Nazi movement was such an amalgam of contradictory social forces that it was capable of producing neither theory nor practice of any realistic new social construct. It was as parasitic as it was predatory.

Where Nazism was ambitious – and extraordinarily so – was in attempting a transformation in subjective consciousness rather than in objective realities.[39] Since the Nazi diagnosis of Germany's problem was in essence one of attitudes, values, and mentalities, it was these which they were attempting to revolutionize psychologically by replacing all class, religious, and regional allegiances by massively enhanced national self-awareness to mobilize the German people psychologically for the coming struggle and to bolster their morale during the inevitable war. Not the revamping of comfortable, small-town, lower-middle-class social views, but the moulding of a people in the image of an army – disciplined, resilient, fanatically single-minded, obedient to death for the cause – was the intention. The idea of the 'national community' was not a basis for changing social structures, but a symbol of transformed consciousness. The attempt to inculcate such values into the German people was in essence a task of propaganda more than of social policy.

These remarks on the character of the Nazi movement and its social aims suggest that notions of social change were – inevitably given its nature, composition, and dominant leadership – negative (destruction of working-class organizations, increased discrimination against minorities); confined to long-term but vacuous, utopian ambitions bearing little relation to reality, or to short-term sectional interests incompatible with war preparation and therefore dispensable; and, finally, rested on conceptions of a revolution of attitudes which, given the strength of previous loyalties to Church, region, or class, were again illusory as a short- or middle-range objective. The nature of the Nazi movement offers pointers towards understanding the impact of Nazism on specific social groups; the widespread disillusionment and disappointments during the Third Reich; the compensatory mechanism of the 'selection of negative stereotypes'[40] as the victims of ever more vicious discrimination; and the difficulty of regarding Nazism as capable of bringing about a 'social revolution' on its own terms.

Understanding of what Schoenbaum dubbed 'objective reality' – actual changes in the class structure and social formations in Germany during the Third Reich – has been greatly advanced by much valuable empirical research in the past few years. The findings of this research have pointed unequivocally in the direction of Winkler's conclusion, 'that there can be no question in real

---

[39] Broszat, *Hitler State* (ch. 2 note 39), p. 18.
[40] Broszat, 'Soziale Motivation', p. 405.

terms of a revolutionary transformation of German society between 1933 and 1945'.[41] The notion that the Third Reich had brought about a social revolution was, as Winkler indicates, largely attributable to an over-ready acceptance of the regime's own pseudo-egalitarian propaganda and exaggerated claims, and, partly too, to actual social changes of the post-war era which were cften projected backwards into the Third Reich, though they had little to do with Nazism, even indirectly.[42]

The emphasis in recent research has, therefore, been far more heavily laid upon the essential continuities in the class structure of Nazi Germany, rather than upon incisive changes. Schoenbaum himself had accepted that the social position of the élites remained relatively unscathed down to the last phase of the war. He may, however, have rather exaggerated the extent of the fluidity in social structures and the amount of upward mobility which took place. Of course, it is true that thrusting, energetic, ruthless, and often highly efficient 'technocrats of power'[43] such as Heydrich or Speer pushed their way to the top. And the war certainly accelerated changes in the high ranks of the *Wehrmacht*. But the new political élite co-existed and merged with the old élites rather than supplanting them.[44] Non-Party preserves such as big business, the civil service, and the army recruited their leadership for the most part from the same social strata as before 1933. Education remained over-whelmingly dominated by the middle and upper middle classes. The most important and powerful Party affiliation, the SS, recruited heavily from the élite sectors of society.[45] If the traditional ruling classes had to make some room for social upstarts from lower ranks of society who had gained advance-ment through positions of power and political influence, such changes amounted to little more than a slight acceleration of changes already perceptible in the Weimar Republic.

At the other end of the social scale, the working class – deprived of a political voice, its social gains of the Weimar Republic reversed, and exposed in the shadow of mass unemployment to the brutal exploitation of employers backed by the repressive apparatus of the police state – had its living standard reduced in the first years of the Third Reich even from the lowly level of the Depression era.[46] The slight rise in real wages in the later 1930s was a by-product of the armaments boom, and was accompanied by intensified pressure – physical and mental – upon the industrial workforce. The class position of workers remained basically unchanged into the middle of the war – except that the most extreme exploitation now fell upon foreign workers. The most significant

---

[41] Matzerath and Volkmann, p. 103 (comment of H.A. Winkler).

[42] Matzerath and Volkmann, p. 102 (comment of H.A. Winkler). See also Winkler, 'Vom Mythos der Volksgemeinschaft', p. 490.

[43] Broszat, 'Zur Struktur der NS-Massenbewegung', p. 67.

[44] See Noakes, 'Nazism and Revolution', pp. 80–5, and also Hans Mommsen, 'Zur Verschrän-kung traditioneller und faschistischer Führungsgruppen in Deutschland beim Übergang von der Bewegungs- zur Systemphase', in Schieder, *Faschismus als soziale Bewegung* (ch. 2 note 29), pp. 157–81.

[45] See e.g. Gunnar C. Boehnert, 'The Jurists in the SS-Führerkorps 1925–1939', in Hirschfeld and Kettenacker (ch. 1 note 23), pp. 361–74, and 'The Third Reich and the Problem of "Social Revolu-tion": German Officers and the SS', in Volker R. Berghahn and Martin Kitchen, eds., *Germany in the Age of Total War* (London, 1981), pp. 203–17; and Bernd Wegner, *Hitlers Politische Soldaten: Die Waffen-SS, 1933–1945* (Paderborn, 1982), ch. 15, esp. pp. 222–6.

[46] See Mason, *Sozialpolitik* (ch. 4 note 62), esp. ch. 4.

changes in the nature and composition of German labour occurred in the last phase of the war and were, in the main, the consequence of military service, losses at the Front, destruction of industries, dislocation of the workforce, evacuation and homelessness, and ultimately foreign conquest.[47] Whatever changes had taken place by 1945 were, therefore, a product of Nazism's collapse more than of its policies while in power.

Studies of middle-class groups in the Third Reich have also stressed how for all the Nazis' archaic rhetoric and anachronistic legislation – such as the Entailed Farm Law of 1933 – such change as took place was the product of industrial recovery and accelerated development in a capitalist economy.[48] Continuity rather than dramatic change was the hallmark down to the mid-war period. Before then, there was some decline in the number of small retail and craft concerns, but no fundamental threat to their position. The number of white-collar employees, the service sector, and the bureaucracy expanded, as in all contemporary capitalist societies, if at a somewhat faster rate. There was no major shift in the pattern of landholding, despite the Entailed Farm Law, and after the early promises of a new deal, peasants found themselves yet further victims of the armaments economy, their labourers drained away to the higher wages of industry and the better living conditions of the city. Again, whatever major shift took place in the social position of the *Mittelstand* and peasantry was a consequence of the extreme disruption and dislocation of the final phase of the war and – especially in the eastern zone – of the immediate post-war era.

Finally, research on the position of women and the structure of their employment has illustrated both the extent to which Nazi anti-feminism corresponded to the traditions and patterns of bourgeois anti-feminism in a capitalist society, and at the same time the contradictions within the Nazi system, where the increased need for female labour forced concessions to the point of ultimate reversal of ideological prerogatives by the middle of the war.[49] Once more, the continuities in social structures under Nazism greatly outweigh the change which, far from being revolutionary, was simply that of an advanced capitalist economy, if one with an unusual degree of state intervention,[50] and one which

[47] See the works of Salter and Werner (ch. 4 note 68). On labour policy during the war, Marie-Louise Recker, *Nationalsozialistische Sozialpolitik im Zweiten Weltkrieg* (Munich, 1985) offers a further important contribution.

[48] See esp. von Saldern (note 20, this chapter); Heinrich August Winkler, 'Der entbehrliche Stand. Zur Mittelstandspolitik im "Dritten Reich" ', *AfS* 17 (1977), pp. 1–40; and, more recently, the valuable study by Michael Prinz, *Vom neuen Mittelstand zum Volksgenossen* (Munich, 1986).

[49] See Dörte Winkler (ch. 4 note 74); Mason; 'Women' (ch. 4 note 74); Jill Stephenson, *Women in Nazi Society* (London, 1975); Stefan Bajohr, *Die andere Hälfte der Fabrik* (Marburg, 1979); Gisela Bock, 'Frauen und ihre Arbeit im Nationalsozialismus', in Annette Kuhn and Gerhard Schneider, eds., *Frauen in der Geschichte* (Düsseldorf, 1979), pp. 113–49; Frauengruppe Faschismusforschung, ed., *Mutterkreuz und Arbeitsbuch* (Frankfurt am Main, 1981); Dorothee Klinsiek, *Die Frau im NS-Staat* (Stuttgart, 1982). Recent work on women in the Third Reich has concentrated more on Nazi biological than economic policy: see, for example, Gisela Bock, *Zwangssterilisation im Nationalsozialismus* (Opladen, 1986); Renate Bridenthal, Atina Grossman, and Marion Kaplan, eds., *When Biology became Destiny. Women in Weimar and Nazi Germany* (New York, 1984); Claudia Koonz, *Mothers in the Fatherland. Women, the Family, and Nazi Politics* (New York, 1986).

[50] See Overy, 'Göring's "Multi-National Empire" ' (ch. 3 note 41) and – still fundamental as an analysis of the Nazi economy – Neumann, *Behemoth* (ch. 2 note 5).

long before the war was extraordinarily lopsided in its concentration on armaments production, and spinning rapidly out of control.

In Schoenbaum's view, it was above all 'interpreted social reality' – attitudes, values, mentality, subjective consciousness – which underwent a transformation in the Third Reich. His assertions in this area, however, were highly speculative and impressionistic. In the nature of things, evaluation of changes in subjective attitudes and consciousness is fraught with difficulties, the evidence full of pitfalls, conclusions necessarily tentative. Recent research, however, which paints an extremely complex picture of social behaviour and attitudes in the Third Reich, suggests strongly that it is easy to exaggerate the nature of changes in values and attitudes under Nazism, and that here too there can be no suggestion of Nazism having effected a social revolution.[51]

The most continuous, and usually the most dominant, influence upon the subjective perception by differing social groups of their own socio-economic position during the Third Reich was, it seems, formed by the material conditions which directly affected the everyday lives of the population. And here, the acute perception of social injustice, the class-conscious awareness of inequalities, and the persistent feelings of exploitation appear to have changed little in the period of the Dictatorship. The alienation of the working class, the ceaseless expression of sectional grievances by middle-class groups and farmers, the massive disillusionment and discontent in most sections of the population deriving from their actual daily experience under Nazism is scarcely compatible with Schoenbaum's view that 'interpreted social reality . . . reflected a society united like no other in recent German history' and a status revolution amounting to a 'triumph of egalitarianism'.[52]

In Nazi eyes, the greatest need to reshape status awareness and to replace class by national consciousness was with regard to the industrial working class. Yet here especially, for all that there *was* some penetration of Nazi values and attitudes, the regime's social propaganda made little serious dent in traditional class loyalties, particularly among older industrial workers. It would appear that Dahrendorf equally overestimated the extent of the breakdown in traditional loyalties to the Christian Churches. The decline in Church membership was trivial during the 1930s, while religious observance and attendance at services increased sharply during the war years. Defence of Church traditions and institutions against piecemeal Nazi attacks was extensive, and partly successful. The hold of the Church and clergy over the population, especially in country areas, was often strengthened rather than weakened by the 'Church struggle'. And, finally, the Churches as institutions recovered enormous social power and political influence in post-war West Germany. Everything points to the conclusion that Nazi policy failed categorically to break down religious allegiances. Even in their attempt to inculcate the German people with racial, eugenic, and social darwinist values – the core of their ideology – the Nazis, it appears, had

---

[51] I have attempted to argue this case in full in my *Popular Opinion and Political Dissent* (ch. 2 note 44). See also Peukert (ch. 2 note 45), and the contributions to Peukert and Reulecke (ch. 4 note 58) for some of the best recent research in this field, of which a perceptive survey is provided by Richard Bessel, 'Living with the Nazis: Some Recent Writing on the Social History of the Third Reich', *European History Quarterly* 14 (1984), pp. 211-20.
[52] Schoenbaum, pp. 273, 286.

only limited success.[53] Enhancement of existing prejudice against Jews and other racial minorities and 'social outsiders' unquestionably occurred, and within the SS in particular – but also to some extent within the *Wehrmacht* – indoctrination with a new value-system was effected.[54] But the growing protest against the 'euthanasia action' and the regime's perception of the need for utmost secrecy in the 'Final Solution' are indirect testimony that exposure to Nazi race values had come nowhere near completely eradicating conventional moral standards.

Much suggests that the Nazis made their greatest impact on young Germans, and that there was a pronounced generation gap between those who had reached adulthood in the Imperial or Weimar eras and those who had experienced little else other than Nazism. The rejection of the old bourgeois world and idealistic notions of a new and more mobile and egalitarian society were the basis of the Nazis' dynamic mobilization of youth. But even here, the regime had only partial success. Hitler's own view, as it was recorded in 1945, was that it would have taken 20 years to produce an élite which would have imbibed Nazi values like its mother's milk. The illusory nature of such hopes was demonstrated by his further comment that he could not afford to wait so long: time was now as always against Germany.[55] In fact, as recent studies have demonstrated, signs of conflict, tension, and opposition within certain sections of German youth were already apparent by the later 1930s and increased in the war years, suggesting that the Nazis had been only temporarily successful in winning over, mobilizing, and integrating young Germans.[56]

Finally, though it is still an under-researched and difficult subject, there is no evidence to suggest that family structures came anywhere near breaking down under Nazism, despite the undoubted accentuation of generational con-

---

[53] See my essay, 'The Persecution of the Jews and German Popular Opinion in the Third Reich', *Yearbook of the Leo Baeck Institute* 26 (1981), pp. 261–89; Otto Dov Kulka, ' "Public Opinion" in Nazi Germany and the "Jewish Question" ', *The Jerusalem Quarterly* 25 (1982), pp. 121–44, and ' "Public Opinion" in Nazi Germany: the Final Solution', *The Jerusalem Quarterly* 26 (1983), pp. 34–45; and Sarah Gordon, *Hitler, Germans, and the 'Jewish Question'* (Princeton, 1984). The most thorough assessment of attitudes towards the Jews in Germany after 1933 is now provided by David Bankier, *The Germans and the Final Solution. Public Opinion under Nazism* (Oxford, 1992).

[54] See the works of Wegner (note 45, this chapter); Streit (ch. 5 note 70); and see now especially Omer Bartov's studies, *The Eastern Front 1941–45. German Troops and the Barbarisation of Warfare* (London, 1985); and *Hitler's Army. Soldiers, Nazis, and War in the Third Reich* (Oxford, 1991).

[55] F. Genoud, ed., *The Testament of Adolf Hitler* (London, 1961), pp. 58–9.

[56] See Lothar Gruchmann, 'Jugendopposition und Justiz im Dritten Reich', in Wolfgang Benz, ed., *Miscellanea. Festschrift für Helmut Krausnick zum 75. Geburtstag* (Stuttgart, 1980), pp. 103–30; Matthias von Hellfeld, *Edelweißpiraten in Köln* (Cologne, 1981); Arno Klönne, *Jugend im Dritten Reich. Die Hitler-Jugend und ihre Gegner* (Düsseldorf, 1982); Heinrich Muth, 'Jugendopposition im Dritten Reich', *VfZ* 30 (1982), pp. 369–417; Detlev Peukert, 'Edelweiß-piraten, Meuten, Swing. Jugendsubkulturen im Dritten Reich', in Gerhard Huck, ed., *Sozialgeschichte der Freizeit* (Wuppertal, 1980), pp. 307–27, and 'Youth in the Third Reich', in Richard Bessel, ed., *Life in the Third Reich* (Oxford, 1987), pp. 25–40. By the latter half of 1943, the SD was reporting widespread negative attitudes towards the party and other aspects of Nazi rule among youth and schoolchildren: see SD-Berichte zu Inlandsfragen, 12 Aug., 22 Nov. 1943, Heinz Boberach, ed., *Meldungen aus dem Reich* (Herrsching, 1984), vol. 14, pp. 5603–7, vol. 15, pp. 6053–5. It is important, however, not to exaggerate the scale and oppositional significance of youth nonconformity, a point stressed by Gerhard Rempel, *Hitler's Children* (Chapel Hill/London, 1989).

flicts between children and parents fostered by Nazi youth organizations. There were, in fact, signs in the Third Reich itself of a reaction against the release of youth from the close shackles of adult authority in the school, the parental home, and elsewhere, and the reaction had not inconsiderable success, particularly in the post-war era.

It seems clear, then, that Nazism did not produce a 'social revolution' in Germany during the period of the Third Reich – whether one of 'objective' or of 'interpreted social reality'. As we noted earlier, the nature of the Nazi movement and the character of its social aims make it possible to go further and to argue that it was in any case incapable of bringing about a complete and permanent social revolution, short of attaining total and final victory in a war which was itself an intrinsically vain gamble to secure German domination. Nazism's intentions were directed towards a transformation of value- and belief-systems – a psychological 'revolution' rather than one of substance – and could only have been effected through the attainment of long-term goals which were themselves illusory, contradictory, and thus innately destructive and self-destructive.

Once the misleading notion that German society was changed in revolutionary fashion during the Third Reich is removed, it seems possible to argue both that during the period of its rule Nazism substantially bolstered the existing class order of society, and that, above all through its destructive dynamism, it paved the way for a new start after 1945.

On the one hand, obvious though it is, the point deserves emphasis that Nazism was not the product of 'pre-modern' society, but emerged in an advanced industrial state whose fragile political system was, in an unprecedented crisis of capitalism, severely wracked by class conflict. The Nazi regime's initial objective function was to re-establish the socio-economic order and the threatened position of the ruling élites by ruthlessly crushing the labour movement. The fateful political intervention of Nazism in 1933 has, therefore, to be seen in one sense as a decisive step in the struggle between capital and labour in an advanced industrial economy. And, indeed, Nazism in power was the most ruthless and exploitative form of industrial class society encountered – one which from a contemporary working-class perspective made the Kaiser's Germany seem in retrospect like 'a heaven of freedom'.[57] The new ordering of class relations in 1933 reversed in violent fashion the advances made by the working class not only since 1918, but since Bismarck's era, strengthened the weakened position of capitalism, and upheld – at least initially – the reactionary forces in the social order.

It is, however, insufficient to leave it at that, and to deny Nazism any motive force for social change of a long-term nature – even if this was in the main a 'negative' feature emanating from the destructive force of the regime. It has been suggested, for instance, that the necessary individualization of the working-class struggle within the Nazi system to gain maximum benefit from the armaments boom had lasting effects in weakening worker solidarity and paving the way for 'a new, more individualistic, performance-orientated, "sceptical" type of worker as described by sociologists in the fifties'.[58]

---

[57] Archiv der Sozialen Demokratie, Bonn, Bestand Emigration Sopade, M32, report of the Border Secretary of Northern Bavaria, Hans Dill, of 18 Nov. 1935.
[58] Peukert, *Volksgenossen* (ch. 2 note 45), pp. 136, 140.

Whether this is to project back into the Third Reich behavioural patterns which were largely a product of the post-war conditions of economic recovery and the 'economic miracle' itself is difficult to estimate. Also speculative, though inherently not unlikely, is the suggestion that Nazi atomization of society led to a 'retreat into the private sphere' which had lasting implications for 'depoliticized' popular culture – a part of the basis of the consumer and 'achievement' society of the 'economic miracle' era.[59]

The extent to which this can be linked to or explained by concepts of 'modernity' or 'modernization' seems debatable. As conventionally deployed in sociological and historical writing, 'modernization' implies long-term change spanning centuries and transforming 'traditional' society based on agricultural and artisanal production, personal relations of dependance, local loyalties, rural cultures, rigid social hierarchies, and religious world-views, into industrial class society with highly developed industrial technologies, secularized cultures, 'rational' bureaucratic impersonal socio-political orders, and political systems of mass participation. Some form of applied modernization theory seems an essential component in explaining long-term historical change. But in such a process, the Nazi era is a mere flash in time. And while 'traditional' value systems and social structures were undoubtedly in certain ways more resistent to the changes of industrialization in Germany than, for example, in Britain, their 'traditionality' can be overplayed and the emphasis on anti-modernization as the secret of Nazism's appeal can easily be greatly exaggerated. On the contrary: though Nazism contained obvious archaic and atavistic elements, they often served as propagandistic symbols or ideological cover for wholly 'modern' types of appeal offering social mobility, a society of equal chances where success came from merit and achievement, and new opportunities to thrive and prosper through letting youth and vigour have its head at the expense of the old, the sterile, the rigid, and the decayed.[60] Though vicious and extreme in its form and nature, this darwinistic appeal to the pure 'achieving society' (*Leistungsgesellschaft*) has parallels in other advanced capitalist economies. In evaluating the brief era of the Dictatorship itself, the modernization concept is unhelpful.[61] What change took place was within the context of,

---

[59] Peukert, *Volksgenossen*, pp. 230, 280–8, 294. This argument is backed by the findings of a major oral-history project in the Ruhr: see Lutz Niethammer, ed., *'Die Jahre weiß man nicht, wo man die heute hinsetzen soll'. Faschismuserfahrungen im Ruhrgebiet* (Berlin/Bonn, 1983) and *'Hinterher merkt man, daß es richtig war, daß es schiefgegangen ist'. Nachkriegserfahrungen im Ruhrgebiet* (Berlin/Bonn, 1983).

[60] See Broszat, 'Zur Struktur der NS-Massenbewegung' (ch. 2 note 53).

[61] The recent, valuable study by Michael Burleigh and Wolfgang Wippermann, *The Racial State* (Cambridge, 1991), is, from a different angle, highly critical of modernization approaches to Nazism (p. 2). The authors view such approaches – though this is surely going too far – as worthless even in heuristic terms (p. 307). They nevertheless 'regard the question concerning the modernity or anti-modernity of the Third Reich as among the most crucial problems confronting modern historical research' (p. 1), though their own answer to this question is limited by their lack of a clear definition of 'modernity' and 'anti-modernity'. The core of their argument, which even so in my view has much to commend it, is to see Nazi racial and social policy 'as an indivisible whole', 'merely different sides of the same coin', and simultaneously 'modern and profoundly anti-modern' (p. 4). Nazi racial policies amounted, therefore, they argue, to 'an unprecedented form of progress into barbarism' (dustjacket). How this could arise still, however, leaves open legitimate and important questions in which the 'modernization' issue can scarcely be avoided; such as whether Germany experienced a peculiar form of 'modernization crisis' or, as Detlev Peukert, in his *Die Weimarer Republik* (Frankfurt am Main, 1987; Engl. trans., *The Weimar Republic*,

for its date, an already highly advanced capitalist society. And while some Nazi measures had an archaic tinge to them, more were (in a neutral sense) 'advancing' or 'modern' – though in a fashion little different from those of other contemporary advanced capitalist states. Nor is the counterfactual question wholly misplaced: much of what is frequently dubbed Nazism's 'modernizing push' would, given the nature of the German economy, undoubtedly have taken place under any form of government.

We return, therefore, to what seems to be the crucial point in the question of Nazism's impact on social change: the intrinsic, all-consuming destructive essence of the system. In its drive to attain increasingly irrational goals, Nazism was a parasitic growth on the old social order, neither wanting nor capable of stability. Through the allocation of absolute priorities to rearmament, war, and expansion – goals actively furthered by collaboration from Germany's ruling classes – Nazism produced a maelstrom of destruction which threatened, then inevitably sucked in, the representatives of the existing social order. Hence, the destructive dynamic of the Nazi regime brought down the pillars of the old social order in its own violent end, and paved the way for a drastically revised form of capitalist state in the west and a genuine social revolution in the east. If the notion of a 'zero hour' in the defeat of 1945 marking a complete break with Germany's past – a notion very popular in West Germany after the war – is a fiction masking the many spheres of continuity in socio-economic structures, institutions, and mentalities, then it is nevertheless true that the demise of the German aristocracy, the bankruptcy of the old army leadership and its Prusso-German ideals, the unending columns of refugees from the east, the physical division of Germany, the social demands of reconstruction, and allied 're-education' policy, denoted a caesura with the past beside which the social changes of the Third Reich itself pale into insignificance.

(London, 1991)) argued, a uniquely traumatic 'crisis of classical modernity'. Nor does the correct emphasis upon the drive for racial purification as the essential characteristic of Nazism in itself exclude other perspectives of analysis which look to questions of modernization, such as those which focus upon Nazism's (unwitting), contribution to long-term modernization in Germany and the Third Reich's legacy – however little it was intended – for its successor German states.

8

'Resistance without the People'?

An attempt to evaluate German resistance in the Third Reich has to confront problems which lie at the very heart of any historical understanding of the phenomenon of Nazism. According to one of the foremost authorities, 'the relation between National Socialism and the Resistance is a key to comprehending the Nazi system'.[1] Yet, almost 50 years after the bomb exploded under Hitler's table on 20 July 1944, and despite a scholarly literature running into thousands of works, historians are still unable to agree not only on how to define 'resistance', but even whether a precise definition of 'resistance' should be sought.[2]

The reasons for the complexity of the term 'resistance' will be explored shortly. The contours of the debate can, however, be swiftly outlined. As historians shifted their scope of analysis from the 1944 conspiracy against Hitler to the more humble actions of ordinary Germans who defied the regime in many different ways, responding as best they could to the challenge they encountered in the Nazi system, so the term 'resistance' seemed in need of revision. And the assumption that resistance had lacked popular backing, had been 'resistance without the people',[3] gave way, in the context of revised definitions, if not to the contrary assumption that there had been 'popular resistance',[4] nevertheless to a readiness to accept that differing manifestations of 'resistance' could be located in all sections of German society.

What constituted 'resistance' in Germany itself during the Third Reich is

---

[1] Peter Hoffman, *German Resistance to Hitler* (Cambridge Mass., 1988), p. 3. Hoffmann's major study, *Widerstand, Staatsstreich, Attentat* (4th edn., Munich, 1985), Engl. trans., *The History of the German Resistance 1933-1945* (revised edn., Cambridge, Mass., 1977) is the standard analysis. A more recent wide-ranging study in English is that of Michael Balfour, *Withstanding Hitler in Germany 1933-45* (London, 1988).

[2] See the comments of Peter Steinbach and Hans Mommsen in Jürgen Schmädeke and Peter Steinbach, eds., *Der Widerstand gegen den Nationalsozialismus* (Munich/Zürich, 1985), p. 1122 (henceforth referred to as Schmädeke). This volume is the best and most comprehensive collection of assessments of German resistance to date. Other valuable essay collections are those edited by Klaus-Jürgen Müller (see note 11, this chapter), by Richard Löwenthal and Patrik von zur Mühlen (note 74, this chapter), Christoph Kleßmann and Falk Pingel (note 80, this chapter), and by Hermann Graml, ed., *Widerstand im Dritten Reich. Probleme, Ereignisse, Gestalten* (Frankfurt am Main, 1984). In English, there is now the collection edited by Francis Nicosia and Lawrence Stokes (note 62, this chapter). An extensive survey of an array of publications which appeared between 1979 and 1984 is presented by Gerd R. Ueberschär, 'Gegner des Nationalsozialismus', *Militärgeschichtliche Mitteilungen*, 35 (1984), pp. 141-97.

[3] The term appears to have been first used by Hans Mommsen in his essay, 'Gesellschaftsbild und Verfassungspläne des deutschen Widerstandes', in Walter Schmitthenner and Hans Buchheim, eds., *Der deutsche Widerstand gegen Hitler* (Cologne/Berlin, 1966), pp. 75-6.

[4] A term (*Volkswiderstand*) explicitly rejected by Theo Pirker in Schmädeke, p. 1141.

indeed less easy to define than 'resistance' in those countries occupied by Germany during the War, where it was synonymous with all attempts to oppose and work for liberation from an invader and conqueror. Resistance by the German people to their own state, and for six years while that state was at war, raises quite different problems of analysis.[5] And it is an issue which plainly highlights the political overtones, methodological differences, and above all the moral dimension involved in grappling with the problems of Nazism.

Although many non-Germans have made major contributions to a deeper understanding of the issues, the historiography on resistance to Hitler has been fundamentally shaped by the continuing and changing attempt within Germany itself to wrestle with the legacy of the Nazi past. Particularly for those who have never had to experience life under a terroristic dictatorship, any attempt to deal with the German resistance to Hitler must begin with humility and respect for those who stood out, in whatever ways, against repression and inhumanity. But respect must not hinder the task of rational, critical assessment of the opposition to Nazism – as long as such an assessment keeps in view the historical realities and the determinants of action, bears in mind the 'art of the possible', and does not attempt to pass judgement on the basis of a-historical moral ideals.

To the survivors of those who suffered under Nazism for their courageous stance, the cool deliberations of historians can seem painfully detached and remote in their abstraction and conceptualization of the motives and behaviour of those opposing Nazi rule.[6] Yet a greater historical understanding of resistance can not be attained by elevating 'resistance fighters' into monumentalized heroes but only by placing them squarely in their time and context and trying to understand their actions, in all their fallibility, as part of the wider problem of the relationship of the Nazi regime and German society. This means examining not only resistance, but also collaboration and conformism under the Nazi system; for the boundaries between conflict and consensus were fluid and fluctuating, even for those whose opposition became so fundamental that they were prepared to pay for it with their lives.

Interpretations

In both parts of post-war divided Germany, though in very different ways, the history of the resistance to Nazi rule played a pivotal role in the portrayed self-image of the new states and in the attempt to mould the political consciousness and values of the population.

The overt political function of the interpretation which prevailed from beginning to end of the German Democratic Republic could scarcely be more plainly stated than on the dustjacket of the most widely-used textbook on resistance: 'The German anti-fascist resistance movement, especially the KPD and the forces allied with it, embodied the progressive line of German policy. The most consistent political force of this movement, the KPD, carried out from the first day of the fascist dictatorship, organized, and centrally directed, the struggle

---

[5] See the comments of Hans Rothfels, *The German Opposition to Hitler* (London, 1961), p. 8.
[6] See Schmädeke, p. xviii (the comments of Wolfgang Treue), and p. 1155 (the remarks of Frau Meyer-Krahmer, the daughter of Carl Goerdeler).

against imperialism and preparation for war in which it was supported by the Communist International and the other fraternal parties and in which it constantly sought to incorporate new allies. The anti-fascist, democratic programme worked out by the KPD with help of the Communist International represented a true alternative to fascist barbarism and war. . . . The expression of the victory of the resolute anti-fascists after the smashing of fascism by the Soviet Union and the other states of the Anti-Hitler Coalition and the defeat of German imperialism is the existence of the GDR in which the legacy of the best of the German people who gave their lives in the anti-fascist struggle was realized'.[7] As these lines make clear, the GDR as a state and, within it, the successor party to the KPD, the SED (Socialist Unity Party), based their claim to legitimacy on the legacy of the communist anti-fascist resistance. But a historiography condemned to function so blatantly in the service of the state was bound to be fatally flawed.[8]

Though this interpretation was to become more nuanced and differentiated, the near-exclusive emphasis within it upon the heroic underground resistance of the KPD meant the deliberate down-playing of all other forms of opposition.[9] Social Democratic resistance was dealt with only briefly, schematically, and critically, while the opposition of the Christian Churches and that of national-conservative and élite groups was either largely ignored or portrayed in a negative light.

Certain members of the 'bourgeois' resistance, such as Claus Graf Schenk von Stauffenberg and Adam von Trott zu Solz, the 'White Rose' Munich students' group, and some individual Catholics or Protestants, were regarded more favourably, as – for their class or position – relatively 'progressive' and part of a 'popular front' against the Hitler regime. But this did not alter the basic tenor of resistance historiography in the GDR, which remained – for all the research uncovering the bravery of ordinary communist workers (who suffered more grievously than any other political grouping for their opposition) – limited in perspective, sterile in approach, and misleadingly monolithic in interpretation.

Resistance historiography in West Germany has been less monolithic, and its central emphases, approaches, and interpretations have changed considerably since 1945 – largely in accordance with changes in the political, cultural, and intellectual climate of the Federal Republic. Though less overt than in the GDR, the political function of the resistance to Hitler as a basis of legitimacy for the Federal Republic is nonetheless plainly visible. This was particularly the case during the period of consolidation of the West German state and in the context of the Cold War, when the research historiography in the Federal Republic

---

[7] Klaus Mammach, *Die deutsche antifaschistische Widerstandsbewegung 1933–1939* (East Berlin, 1974). Mammach added similar comments to his contribution, 'Zum antifaschistischen Kampf der KPD', in Dietrich Eichholtz and Kurt Gossweiler, eds., *Faschismusforschung. Positionen, Probleme, Polemik* (East Berlin, 1980), pp. 323–54, here esp. pp. 353–4.

[8] For a summary of GDR historiography on resistance, see Andreas Dorpalen, *German History in Marxist Perspective. The East German Approach* (Detroit, 1985), pp. 418–28.

[9] An example from many of an informative, but one-sided and ideologically weighted account of resistance within a specific region is *Der antifaschistische Widerstandskampf unter Führung der KPD in Mecklenburg 1933 bis 1945*, edited by the Bezirkskommission zur Erforschung der Geschichte der örtlichen Arbeiterbewegung bei den Bezirksleitungen Rostock, Schwerin und Neubrandenburg der Sozialistischen Einheitspartei Deutschlands (Rostock, 1970).

formed almost an exact mirror image of the picture portrayed in the GDR.

Not surprisingly, the earliest works on resistance to Nazism to appear in Germany – including memoirs of those who had been connected with the 1944 plot[10] – were advanced as a counter to the crude 'collective guilt' notions of the victorious Allies. The conscious objective was to emphasize, to the German people themselves as well as to their former enemies, the existence of 'the other Germany', and to demonstrate that those resisting Nazism had acted as patriots, not as traitors.[11] The Stauffenberg bomb plot formed the obvious focal point of early works, with an emphasis already, therefore, on conservative resistance from élite, bourgeois, or military figures. In this phase, nevertheless, other forms of resistance – from the churches, but also from the socialist and even communist Left – were not excluded.[12]

In the wake of the onset of the Cold War, and with the division of Germany sealed, the tone of West German resistance historiography altered. The 'totalitarianism' concept which bracketed together National Socialism and Communism as twin evils, and which saw the fight of western democracies against the Soviet menace as the paramount political objective, meant that the initial recognition of communist resistance now disappeared. Instead, resistance became portrayed as the shining beacon of freedom and democracy in the darkness of the totalitarian state. Resistance was understood as legitimate only in the extreme conditions of tyranny and with the aim of restoring the original legal order. A revolutionary challenge to the social order did not accord with this notion of resistance.[13] In this picture, communist resistance had no place and resistance of the Social Democrats only a marginal one. Resistance was essentially bourgeois, Christian, and individual. It arose from a moral-ethical choice of the individual to uphold, whatever the cost, the values of freedom and democracy in the face of tyranny. The classic study of this kind was Gerhard Ritter's biography of Carl Goerdeler.[14] It was only natural that such an emphasis should lead to what has been called a 'heroicization' and 'monumentalization' of resistance,[15] and to placing the conservative resistance to Hitler squarely in the service of Adenauer's Federal Republic.

In resistance historiography, as in approaches to many other aspects of the Third Reich, the 1960s ushered in significant change. A more solid basis of sources was by now available; a younger generation of historians looked more critically at the motives and aims of the resistance (as they did at the way the memory of the resistance was being used in Adenauer's Germany); the backwash of the 'Fischer Controversy' encouraged in general a more critical approach to the recent past (not just to the Third Reich), and a consideration, too, of the broader social and political framework conditioning the actions and

[10] For example, Fabian von Schlabrendorff, *Offiziere gegen Hitler* (Zürich, 1946); Hans Bernd Gisevius, *Bis zum bitteren Ende* (Zürich, 1946); and Ulrich von Hassell, *Vom anderen Deutschland* (Zürich/Freiburg i.B., 1946).
[11] See Klaus-Jürgen Müller and Hans Mommsen, 'Der deutsche Widerstand gegen das NS-Regime. Zur Historiographie des Widerstandes', in Klaus-Jürgen Müller, ed., *Der deutsche Widerstand 1933–1945* (Paderborn, 1986), pp. 13, 221 note 6.
[12] See Müller and Mommsen, p. 13; Günter Plum, 'Widerstand und Resistenz', in Martin Broszat and Horst Möller, eds., *Das Dritte Reich. Herrschaftsstruktur und Geschichte* (Munich, 1983), p. 250.
[13] See Plum, pp. 251–2.
[14] Gerhard Ritter, *Carl Goerdeler und die deutsche Widerstandsbewegung* (Stuttgart, 1954).
[15] Müller and Mommsen, p. 15.

intentions of individuals; and, finally, the stuffy Adenauer years were giving way to political and intellectual restlessness and sharpened ideological conflict, frequently focusing upon the legacy of Nazism. In this context, the roles of the non-Nazi élites and the Left came to be viewed in a different light.

A major breakthrough in the analysis of the national-conservative resistance occurred in 1966 with two pioneering articles by Hans Mommsen and Hermann Graml.[16] These analysed the social, constitutional, and foreign-policy ideas of the individuals and groups associated with the 1944 bomb-plot, and marked a conscious move away from a concentration on the moral motivation of élite resistance and to a critical analysis of the historical development of political aims and objectives.

Mommsen was able to show convincingly that the ideas and plans of the national-conservative resistance could hardly be seen as the font of post-war liberal-democratic thinking. The 'men of the 20 July' could not, he argued, be regarded as inspired by the sentiments which were to be enshrined in the Federal Republic's 'Basic Law'. Rather, their ideas were rooted in the pre-Nazi era, in the search for alternatives – non-Nazi, but also non-liberal-democratic – to the Weimar parliamentary system.[17] Strikingly, hardly anyone who had played a prominent role in Weimar was involved in the resistance.[18] The fateful experience of Nazism confirmed for conservative resistance groups their distrust of mass democracy, since they saw the Third Reich as the logical culmination of the plebiscitary, populist, and demagogic potential of the party-political system. Though there were significant differences of emphasis, their political ideas were essentially oligarchic and authoritarian, resting heavily on corporatist and neo-conservative notions advanced in the Weimar Republic, envisaging self-governing communities, limited electoral rights, and renewal of Christian family values. The national-conservative resistance even sought to incorporate what they saw as the 'right' ideas – such as the attainment of a true 'national community' – represented by National Socialism, but perverted by the corruption and incompetence of the Third Reich's leaders and functionaries.[19] With the important exception of the leading figures in the Kreisau Circle,[20] there was no wish to entertain fundamental social change. Even

---

[16] Hans Mommsen, 'Gesellschaftsbild und Verfassungspläne des deutschen Widerstandes', and Hermann Graml, 'Die außenpolitischen Vorstellungen des deutschen Widerstandes', both first appeared in Schmitthenner and Buchheim (see note 3, this chapter). All subsequent references are to the English versions in Hermann Graml *et al.*, *The German Resistance to Hitler* (London, 1970). For a more recent comment by Mommsen on the same theme, see his essay 'Verfassungs- und Verwaltungsreformpläne der Widerstandsgruppen des 20 Juli 1944', in Schmädeke, pp. 570-97.
[17] More recent excellent studies of national-conservative and military resistance are those of Klaus-Jürgen Müller and Manfred Messerschmidt in Müller, ed., *Der deutsche Widerstand*, pp. 40-59 and 60-78. In English, there is Klaus-Jürgen Müller, 'The Structure and Nature of the National Conservative Opposition in Germany up to 1940', in H.W. Koch, ed., *Aspects of the Third Reich* (London, 1985), pp. 133-78.
[18] Mommsen, in *The German Resistance to Hitler*, p. 60; see also his essay 'Der Widerstand gegen Hitler und die deutsche Gesellschaft', in Schmädeke, p. 9 (Engl. trans., 'German Society and the Resistance to Hitler', in Hans Mommsen, *From Weimar to Auschwitz. Essays in German History* (Oxford, 1991), pp. 208-23).
[19] See Mommsen in Schmädeke, p. 11.
[20] The standard work on the Kreisauer Circle is Ger van Roon, *Neuordnung im Widerstand. Der Kreisauer Kreis innerhalb der deutschen Widerstandsbewegung* (Munich, 1967). See also van Roon's essay, 'Staatsvorstellungen des Kreisauer Kreises', in Schmädeke, pp. 560-9.

within the Kreisau Circle, the significance attached to a specifically German cultural heritage signalled the distance from western liberal democracy in the group's thinking.

Foreign-policy ideas within the German resistance not surprisingly fluctuated during the course of the war. Before the middle of the war, however, the views of the national-conservatives (those in the group led by Ludwig Beck, Carl Goerdeler, Ulrich von Hassell, Johannes Popitz, and the *Abwehr* Circle centring on Hans Oster), unmistakably bore some partial affinities with Nazi expansionist aims. Nazi methods and practices were utterly rejected as barbaric. But what was wanted was the restitution of the major-power status of the German Reich, while German dominance over central and eastern Europe was taken for granted. With Hitler at the height of his power, in late 1941 and early 1942, Carl Goerdeler, the projected Reich Chancellor in the event of a successful coup, saw the prospect of 'a European federation of states under German leadership within 10 or 20 years' if the war could be ended and a 'sensible political system' restored.[21] There was a reluctance to accept anything less than the borders of 1914. Von Hassell hoped to salvage 'at least the rudiments of Bismarck's Reich'.[22] Goerdeler, from fear of Bolshevism, wanted to retain the 1938 eastern border of Poland.[23] Even Adam von Trott zu Solz, associated with the more 'progressive' Kreisau Circle, echoed Germany's claim to the Sudetenland and parts of West Prussia in any post-war settlement.[24] However, generally within the Kreisau Circle (whose leading figures, apart from Trott, included Graf Helmut von Moltke, Paul Graf Yorck von Wartenburg, and the Jesuit Pater Alfred Delp), it was recognized that significant territorial sacrifices were inevitable.[25] And with the growing realization that there could be no turning back of the clock, the national-conservatives, too, from 1942–3 onwards acknowledged that Germany's role in a future Europe would be more modest than they had initially desired.[26]

As Hermann Graml aptly put it, the thinking of the resistance leaders 'was centred on the Reich. Led astray, not by Hitler but by the unfolding power of Germany, they saw the Reich for a brief moment of time as the power that would lead and order Europe'.[27]

For more than 20 years, West German resistance historiography was largely preoccupied – apart from works on the 'Church struggle' and the 'White Rose' student resistance in Munich in 1942–3 – with élite opposition by conservative and bourgeois groups and individuals.[28] The plot of July 1944 was the focal-

---

[21] Graml, in *The German Resistance to Hitler*, p. 27.

[22] Graml, in *The German Resistance to Hitler*, p. 43.

[23] Graml, in *The German Resistance to Hitler*, p. 45.

[24] Graml, in *The German Resistance to Hitler*, p. 50.

[25] Graml, in *The German Resistance to Hitler*, p. 52.

[26] Graml, in *The German Resistance to Hitler*, p. 49.

[27] Graml, in *The German Resistance to Hitler*, p. 54. See also Klaus Hildebrand, 'Die ostpolitischen Vorstellungen im deutschen Widerstand', *Geschichte in Wissenschaft und Unterricht*, 29 (1978), pp. 213–41.

[28] One of the most influential early works on opposition within the Catholic Church was Johann Neuhäusler, *Kreuz und Hakenkreuz*, 2 vols. (Munich, 1946). An extensive documentation of the Evangelical Church's conflict with the regime was Heinrich Hermelink, ed., *Kirche im Kampf. Dokumente des Widerstands und des Aufbaus in der Evangelischen Kirche Deutschlands von 1933 bis 1945* (Tübingen, 1950). On the 'White Rose', the most influential early work was the account

point of all analyses;[29] and the premiss was that resistance under Hitler had been, of necessity, 'resistance without the people', that in the context of a totalitarian state there had been no popular resistance as such.

In the 1960s, this perspective began to change, and in the 1970s and 1980s became substantially transformed. In part, this was a consequence of the altered political climate in West Germany (where from 1969 the SPD formed a party of government, for the first time since 1930), the mood of protest expressed in the student demonstrations of 1968, and the generational shift. During the 1960s, however, the notion of the Third Reich as a totalitarian monolith had itself begun to break down as numerous works on the governmental system exposed polycratic structures fitting uneasily into unrefined versions of the totalitarianism model. The growth of interest in social history, in the behaviour of non-élite groups, also started to show through in the historiography. Initially, this was reflected for the most part in institutional or organizational histories of underground resistance activities of working-class groups. Then, more and more, the shift became visible in increasingly subtle and well-researched analysis of the actions, attitudes, and behaviour of the mass of ordinary Germans.

The trend away from concentration on élite resistance encouraged revised approaches to resistance under Hitler. Tentatively, the view emerged that there had not, after all, been 'resistance without the people'. Increasingly, it became possible even to conceive of 'resistance *by* the people'.

Already in 1953, the writer Günther Weisenborn, had attempted, on the basis of a great deal of unpublished police material, to reconstruct 'the resistance movement of the German people 1933–1945'. Weisenborn claimed that, as a consequence of the concentration on the events of the 20 July 1944, the extent and significance of any wider resistance was little known about since (for political reasons, he implied) it had been deliberately suppressed.[30] When it first appeared, the book ran counter to the historiographical trend of the day, and remained uninfluential. It was re-published, in a very different climate, in 1974 by the left-wing publishing house Röderberg, which by then had published over 20 volumes in its 'Library of Resistance' – mostly local studies of working-class opposition written by former members of the communist resistance. These brought a new dimension to resistance historiography, even if the volumes were frequently distinguished more by political engagement and moral fervour than by scholarly precision and judgement.[31]

Of a different calibre were a number of studies, appearing in the late 1960s

---

by Hans and Sophie Scholl's surviving sister: Inge Scholl, *Die Weiße Rose. Der Widerstand der Münchner Studenten* (Frankfurt am Main, 1952). Important documentation is now provided in Anneliese Knoop-Graf and Inge Jens, eds., *Willi Graf. Briefe und Aufzeichnungen* (Frankfurt am Main, 1988) and Inge Jens, eds., *Hans und Sophie Scholl. Briefe und Aufzeichnungen* (Frankfurt am Main, 1989). See also Heinrich Siefken, ed., *Die Weiße Rose. Student Resistance to National Socialism 1942/1943* (Nottingham, 1991).

[29] Characteristic in this respect was the first general analysis of resistance, by Hans Rothfels, *The German Opposition to Hitler* (Hinsdale, Illinois, 1948).

[30] Günther Weisenborn, *Der lautlose Widerstand. Bericht über die Widerstandsbewegung des deutschen Volkes 1933–1945* (Hamburg, 1953; republished Frankfurt am Main, 1974), p. 8.

[31] For a useful survey of these volumes (and other studies of working-class resistance) see Reinhard Mann, 'Widerstand gegen den Nationalsozialismus', *Neue politische Literatur*, 22 (1977), pp. 425–42.

and early 1970s under the auspices of the Friedrich-Ebert-Stiftung, which meticulously researched the history of working-class resistance in the Ruhr cities of Dortmund, Essen, and Duisburg.[32] Other local studies now also concentrated on the previously under-researched working-class resistance in big cities, and were able to build up a compelling and differentiated picture of the extent of such illegal activity.[33]

Some of the literature on worker resistance amounted to works of piety and monumentalization. But depicting the misery of the daily struggle against hopeless odds, the isolation and ineffectiveness of the illegal work, the enormity of the risks, the almost inevitable penetration of resistance groups by infiltrating Gestapo informers, the reprisals and suffering, all brought closer to home the social world of people from working-class backgrounds who had become involved in resistance because of political convictions shared by millions rather than from lofty moral and ethical considerations apparently beyond the capability of ordinary mortals.

The next step was to turn from what was nevertheless still an exceptional minority, prepared to go to such lengths for their political beliefs, and to alter the perspective – from resistance as the *organization* of illegal activity, to the ordinary mortals themselves and how they coped in their everyday lives with more partial, but also more *normal*, forms of opposition to those aspects of the Nazi regime which affected them most directly.

The direction in which resistance historiography was moving was part of a wider trend, extending far beyond the history of the Third Reich, towards 'social history from below' (*Sozialgeschichte von unten*), 'the history of experience' (*Erfahrungsgeschichte*), and – as the genre in general came above all to be called – 'the history of everyday life' (*Alltagsgeschichte*). The 'everyday history' approach helped open up new questions and problems about the character and extent of resistance in the Third Reich; not least, the question of how that resistance was to be defined.[34]

A crucial role was played in this development by the 'Bavaria Project', on 'Resistance and Persecution in Bavaria 1933–1945', launched by the Institute of Contemporary History in Munich in 1973. In this project, the conception of what constituted 'resistance' was not only widened far beyond any earlier usage of the term, but also removed from the previously prevailing linkage with ethical motivation and organizational framework.[35] Instead, the emphasis was placed upon the impact of the Nazi regime on all areas of 'everyday life', allowing a multifaceted picture of spheres of conflict between rulers and ruled to emerge.

[32] Kurt Klotzbach, *Gegen den Nationalsozialismus. Widerstand und Verfolgung in Dortmund 1930-1945* (Hanover, 1969); Hans-Josef Steinberg, *Widerstand und Verfolgung in Essen 1933-1945* (Hanover 1969); and Kuno Bludau, *'Gestapo-geheim!'. Widerstand und Verfolgung in Duisburg 1933-1945* (Bonn-Bad Godesberg, 1973).

[33] For an excellent evaluation of working-class resistance and survey of historiography, see Detlev Peukert, 'Der deutsche Arbeiterwiderstand 1933-1945', in Müller, *Der deutsche Widerstand*, pp. 157-81. In English there is now his contribution, 'Working-Class Resistance: Problems and Options', to David Clay Large, ed., *Contending with Hitler. Varieties of German Resistance in the Third Reich* (Cambridge, 1991), pp. 35-48.

[34] For the potential of the 'history of everyday life' approach, see above all Detlev Peukert, *Volksgenossen und Gemeinschaftsfremde* (Cologne, 1982); Engl. trans., *Inside Nazi Germany* (London, 1987).

[35] See Plum, pp. 263-4.

The definition of 'resistance' deployed at the start of their work by the team of archivists from the Bavarian State Archives involved in the project clearly indicates this widening and new emphasis: 'Resistance is understood as every form of active or passive behaviour which allows recognition of the rejection of the National Socialist regime or a partial area of National Socialist ideology and was bound up with certain risks'.[36]

In an original, if highly abstract, theoretical exposition of the concept of resistance soon after the project began, its first director, Peter Hüttenberger – after a short time this role was assumed by Martin Broszat – defined resistance as 'every form of rebellion against at least potentially total rule within the context of asymmetrical relations of rule'.[37] This definition was based upon concepts of changing mechanisms of rule and social reactions to that rule. It was premissed upon an understanding of 'rule' or 'domination' (*Herrschaft*) as a process of balancing the aims, interests, standards, and norms of rulers with those of the ruled. For Hüttenberger, 'symmetrical' rule obtains where such a 'bargain' is struck. In such a system, most notably in a democracy, there is no 'resistance', but merely rivalry and conflict within, and immanent to, the system. Even where the system is breaking down, it is misleading to speak of 'resistance'. (Hence, the political struggle between the NSDAP and the KPD during the Weimar Republic could not, under such a definition, be regarded as communist resistance against Nazism; such resistance was a creation of the altered relations of rule from 1933 onwards.)[38] Resistance, therefore, arises only under 'asymmetrical' relations of rule; that is, where the attempt is made to erect a total or complete system of domination, and thereby to destroy the 'bargain' or 'balance' of the 'symmetrical' system.

In simple terms, this means that resistance is a product and reflection of the system of rule itself; the nature of that rule determines the nature of resistance. And it then follows that the more comprehensive the claim to rule, the *more*, not the less, resistance there will be, since the regime itself turns behaviour and actions into resistance which would not be so in the 'symmetrical' rule of a pluralist democratic system. Hüttenberger's definition, it can be seen readily, widens enormously the scope for categorizing and appraising resistance.

As these definitions clearly implied, the 'Bavaria Project' by no means confined itself to dealing with fundamental, principled, and total resistance to Nazism. Rather, it embraced all forms of limited and partial rejection, whatever the motives, of specific aspects of Nazi rule. Instead of dealing in images of black and white, resistance was portrayed in shades of grey; as a part of the everyday reality of trying to adjust to, and cope with, life in a regime impinging on practically all aspects of daily existence, posing a total claim on society, but – as a direct consequence – meeting numerous blockages and

---

[36] Harald Jaeger and Hermann Rumschöttel, 'Das Forschungsprojekt "Widerstand und Verfolgung in Bayern 1933-1945" ', *Archivalische Zeitschrift*, 73 (1977), p. 214.

[37] Peter Hüttenberger, 'Vorüberlegungen zum "Widerstandsbegriff" ', in Jürgen Kocka, ed., *Theorien in der Praxis des Historikers* (Göttingen, 1977), p. 126.

[38] In contrast, Leonidas E. Hill, 'Towards a New History of German Resistance to Hitler', *CEH*, 14 (1981), pp. 369, 395 sees 'resistance before 1933' as a desideratum of research. Hill's paper was one of five contributions – the others were by Harold C. Deutsch, Peter Hoffmann, Klemens von Klemperer, and Robert O. Paxton – to a 'Symposium: New Perspectives on the German Resistance against National Socialism' (pp. 322-99), which was concerned for the most part with military and conservative resistance.

restrictions in its attempt to make good this claim.

The six volumes arising from the project between 1977 and 1983 mark a milestone in the social history of the Third Reich as well as in resistance historiography.[39] In line with the open brief of the 'conflict' approach and methodology, contributions ranged over themes and aspects of reactions to Nazi rule which had not previously been contemplated as falling under the rubric of 'resistance'. Alongside those exploring the actions of organized social-democratic and communist groups (which, more than in earlier works, were located firmly in their social milieu), stood descriptions of numerous forms of 'civil disobedience', such as refusal to give the 'Heil Hitler' greeting; insistence on hanging out the church flag instead of the swastika banner; objections by peasants to farm legislation; public criticism of anti-church measures by Catholic priests; continued trafficking with Jewish cattle-dealers; or fraternization with foreign workers. The detailed attention to the social milieu uncovered motives which frequently fell short of the 'heroic' image of 'resistance fighters' but offered a far more human and approachable picture of individuals acting at times out of social resentment, economic misery, blind protest, fury at Nazi treatment of family and friends, religious conservatism, as well as principled struggle for a better social and political order or a 'dictatorship of the proletariat'.[40]

Instead of a concentrating on motive, intention, and goal, therefore, the 'Bavaria Project' dealt with 'resistance' as defined in the broadest sense, and placed a methodological emphasis upon the actions themselves (however limited) and their 'effect' (*Wirkung*) in blocking or partially restricting Nazism's societal penetration. In reality, therefore, the project had turned away from resistance as an intended fundamental challenge to the regime, to the capacity of specific pre-existing social or political groups, institutions (such as the churches, bureaucracy, and the army), subcultures, and individuals in certain social milieus, to offer a level of relative immunity to the 'total claim' of the Nazi regime and its ideology.

In embracing this 'functional' rather than 'intentional' approach to societal conflict with Nazism, the director of the 'Bavaria Project', Martin Broszat, introduced a wholly new – and much disputed – concept into the terminology and historiography of resistance: the concept of '*Resistenz*'.[41] Broszat explained *Resistenz* as a structural concept, and a morally neutral one, which – implying 'immunity' as used in medicine, or 'resistance' as deployed in physics – could help in examining the actual effects of actions limiting the penetration of Nazism and blocking its total claim to power and control.[42] 'In every socio-political system', he stated, 'and particularly under a form of

[39] Martin Broszat et al., eds., *Bayern in der NS- Zeit*, 6 vols. (Munich/Vienna, 1977–83).
[40] See Plum, pp. 270–1.
[41] Martin Broszat, 'Resistenz und Widerstand', *Bayern in der NS-Zeit*, iv., pp. 691–709, reprinted in Martin Broszat, *Nach Hitler* (Munich, 1986), pp. 68–91. The obvious linguistic difficulties of translating '*Resistenz*', given its proximity to 'resistance' in English and several other languages, have been pointed out by numerous commentators and provide in themselves an evident obstacle to any non-German usage of the term. For a valuable appraisal of Broszat's contribution to scholarship on German resistance, see Hans Mommsen, 'Widerstand und Dissens im Dritten Reich', in Klaus-Dietmar Henke and Claudio Natoli, eds., *Mit dem Pathos der Nüchternheit* (Frankfurt am Main, 1991), pp. 107–18.
[42] *Bayern in der NS-Zeit*, i., pp. 11; iv., pp. 697–9.

political domination such as that of National Socialism, what counts politically and historically is above all what was *done* and *accomplished* (*bewirkt*), not just *desired* or *intended*'.[43]

Broszat distinguished '*Resistenz*' from 'resistance' (*Widerstand*) , which, he argued, frequently obscured the actual social and political impact, effect, and consequences of actions through an over-concentration on subjective motivation, organization, and the moral-ethical framework of action. By contrast, the 'behavioural' concept of *Resistenz* allowed partial opposition (coexisting with partial approval of the regime) to be included within the broad framework of 'resistance' (*Widerstand*), which had conventionally excluded non-fundamental opposition to the regime. The *Resistenz* concept, therefore, made possible both a deepened understanding of the social base of conflict with the regime, and a more nuanced explanation of the spheres of underlying consensus with aspects of Nazi rule.

For the investigation of grass-roots behaviour of the German population under Nazism, the change of emphasis was enormously fruitful. But both the '*Resistenz*' concept itself, and the adequacy of the new approach for tackling the issue of '*resistance*' to Nazism, provoked controversy and criticism.[44] The sharpest criticism was formulated by the Swiss historian Walter Hofer: 'The concept of *Resistenz* leads to a levelling down of fundamental resistance against the system on the one hand and actions criticizing more or less accidental, superficial manifestations on the other: the tyrannicide appears on the same plane as the illegal cattle-slaughterer'. Hofer went on to decry the tendency to 'disqualify' as a 'moralizing perception of history' those interpretations which placed the emphasis on the 'moral quality' and 'political content' of resistance; he did not accept the alleged 'monumentalization' of the German resistance which was under attack; and he claimed that to remove 'moral-political judgements' from the debate on resistance was to fall victim to a false sense of objectivity. He concluded that there was little point anyway in moving from 'behaviour' to 'effect' in assessing resistance, since it was plain 'that the oppositional stances subsumed under the new concept of *Resistenz* had very little relevant effect on the ruling totalitarian regime or none at all'.[45]

The responses to Hofer's comments illustrate the difficulties with any concept of 'resistance' and the lack of agreement, and variety of interpretations, possible among leading experts. Lending support to Hofer's position, the concept of '*Resistenz*' was attacked by Marlis Steinert as too broad, inclusive of 'silent acceptance, resignation, apathy' and not even excluding 'integration in the regime'. The notion of 'resistance' (*Widerstand*), it was further argued by Klaus-Jürgen Müller, ought to carry the meaning of 'will to overcome the system' (though such a statement was still not free of ambiguities about the

---

[43] *Bayern in der NS-Zeit*, iv, p. 698.

[44] Even those engaged on the 'Bavaria Project' found '*Resistenz*' a difficult concept to deploy, and had misgivings about the term. See Plum, pp. 264–5.

[45] Schmädeke, pp. 1121–2. In seeking definitional clarity, Hofer argued (pp. 1120–1) that, since the term '*Resistenz*' involved at least partial *collaboration*, it either meant the opposite of 'resistance' (as, for example, used by the French), or was at least guaranteed to produce confusion. Others have remarked on its ambivalence, even in medicine, where it can mean being resistent to antidotes offering recovery from illness, and, perhaps more compellingly, have stressed that it is essentially a *passive* concept whereas 'resistance' conveys an *active* meaning. See Plum, pp. 264–5.

nature and extent of the actions arising from such a 'will', nor how comprehensive and all-embracing-such a 'will' needed to be).[46]

Another response (that of Heinz Boberach and Manfred Messerschmidt) defined 'resistance' from the perspective of the wielders of power in the Nazi state.[47] There is, in fact, much to be said for this interpretation (which accords with Hüttenberger's theoretical 'asymmetrical rule' approach). In posing their total claim on society, the Nazis were not willing to grant any institutional or organizational space which they themselves did not control. Though their total claim was not realized, there can be no doubting the intent to control all aspects of society. Thus, many forms of behaviour which would not even be noticed, or would be regarded as harmless, in a liberal democracy – for example, youth groups aping western clothing styles and listening to swing music or musicians playing jazz – were politicized and criminalized in the Nazi police state and interpreted as a threat to the system.[48]

Other experts (Peter Steinbach, Hans Mommsen) warned against the use of an inflexible definition, since this would be incapable of embracing the variety of 'resistance practice' (*Widerstandspraxis*), the differing challenge to each individual, the myriad ways in which persons came to be involved, and the fact that resistance ought to be perceived not in static or absolute terms but as a 'process', subject to gradual increase in radicalism over time.[49] What was meant by 'resistance as process' was that many of those who ultimately came to be involved in absolute resistance to Hitler (including many of the 1944 conspirators) had initially, often even for a long time, approved of much that Nazism had to offer, and had indeed been part of the system.

One of the central figures in the resistance, Carl Goerdeler, for instance, had served as Hitler's Price Commissioner in the early years of the regime, favoured racial legislation, and initially saw National Socialism as offering the best chance of securing Germany's rights as a nation on a basis of internal unity, *völkisch* principles, and moral leadership. But by 1935 he was already in fundamental disagreement with the development of economic policy; thereafter, his disillusionment with the Nazi abuse of power mounted sharply; in 1937 he resigned his office as Lord Mayor in protest at the removal of the statue of Mendelssohn from the Leipzig town centre; and, increasingly alienated and worried by the likely disastrous consequences of Hitler's foreign and economic policy, he rapidly thereafter took the path into outright opposition which made him, following the outbreak of war, the fulcrum of bourgeois resistance to Hitler.[50]

[46] Schmädeke, p. 1122.
[47] Schmädeke, p. 1122.
[48] See Detlev Peukert, *Die Edelweißpiraten. Protestbewegungen jugendlicher Arbeiter im Dritten Reich* (Cologne, 1980), and the works referred to in ch. 7 note 56; on music see the fascinating accounts by Michael Kater, 'Forbidden Fruit? Jazz in the Third Reich', *American Historical Review* 94 (1989), pp. 11–43; and *Different Drummers: Jazz in the Culture of Nazi Germany* (Oxford, 1992). For an analysis of many of the minor disdemeanours subject to often draconian penalties, see Peter Hüttenberger, 'Heimtückefälle vor dem Sondergericht München 1933–1939', in *Bayern in der NS- Zeit*, iv, pp. 435–526.
[49] Schmädeke. p. 1122.
[50] See Michael Krüger-Charlé, 'Carl Goerdelers Versuche der Durchsetzung einer alternativen Politik 1933 bis 1937', in Schmädeke, pp. 383–404, and the comments of Broszat, *Nach Hitler*, p. 170.

The cases of Goerdeler and the many others who hesitantly found their way from collaboration – sometimes enthusiastic – into fully-fledged resistance to the regime led Hans Mommsen to describe, fittingly, the national-conservative conspiracy as 'a resistance of servants of the state (*Staatsdiener*)',[51] who for the most part came only gradually to recognize the need to undertake the ultimate act of saving the state by the killing of the head of state. Even among the conspirators, he pointed out, 'the boundaries between partial criticism, open opposition, and active resistance were, under the given conditions, necessarily fluid'.[52]

This clash of interpretations demonstrated that, while four decades of writing and research on German resistance to Hitler had produced extensive and impressive empirical findings, they had increasingly – particularly in the wake of the move to what one might call the social history of resistance – given rise to complex and unresolved, interlocking theoretical and intepretational problems.

In the developing historiography, as we have seen, changing views of the effect of resistance have been related to the widening of its definition, and both have reshaped notions of its forms and extent. The following evaluation begins by accepting, initially, the widened definition, and proceeds to assess – on that basis – the *effect* of resistance. The conclusions will prompt an examination of which is analytically more useful, a wider or narrower *definition*. This in turn will lead to a reassessment of the *extent* of resistance among the German people. Finally, I shall ask whether the currently antagonistic interpretations can be reconciled; and whether a synthesis which might assist in reframing approaches to the resistance problem is possible.

Evaluation

I

As we have noted, the widening of the scope of 'resistance' to encompass all forms of 'everyday' conflict with the regime went hand in hand with the shift of emphasis from 'motive' to 'effect' or 'function'. The *Resistenz* concept in particular was premised on the notion that the *effect* or *function* of actions (whatever the motives that lay behind them) in limiting or blocking the regime's penetration was the central concern of enquiry. While, as we saw, some historians have rejected outright the suggestion that *Resistenz* placed any significant limits on the regime's scope for action, others have interpreted 'functional resistance' as a genuine obstacle to the regime in realizing its aims.

Is Hofer correct that '*Resistenz* had very little relevant effect on the ruling

---

[51] Schmädeke, p. 9. Among those to whom Mommsen's description applies were Ulrich von Hassell (German Ambassador in Rome until 1938) and Fritz-Dietlof Graf von der Schulenburg (a higher civil servant and Party member since 1932, who expected Nazism to accomplish Germany's renewal through the inculcation of Prussian virtues). See the new edition of *Die Hassell-Tagebücher*, edited by Friedrich Freiherr Hiller von Gaertringen (Berlin, 1988) and Ulrich Heinemann, *Ein konservativer Rebell. Fritz-Dietlof Graf von der Schulenburg und der 20. Juli* (Berlin, 1990).
[52] Schmädeke, p. 11.

totalitarian regime'?[53] Or ought Broszat's contention, of 'a *Resistenz* which did not, like most active resistance, generally fail, but could be thoroughly effective' in limiting Nazi rule in numerous spheres of activity through 'many minor forms (*Kleinformen*) of civil courage', be upheld?[54]

The apparently irreconcilable conflict of interpretation reflects utterly contrasting approaches, which we might call *fundamentalist* and *societal*. It can, in fact, be argued that both approaches are legitimate; and that a case can be made for each of the opposing answers to the question about the effectiveness of *Resistenz*.

Strictly speaking, the effectiveness of *Resistenz* could only be established by a counter-factual case. What difference would it have made had there been no *Resistenz*? Hofer's answer to this question is simple and straightforward: it would have made no difference at all. No difference to what?, it must be asked. Hofer would answer this question in a *fundamentalist* fashion by emphasizing the aims, ambitions, and intentions of the regime. He might, quite correctly, point to a regime whose 'total claim' was not simply concerned with power for power's sake, but rather as preparation for a war of racial conquest to establish, on the ashes of Bolshevism, the lasting dominance of a racially purified Germany. And he might add that it took the combined might of the Allies to prevent the realization of this aim, concluding – sensibly enough – that the regime's central ambitions were scarcely dented by *Resistenz*.

Broszat's answer would be different because he interprets the question in a different way. For him, the point that Nazism was *not* prevented by *Resistenz* from waging a war of annihilation and perpetrating genocide is so self-evident that it needs no discussion. In using the *Resistenz* concept, his sights are set on a different trajectory of the regime's aims. His interest lies not in questioning the extent to which the regime might have been hindered by *Resistenz* from engaging in war and genocide, but in examining the total penetration of society which it sought, and the degree to which specific social groups and institutions were able to ward off such penetration. His answer to the counter-factual question would vary, therefore, on the aspect of society under consideration. But his general conclusion would certainly be that *Resistenz* did make a difference to the ability of the regime to manipulate society at will.

From this *societal* angle, then, Broszat accepted as the sole criterion for seeing them as *Resistenz* that actions 'had indeed a limiting effect (*einschränkende Wirkung*) on National Socialist rule and ideology'. He listed a variety of actions (including strikes, criticism from the pulpit, non-participation in Nazi meetings, refusal of the 'Heil-Hitler' greeting, ignoring the ban on relations with Jews, and maintaining social contact with former SPD members) as illustrations of the limits of Nazi penetration and control and of the barriers to the implementation of the regime's total claim. And he pointed to the continued existence throughout the Third Reich of institutions which managed to sustain a relative independence of Nazism (the Churches, the bureaucracy, the *Wehrmacht*), and to the persistence of moral and religious norms, and economic, legal, intellectual, or artistic values which Nazi rule failed to erase.[55] From the perspective of society's response to Nazism, and considering the

[53] Schmädeke, p. 1122.
[54] Broszat, *Nach Hitler*, p. 112.
[55] *Bayern in der NS-Zeit*, iv, p. 697.

brakes placed on Nazi penetration as it affected individuals, groups, or institutions, Broszat's argument for effective *Resistenz* – for spheres of relative immunity and for the functional efficacy of nonconformist behaviour in the limited sense in which he intended it to be understood – is correct.

The interpretations of Hofer and Broszat address different problems. Despite appearing to be absolute opposites, they by-pass one another. The difficulty is plainly the confusing and misleading term *Resistenz*. Its linguistic proximity to 'resistance' (in several languages, if not in German) not only misleads, but seems to make it part of a debate on 'resistance', when in reality it is intended to be a conceptual device for attempting to understand channels of conflict and co-operation between regime and society. As such, it can help to explain the context in which 'resistance' did – or more usually did *not* – arise. And in illustrating why, for example, Catholic practices could at times be forcefully and successfully defended while little attempt was made to defend the Jews, this approach can contribute significantly – if more subtly than Hofer is prepared to recognize – to understanding how and why the regime *was* able to implement its central policies, and how ineffective *resistance* was in checking such implementation. Broszat certainly tried to make plain that he was setting *Resistenz* against 'resistance' (*Widerstand*), not offering it as a substitute or alternative term. If he was let down at all in his imaginative and path-breaking approach, then it was by the inadequacy of his own conceptual construct.

A notable weakness of the concept of *Resistenz* is that, despite its claims to deal with '*effect*' instead of motive as a criterion and to treat actions as 'value-free', the term '*Resistenz*' in practice cannot completely separate intention and motivation from the evaluation of action. The attempt is naturally, and correctly, still made to understand *why* individuals inflicted major, sometimes fatal, risks upon themselves, their families, and their friends. If the motives were often less than lofty, heroic, independently reached, and moral-ethical in nature, and were frequently shaped by milieu and circumstance, they were motives nevertheless, and distinguishable from spontaneous outbursts of anger, frustration, or suffering.

Nor is it in practice possible or sufficient to look at an action and its 'effect' in isolation from any moral value or implications it may have had. It is important, for instance, to know that peasants who continued trafficking with Jewish cattle-dealers were doing so out of material self-interest, and that their objections to Nazi measures to eliminate Jewish dealers in the countryside prevented them neither from being anti-semites nor from approving the broad thrust of anti-Jewish policy. It is relevant, for example, to know that among the workers engaged in strikes in 1935 and 1936 were many discontented SA men whose pro-Nazi sympathies were otherwise much in evidence. And it is also necessary to distinguish the principled and consistent refusal from the beginning of the regime by the conservative Ewald von Kleist-Schmenzin to give the 'Heil-Hitler' greeting[56] from the same refusal, for example, by Bavarian farmers who welcomed the internment of 'marxists', approved of the Nuremberg Laws, but were irritated by the inability of the government to do anything about the shortage of rural labour.

The 'functional' argument is at its weakest where its claims are greatest. We

[56] See Schmädeke, p. 5.

outlined in chapter 4 the thesis of Tim Mason (and the substantial criticism it encountered) that workers in the Third Reich were able, in conditions of acute labour shortage, to impose demands on their employers (even without the benefit of trades unions, and in the teeth of a ferociously hostile state) sufficient to limit seriously the potential of the Nazi regime to fight the war for which it was preparing on the terms and at the time it would have wished.[57] The *function* of workers' actions in defence of their own position consequently, in this argument, weakened the regime at its key point – the ability to wage imperialist war.

The same approach was unfolded, on the basis of a detailed investigation of reports by the exiled SPD organization (calling itself the 'SOPADE') about 'resistance in the factories' (*'Widerstand in den Betrieben'*), by Michael Voges.[58] The economic class struggle in the factories, claimed Voges, was, as far as the intention of the workers was concerned, only in part politically motivated. But 'in its function', it had to be seen as 'resistance', in the sense that the regime was forced to deal politically with it, and also because it was perceived to constitute a political threat to the regime.[59]

Such arguments, that the realization of the regime's fundamental aims was threatened by industrial workers often acting without political motivation or intention, went far beyond any claims made by Broszat for the impact of *Resistenz*. And, they expose themselves far more to the type of objection raised by Hofer.

Working-class 'subcultures', as we commented in chapter 7, did remain relatively impervious to Nazi inroads after 1933 as before. And manifestations of discontent and signs of unrest – politically interpreted by the regime, if not necessarily politically motivated – did become increasingly apparent from the mid 1930s onwards, *potentially* endangering the regime's stability and the accomplishment of its aims.

However, it would be as well not to make too much of these in terms of their effect on the functioning of the regime. Collective industrial 'protest' actions through strikes, which reached a numerical peak in the years 1935-7, were small in scale and politically ineffective, even when compared with those in Fascist

---

[57] The argument was expounded at its fullest in Timothy W. Mason, *Arbeiterklasse und Volksgemeinschaft* (Opladen, 1975). Mason developed the argument a stage further in his essay, 'The Workers' Opposition in Nazi Germany', *History Workshop Journal*, 11 (1981), pp. 120-37; a slightly shorter version appeared in German as 'Arbeiteropposition im nationalsozialistischen Deutschland' in Detlev Peukert and Jürgen Reuleck, eds., *Die Reihen fast geschlossen* (Wuppertal, 1981), pp. 293-313. Here, he distinguished explicitly between politically motived actions of workers (which he classed as 'political resistance'), and the refusal of the working class to subordinate itself fully to the Nazi system, as reflected in its economic class struggle (which he called 'opposition'). Though the former, the illegal underground resistance, failed to undermine the regime, the latter touched on its achilles heel, and *was*, therefore, political – because of its own functional implications for the regime, and because the Nazi State itself was the recognizable agent of repression and persecution of workers acting in defence of their own economic interests – 'Arbeiteropposition', pp. 293-5, 309-12.

[58] Michael Voges, 'Klassenkampf in der "Betriebsgemeinschaft". Die "Deutschland-Berichte" der Sopade (1934-1940) als Quelle zum Widerstand der Industriearbeiter im Dritten Reich', *Archiv für Sozialgeschichte*, 21 (1981), pp. 329-84.

[59] Voges, pp. 376-7 (and note 189), 382-3. Voges also used the concept of *Resistenz*, though he did not assist conceptual clarity by suggesting that 'intentionality' should be a necessary component of the concept, whereas Broszat had specifically excluded 'intention' as a criterion of *Resistenz*.

Italy.[60] Research on key groups like miners, and on the war period, has considerably qualified earlier, somewhat idealistic, notions of the pronounced anti-Nazi character of industrial unrest and especially of its impact on the functioning of the regime.[61] Certainly, the regime's leadership did not give the impression in the later 1930s that it was *politically* worried about the industrial working class.

The view that working-class opposition and the class struggle put the Nazi regime under pressure (and, in so doing, contributed ultimately in a significant, if indirect, way to its defeat) was attractive and, in a sense, reassuring. But it has rightly given way to a more sober and pessimistic view of a working class that was neutralized, contained, resigned, demoralized, at best only partially integrated – but neither rebellious, nor posing a serious threat to the regime.[62] The period when the regime's perception of industrial unrest most influenced its decision-making was, arguably, not immediately prior to the war, in 1938-9, but in the years 1935-6.[63] But even then it only pushed the regime's leaders in the direction they wanted to go.

A number of conclusions can be drawn from the preceding discussion. We have seen, firstly, that Hofer's objections to the *Resistenz* concept are largely irrelevant, since they ignore or overlook the intended meaning of the term and its central thrust as a conceptual device. What *Resistenz* – despite sounding so similar (which is part of the problem) – was *not* intended to do, was to confront the problem of 'resistance'. Martin Broszat and the research team on the 'Bavaria Project' accepted it as axiomatic that German resistance to Hitler was a tragic failure. There was no suggestion that the realization of the regime's fundamental aims was blunted by *Resistenz*. But for all the undeniable weaknesses of the term, the functional approach which it characterises is, we have argued, a valid and fruitful one in looking at *conflict* spheres in the relationship between rulers and ruled in the Third Reich.

[60] A point recognized by Tim Mason in his article 'Arbeiter ohne Gewerkschaften. Massenwiderstand im NS-Deutschland und im faschistischen Italien', *Journal für Geschichte* (Nov. 1983), pp. 28–36.
[61] Klaus Wisotzky, *Der Ruhrbergbau im Dritten Reich* (Düsseldorf, 1983); Wolfgang Werner, '*Bleib übrig! Deutsche Arbeiter in der nationalsozialistischen Kriegswirtschaft* (Düsseldorf, 1983); Stephen Salter, 'The Mobilisation of German Labour, 1939-1945', unpubl. D.Phil. thesis, Oxford, 1983.
[62] This was acknowledged in a later essay, reflecting a shift of emphasis in his own work, by Tim Mason, 'Die Bändigung der Arbeiterklasse im nationalsozialistischen Deutschland', in Carola Sachse *et al.*, *Angst, Belohnung, Zucht und Ordnung* (Opladen 1982), pp. 11–53. More recent research, particularly that of Gunther Mai, has placed far more emphasis than did Mason on the integrative role played by the German Labour Front, functioning as a type of surrogate trade union. See Gunther Mai, 'Die Nationalsozialistische-Betriebszellen-Organisation. Zum Verhältnis von Arbeiterschaft und Nationalsozialismus', *VfZ*, 31 (1983), pp. 573–613; Gunther Mai, ' "Warum steht der deutsche Arbeiter zu Hitler?" Zur Rolle der Deutschen Arbeitsfront im Herrschaftssystem des Dritten Reiches', *GG*, 12 (1986), pp. 212–34; Gunther Mai, 'Arbeiterschaft zwischen Sozialismus, Nationalismus und Nationalsozialismus', in Uwe Backes *et al.*, eds., *Die Schatten der Vergangenheit. Impulse zur Historisierung des Nationalsozialismus* (Frankfurt am Main/Berlin, 1990), pp. 195–217. On the Labour Front, see also the study by Ronald Smelser, *Robert Ley. Hitler's Labour Front Leader* (New York/Oxford, 1988). A balanced appraisal of the state of research on the working class in Nazi Germany is provided by Ulrich Herbert, 'Arbeiterschaft im "Dritten Reich". Zwischenbilanz und offene Fragen', *GG*, 15 (1989), pp. 320–60.
[63] I try to support this claim in my article 'Social Unrest and the Response of the Nazi Regime 1934-1936', in Francis R. Nicosia and Lawrence D. Stokes, eds., *Germans against Nazism* (Oxford, 1990), pp. 157–74.

Secondly, we have suggested that the 'functional' argument – now detached from the *Resistenz* concept – has been pushed too far in claims that the non-politically motivated actions by the industrial working class significantly weakened the regime in the realization of a fundamental aim – the preparation for and pursuit of war. We have, in addition, pointed out that motive and effect cannot be separated and held apart as clinically as the functional approach presumes.

Finally, and most important of all, we have suggested that a 'moderate functional' approach, along the lines which Broszat intended the *Resistenz* concept to serve, is not only sensible, but necessary, in order to explain the social context within which *resistance* – that is, fundamental forms of opposition to Nazism – could develop. Moreover, the insights into relations between regime and society prompted by the *Resistenz* concept help to reveal the immense difficulties facing those engaging in such fundamental opposition. For, in the way Broszat used it, and for all its weaknesses, *Resistenz* opened up new perspectives not only on opposition but – perhaps even more importantly – on the wide areas of underlying social consensus in major aspects of Nazi rule. It was, in good measure, this underlying consensus, which allowed the regime to function, and to further its central aims. Hüttenberger's pessimistic assessment rings true: 'Whatever the perceptible reserve and discontent of the workers, sections of the middle class, and the peasantry, the fact cannot be ignored that the leadership of the Third Reich largely succeeded in producing such a degree of conformity, indeed readiness to collaborate, that its plans, especially preparation for war, were not endangered from within'.[64] With this statement, Hüttenberger, who was a key figure in conceptualizing the objectives of the 'Bavaria Project', rightly saw the problem of conformity and collaboration, rather than that of 'resistance', as posing the most demanding questions for future research.

II

We have seen that at the root of the conflicting usages of the term 'resistance' lie two quite distinct methods and approaches – each legitimate – bound together only by the same emotive term 'resistance' (*Widerstand*). What we have called the *fundamentalist* approach concerns itself with organized attempts to combat Nazism and especially with the moral courage of high-risk political action challenging the regime as a whole. Exponents of this approach focus of necessity heavily upon the *élite groups capable of undertaking such exceptional action*. The contrasting *societal* approach takes as its starting-point the total claim of the Nazi regime and uses this to explore a multiplicity of points of *conflict with ordinary citizens*. Although its exponents, from their regime-centred definition, conceive of a spectrum running from the most minor to the most fundamental types of nonconformist behaviour, they are essentially little concerned with high politics and the conspiratorial opposition of élite groups. Anthologies of works on resistance show in practice how little the two approaches have in common with each other.

We noted earlier the pleas, rejecting Hofer's 'fundamentalist' position, to

[64] Peter Hüttenberger, 'Nationalsozialistische Polykratie', *GG*, 2 (1976), p. 440.

avoid an inflexible definition of 'resistance', and the claim that a narrow definition – or perhaps any definition – would fail to do justice to the fluid boundaries between criticism, opposition, and active resistance. Resistance, it was
argued, ought to be seen as a 'process', not a closely definable fixed entity.

Correct though these observations are, they do not in themselves obviate the
need for clear definition. It might be added that the notion of resistance as a
'process', while correct in many instances when applied to the conservative
resistance, is not always valid even there,[65] and scarcely applies to the experiences of many in the working-class resistance groups,[66] who offered fundamental and principled opposition from the beginning. Nor, it could be
objected, did the 'minor' kinds of nonconformist behaviour subsumed under
the '*Resistenz*' concept usually form an early stage of opposition for those who
eventually became involved in fundamental resistance to the regime.[67]

In fact, on common-sense grounds, the fact that most historians of resistance
who favour a wide-ranging definition (or none at all) find it necessary to use
some term such as 'fundamental resistance' to speak of attempts to bring down
the regime, or opposition against the system as a whole, suggests that, however
fluid the boundaries between different manifestations of rejection of Nazism,
definitions of what constituted resistance narrower than those which cover the
entire gamut of nonconformist behaviour are both possible and necessary.

An all-embracing definition of resistance such as that favoured in the
'Bavaria Project' (embracing passive as well as active behavioural forms, partial as well as total rejection of the regime)[68] opens up, in comparison with the
earlier restricted moral-ethical usage of 'resistance', new avenues for understanding how people behaved during the Nazi dictatorship, how they com-
promised with the regime but also where they drew the line – sometimes
successfully – at the regime's attempts at interference, penetration, and control. It lends itself readily, therefore, to social history and to 'history of everyday life' approaches. It demythologises resistance to a large extent, taking it out

[65] A point implicitly accepted by Mommsen, in Schmädeke, p. 5, with reference to Ewald von
Kleist-Schmenzin.
[66] This was emphasized by Hermann Weber in response to Steinbach and Mommsen in
Schmädeke, p. 1123.
[67] In the case of the partisan-like activities in 1944 of the Cologne-Ehrenfeld '*Edelweißpiraten*',
who engaged in armed combat with Party, Hitler Youth, and Gestapo and were eventually publicly
hanged without trial, the passage from 'everyday' resentment and protest to outright and fully-
fledged resistance was, however, complete. The example shows, it has been said, that it is correct
to speak of 'involuntary resistance within a system which tolerated no nonconformist behaviour,
however trivial, and thereby created for itself a considerable resistance potential'. Matthias von
Hellfeld, *Edelweißpiraten in Köln* (2nd edn., Cologne, 1983), p. 7. Von Hellfeld was concerned to
portray an aspect of what he regarded as the 'resistance of the people' (*Widerstand des Volkes*).
His book had a political as well as a scholarly purpose: to achieve the 'rehabilitation' of the Cologne
Edelweißpiraten as 'resistance fighters' in the light of the continued readiness of the Cologne police
authorities to accept the Gestapo classification of their activities as 'criminal'. See pp. 112-21. For
other works showing the ways in which youth revolt against the straitjacket of Nazi control could
merge into political opposition, see the references in note 48, this chapter, and in ch. 7 note 56.
[68] See note 36, this chapter for the reference. This definition is distinguishable from the concept
Resistenz, used in the 'Bavaria Project', which has just been discussed. See also Broszat's 'revised
definition of resistance', extended to include 'the less heroic cases of partial, passive, ambivalent,
and broken opposition', in his essay 'A Social and Historical Typology of the German Opposition
to Hitler', in Large, ed., *Contending with Hitler*, pp. 25-33, here esp. p. 25.

of the realms of unreachable heroics down to the level of ordinary people.[69] It provides, too, the possibility of seeing *scales* of behaviour, and highlights the real situation of ordinary people, in which confusion, dilemmas of choice, and uneasy compromises were commonplace.

Its benefits in furthering empirical research on the social history of the Third Reich have been enormous. A substantial drawback, however, as it has been rather pointedly put, is 'its tendency to expand the concept of resistance until it covers anything short of positive enthusiasm for the regime'.[70]

Partly because of the difficulties with an open-ended definition, attempts have been made to develop a typology of resistance. The suggested typologies differ in detail. But each works on the basis of a broad pyramidal categorization of 'nonconformist' or 'deviant' behaviour; and each accepts the need to distinguish between essentially private and more public forms of behaviour, between organized and spontaneous actions, and between more fundamental and more partial types of anti-regime behaviour. Detlev Peukert, for example, sketched a pyramidal model building from a broad base of 'nonconformity' (much of it private, and mainly partial criticism), through 'refusal of co-operation' (*Verweigerung*), to 'protest', to, finally, the narrow peak of 'resistance' proper (*Widerstand*), which he sensibly restricted to forms of behaviour 'in which the National Socialist regime in its entirety was rejected'.[71]

There are, however, obvious difficulties about watertight definitions of the various stages involved. Furthermore, they are not – perhaps as the pyramid suggests – easily understandable as upward projections. The third level ('protest') is not necessarily qualitatively different from the second and first levels. But the fourth level ('resistance' itself in the fundamental sense) *is* surely qualitatively different from all other levels.

Some of the definitional problems involved are suggested by another typology, offered by the Austrian historian Gerhard Botz.[72] In his base-level category of 'deviant behaviour' he includes, for example, workers' absenteeism and farmers' illegal slaughtering of livestock; 'social protest', his next stage, comprises among other things the maintenance of contact with ex-party comrades, jokes about the Führer, sermons critical of anti-Church policy, the spreading of rumours, and listening to foreign broadcasts; the highest category, 'political resistance', embraces conspiracy, sabotage, the spreading of information, and the distribution of oppositional broadsheets. Botz distinguishes between actions which were *defensive* in intent, and those which were more *offensive*. Again, in this typology, precisely where the divide between intention of the individual and politicization of behaviour by the regime as a criterion of categorization occurs is not altogether clear. 'Deviance' seems, if anything, too narrow in scope, while 'social protest' includes actions which were arguably not protest at all; and 'political resistance' brackets together strikes, petitions, and

[69] See the comments by Broszat in his *Nach Hitler*, pp. 110–13, 170–1.

[70] Richard J. Evans, 'From Hitler to Bismarck: 'Third Reich' and Kaiserreich in Recent Historiography: Part II', *The Historical Journal*, 26 (1983), p. 1013.

[71] Detlev Peukert, *Volksgenossen*, p. 97. See also Peukert, 'Working-Class Resistance', in Large, ed., *Contending with Hitler*, pp. 36–7.

[72] Gerhard Botz, 'Methoden- und Theorieprobleme der historischen Widerstandsforschung', in Helmut Konrad and Wolfgang Neugebauer, eds., *Arbeiterbewegung-Faschismus-Nationalbewußtsein* (Vienna/Munich/Zürich, 1983), pp. 145ff. See also Botz's article 'Widerstand von Einzelnen' in *Widerstand und Verfolgung in Oberösterreich 1934–1945. Eine Dokumentation*, vol. 2 (Vienna, 1982), pp. 341–63.

episcopal pastoral letters with distribution of illicit flysheets, sabotage, partisan activity, and bomb conspiracies.

A great deal – not least common sense – speaks in favour of distinguishing the distinguishable and separating the non-identical in the various nonconformist actions. Putting a bomb under Hitler's table was certainly a demonstration of political nonconformity. But it was a very different manifestation of nonconformity to the criticisms of a peasant about the high-handedness of a local Party leader, or of a priest about interference with a Corpus Christi Day procession. The typologies of Peukert and Botz recognize the essential difference, but use the same umbrella term – 'resistance' – to define all categories of nonconformity. 'Resistance' is asked to serve, therefore, at one and the same time as an ordering concept for the whole span of behaviour, and as a definition (with all its moral, ethical, and politically normative connotations) for a narrow range of actions qualitatively different from the rest.

Methodologically, the 'fundamentalist' approach is less innovative, and its findings relatively familiar and less challenging, except in the emotional sense of awakening admiration for quite extraordinary courage and humanity amid such barbarity. But, for all its limitations, this approach does unquestionably deal with methods of combating the Nazi regime at its core, in its essence, and in its entirety – with *resistance*. The 'societal' approach is methodologically richer; touches behavioural patterns which reveal much of life in such a brutal police state; and contain implications transcending the history of Nazi Germany. But it is an approach which, whatever its claims, is essentially concerned with *conflict* rather than resistance. It is best regarded not as a sub-variety (let alone an inferior kind) of resistance research – at which Hofer's remarks perhaps hinted – but as dealing with a quite differently conceived set of issues.

However desirable, it is impossible to break away from normative usages in order to create a clean analytical concept out of 'resistance' (*Widerstand*). We might as well, therefore, accept the implications of the term and restrict it to behaviour in which the rejection of Nazism is fundamental: unless the meaning of 'resistance' is to be wholly diluted, it should be restricted to the description of active participation in *organized* attempts to work against the regime with the conscious aim of undermining it or planning for the moment of its demise. Another, less emotive and morally laden, term ought ideally to be deployed instead of 'resistance' as the 'umbrella concept' to cover *all* forms of behaviour which deviated from the norms demanded by the regime and opposed – perhaps even restricted – its total claim.[73] Whatever the term used, it is bound to be more porous than watertight, and less than clinically precise.

'*Resistenz*', for the reasons already given, is less than satisfactory. '*Opposition*' seems a less emotive and more satisfactorily descriptive concept comprising all forms of *action*, including the fundamental actions classed as 'resistance', but also embracing many forms of actions with partial and limited aims, not directed against Nazism as a system and at times deriving from individuals or groups at least partially sympathetic towards the regime and its ideology. A third, and even broader, concept which seems better able to serve

[73] I part company at this point from Peter Steinbach, who retains a definition so wide that it is inclusive of 'any form of opposition to a regime that attempted to control every aspect of political, cultural, religious, and social life'. Peter Steinbach, 'The Conservative Resistance', in Large, ed., *Contending with Hitler*, pp. 89–97, here esp. p. 89.

as an umbrella term to cover the *passivity* of 'oppositional' feeling which did not necessarily result in any action, and the voicing of attitudes, often spontaneous, at all critical of any aspect of Nazism is *dissent*. 'Dissent' could become 'opposition', but did not necessarily do so; while 'resistance' was 'opposition' of a distinctive and fundamental kind, separable from all other forms.[74]

However fluid the overlaps between these types of behaviour, this conceptualization does offer a degree of clarity often lacking in the 'resistance' debate, and corresponds to the historiographical distinction between works on resistance in its narrow sense and works on regime-society conflict spheres. It does not mean artificially sealing off 'resistance' from other forms of dissident behaviour. But it perhaps means that if we want to picture the overlaps in graphic form, we ought to imagine, instead of the rising scale suggested by Peukert, concentric circles blurring into each other: a wide 'soft pulp' of 'dissent'; a narrower, though still wide, band of 'opposition'; and at the core, a small circle of fundamental 'resistance'. Even this spatial metaphor would demand the inclusion of a broad gulf separating the band of 'opposition' from the inner core of 'resistance' proper – the crossing of which amounted to a 'quantum leap' in attitude and behaviour.

III

There remains the important issue of the *extent* of resistance in the Third Reich. Was resistance to Hitler indeed 'resistance without the people'? How popular *was* opposition to Nazism? Such questions go to the heart of the problem of the relationship between German society and the Nazi regime.

Obvious though they may be, a number of points nevertheless warrant renewed emphasis. First, the Nazi regime *was* a terroristic dictatorship – in a literal sense, a terrifying regime – which knew no bounds in the repression of its perceived enemies. 'Keep quiet or you'll end up in Dachau' was a common sentiment indicating an all-pervasive fear and caution sufficient to deter most people from challenging the regime in any way. Passivity and co-operation – however sullen and resentful – were the most human of responses in such a situation. Secondly, once the Nazis had taken a hold on power, the only realistic potential for a challenge capable of ousting Hitler came from *within* the regime's own power élites. Since, apart from the SA (before the 'Night of the Long Knives' in 1934) and the SS, only the *Wehrmacht* had recourse to the weaponry and force needed to sustain a coup, the sole possibility of a successful putsch had to arise from within the army leadership (apart, that is, from a one-man assassination attempt on Hitler's life, such as that of the Swabian joiner Georg Elser in 1939).[75] Fundamental resistance was far from confined to the élites. But, realistically, élite resistance offered the only chance of toppling the regime from within. Thirdly, if we discount the armed forces, outside the Christian Churches no mass institutions survived which were capable of articulating

[74] I tried to operate on the basis of such a distinction in my *Popular Opinion and Political Dissent in the Third Reich. Bavaria 1933–1945* (Oxford, 1983); see in particular p. 3 and note 7.

[75] See Anton Hoch, 'Das Attentat auf Hitler im Münchner Bürgerbräukeller 1939', *VfZ*, 17 (1969), pp. 383–413; and Lothar Gruchmann, ed., *Autobiographie eines Attentäters. Johann Georg Elser* (Stuttgart, 1970).

and organizing opposition. Spheres of conflict and opposition remained, as a consequence, largely isolated from each other, while critical opinion fragmented into its component parts. Fourthly, many aspects of Nazism enjoyed a popularity extending far beyond Party die-hards. Economic recovery, the destruction of 'marxism', the rebuilding of a strong Germany, territorial expansion, foreign policy and military successes, all seemed, before the middle of the war, stunningly impressive to millions. They found their embodiment in Hitler's personal popularity, boosted by propaganda into a leadership cult of great potency. The apparent 'achievements' of the regime both disarmed criticism and created an atmosphere in which opposition was unable to reckon with any broad base of growing popular disaffection that could have proved dangerous to the regime.

In such a context, the extent of defiance – however ineffective – on the part of ordinary Germans cannot fail to impress. By 1939, around 150,000 Communists and Social Democrats had suffered internment in concentration camps; 40,000 Germans had fled the country for political reasons; 12,000 had been convicted of high treason; and a further 40,000 or so had been imprisoned for lesser political offences over the same period. During the war, when the number of offences punishable by death rose from 3 to 46, some 15,000 death sentences were handed out by German civilian courts.[76] One jail alone, the Steinwache prison in Dortmund, has records of the imprisonment for political 'delicts' of 21,823 Germans during the Nazi dictatorship, the overwhelming majority of whom (in cases where occupations are given) were industrial workers.[77] In the Rhine-Ruhr area, a total of 523 mass trials, involving 8,073 persons, resulted in 97 instances in the death penalty and in the imposition of a total of 17,951 years of imprisonment on the convicted members of the worker resistance groups. It is reckoned that over 2,000 working-class members of illegal resistance organizations in this region lost their lives to Nazi terror.[78]

It amounts to a moving testimony to bravery, dignity, and suffering. Yet everything points to the tragic conclusion that this working-class resistance was not only without resonance among other social groups, but was increasingly isolated even from its own mass base in the working class. A legacy of bitterness among the left-wing parties and disillusionment with the failings of the leadership, major Nazi successes and the hopelessness of any confrontation with such might and unrestrained brutality, growth in earnings opportunities for industrial workers during the armaments boom, and, not least, the seemingly unlimited ability of the regime to infiltrate and crush illegal groups, reinforced such isolation. Growing militancy in industrial relations in the later 1930s, particularly the struggle for higher wages, was largely detached from the political resistance, which had relatively little success in exploiting material discontent.

[76] Figures from Martin Broszat, 'The Third Reich and the German People', in Hedley Bull, ed., *The Challenge of the Third Reich* (Oxford, 1989), p. 93. It has been estimated that around a half of the 300,000 members of the KPD in 1932 were arrested during the Third Reich. Richard Löwenthal and Patrik von zur Mühlen, eds., *Widerstand und Verweigerung in Deutschland 1933 bis 1945* (Berlin/Bonn, 1984), p. 83.

[77] Klotzbach, pp. 242–5.

[78] Detlev Peukert, *Ruhrarbeiter gegen den Faschismus. Dokumentation über den Widerstand im Ruhrgebiet 1933–1945* (Frankfurt am Main, 1976), p. 347. A glimpse into the wide spectrum of oppositional forms of behaviour in this region is provided by Heinz Boberach, 'Widerstand an Rhein und Ruhr', in Walter Först, eds., *Leben, Land und Leute* (Cologne, 1968), pp. 130–42.

In the early war years, despite widespread reports of poor work discipline, low morale, and anti-regime feeling among workers, only around 5 per cent of political arrests reported by the Gestapo fell under the rubric 'communism/marxism', while the rest – covering a wide variety of 'offences' – bore no sign of any link with illegal groups or any organized background.[79] Even once the invasion of the Soviet Union had ended the most difficult phase for the KPD, internal KPD appraisals indicated how isolated the underground resistance remained, and how its energies had to be poured almost entirely into sustaining the organization itself in preparation for the day when the regime would be destroyed from without.

One perceptive report in May 1942 from Wilhelm Knöchel, a leading member, based in Berlin, of the KPD's Central Committee, took issue with the continued overestimation by Moscow of the resistance potential of the KPD: 'With the horror of a military defeat staring them in the face, the great majority of our people would like to see the back of the Hitler government now rather than later. Even so, Hitler appears to them as the lesser evil and they hope for a victory, the possibility of which they gravely doubt'. Only a small minority, it was added, put their hopes in a victory of the Red Army. The situation, it was said, was ripe for political propaganda, but there was no hope of a mass movement against the regime similar to that which had emerged in 1918. The report not only pointed to the inroads made by Nazi ideology and propaganda, but indicated the phenomenon of social disintegration under Nazism which had atomized workers, robbed them of their traditional framework of solidarity, organized them in rival cliques and patronage clienteles, and disrupted channels of communication and levels of expertise; these in turn had assisted Nazi inroads and hindered solidarity through rivalries, distrust, and denunciation.[80]

In the very last phase of the war, with defeat imminent, the prospect of suffering immediate and massive retaliation for action against a regime patently in its death-throes was sufficient to persuade most workers to remain aloof from dangerous involvement with resistance groups. Those anti-fascist committees which, in the last days of the war, fought to ensure the intact surrender of their factories and communities had only tangential links with the illegal groups and frequently amounted to new, *ad hoc* formations of workers.[81]

The Christian Churches, never institutionally 'co-ordinated' by the regime, represented large bodies of support for opposition to attempts to undermine Christian practices, institutions, and beliefs. Both major denominations engaged in a bitter war of attrition with the regime, receiving the demonstrative backing of millions of churchgoers. Applause for Church leaders whenever they appeared in public, swollen attendances at events such as Corpus Christi Day processions, and packed church services were outward signs of the popularity of the struggle of both Churches – especially of the Catholic

[79] Detlev Peukert, *Die KPD im Widerstand. Verfolgung und Untergrundarbeit an Rhein und Ruhr 1933 bis 1945* (Wuppertal, 1980), p. 335.

[80] Peukert, *KPD*, p. 345, 354–5. Knöchel had moved to Berlin from Holland in January 1942. Allan Merson, whose book *Communist Resistance in Nazi Germany* (London, 1985) forms the most comprehensive study in English of the KPD's opposition, emphasises (p. 253) that, despite some divergence in line, 'there was no question of a breach between Knöchel and the Moscow leadership'.

[81] Peukert, *KPD*, p. 412 and note 25.

Church – against Nazi oppression. The fight to protect youth organizations and schools was eventually lost. In some cases, such as the struggle to retain the very symbol of Christianity, the Crucifix, in school classrooms, however, the Nazi rulers were forced to concede ground to the protest.[82]

Many acts of extreme courage were carried out by individual Christians, especially by priests and pastors, of both denominations. Alongside the revered figures of Dietrich Bonhoeffer, Martin Niemöller, Pater Alfred Delp, Pastor Grüber, and Father Lichtenberg stood many little known priests and pastors who sacrificed liberty and sometimes life for their opposition to the regime. It has been estimated that around one in three Catholic priests suffered some kind of reprisal during the Third Reich.[83] Around 400 German Catholic priests and 35 Evangelical pastors were incarcerated in the Priests' Block at Dachau alone as a consequence of their unwavering Christian beliefs and principles.[84]

As institutions, nevertheless, the Churches offered something less than fundamental resistance to Nazism.[85] Their considerable efforts and energies consumed in opposing Nazi interference with traditional practices and attempts to ride roughshod over Christian doctrine and values were not matched by equally vigorous denunciation of Nazi inhumanity and barbarism – with the notable exception of Galen's open attack on the 'euthanasia' programme in August 1941.[86] In defence of humanitarian rights and civil liberties, the response of both Churches was muted. What rejection there was from high quarters in the Church of, for example, anti-Jewish policy, came largely by way of private protest letters to government ministers.[87] The public silence following the '*Reichskristallnacht*' pogrom in 1938, in contrast to the courageous denunciation, paid for dearly, by numerous individual priests and pastors, epitomizes this institutionally understandable but morally regrettable reluctance to engage the regime on issues beyond the 'Church struggle'.

The detestation of Nazism was overwhelming within the Catholic Church and grew more extensive within the ideologically as well as theologically divided Evangelical Church. But defiant opposition in the sphere of the 'Church struggle' was compatible in both major denominations with approval of key areas of the regime's policies, above all where Nazism blended into 'mainstream'

---

[82] For valuable surveys and analysis of opposition with the two major Christian Churches, with differing accents and emphases, see Schmädeke, pp. 227–326, 1125–7 (Drinter Teil: 'Kirchen und Konfessionen zwischen Kooperation und Teilwiderstand'); the contributions of Günther van Norden and Ludwig Volk to Christoph Kleßman and Falk Pingel, eds., *Gegner des Nationalsozialismus* (Frankfurt am Main, 1980), pp. 103–49; those of van Norden and Heinz Hürten to Müller, *Der deutsche Widerstand*, pp. 108–56; and those of van Norden, Heinz Gollwitzer, and Walter Dirks to Löwenthal and von zur Mühlen, pp. 111–42. The framework of dissent and opposition in Bavaria is the subject of chs. 4, 5 and 8 of my *Popular Opinion*. Catholic support for Hitler's foreign policy is particularly forcefully dealt with by Guenter Lewy, *The Catholic Church and Nazi Germany* (London, 1964), ch. 7.

[83] Ulrich von Hehl, *Priester unter Hitlers Terror. Eine biographische und statistische Erhebung* (Mainz, 1984), pp. xlii–iii, liii.

[84] Günther van Norden, 'Widerstand in den Kirchen', in Löwenthal/von zur Mühlen, p. 128.

[85] A distinction between individual and institutional opposition is drawn by van Norden in Löwenthal/von zur Mühlen, pp. 111–28.

[86] For Galen's sermon, see Ernst Klee, *'Euthanasie' im NS-Staat. Die 'Vernichtung lebensunwerten Lebens'* (Frankfurt am Main, 1983), pp. 334–5.

[87] See von Norden, in Kleßman/Pingel, pp. 114–6 and note 47a; and in Löwenthal/von zur Mühlen, p. 125.

national aspirations:[88] support for 'patriotic' foreign policy and war aims; obedience towards state authority (except where it was regarded as contravening divine law); approval for the destruction of 'atheistic' marxism (and for the 'crusade' against Soviet Bolshevism); and readiness to accept discrimination against Jews (where traditional Christian anti-Judaism, though distinct from biological *völkisch* racism, nevertheless could offer no bulwark against the dynamic anti-semitism of the Nazis).[89] In all of these areas, the Churches as institutions felt on uncertain ground – a reflection of the fact that popular backing could not be guaranteed, and that such issues fell outside what was regarded as the legitimate sphere of Church opposition, which was correspondingly limited, fragmented, and largely individual.

So lethally dangerous was the Nazi regime to its opponents that, as with the cobra, hitting out at the tail was likely to result only in being destroyed by the head. The only hope of smashing the regime was, therefore, by striking at the head itself. Why did those élite groups with recourse to arms who could have resisted in the most effective way – by destroying the regime from within – fail to mount a serious challenge to the Nazi leadership until military defeat was staring Germany in the face? The failure had far deeper causes than the bad fortune which bedevilled several embryonic plots to remove Hitler.

The substantial complicity of the élites in Nazi rule, and the lack of a popular basis for resistance are among the most important elements of an explanation. And linking both together is the crisis-ridden character of the regime, whose 'cumulative radicalization',[90] destruction of ordered channels of government and administration, fragmentation of lines of communication (other than those prescribed by Goebbels), ruthlessness towards its opponents, and propaganda success in establishing powerful (if often superficial) plebiscitary support focused above all on Hitler, had an important impact on the élites at the same time as it nullified any possibility of a rising from below, or even mass backing for an élite coup.

We have already referred to the early enthusiasm for the Third Reich of most of those who subsequently became involved in the Stauffenberg resistance conspiracy, and to the fact that many of them were important 'servants' of the Hitler state in the upper echelons of the bureaucracy and army. Their gradual estrangement and disillusionment accompanied the increasing radicalization of the regime. But they served the regime long enough to strengthen it to the point where resistance became more difficult, and where they themselves became dispensable; to the point at which the regime's leadership could distance itself from the traditional national-conservative power élites and reduce them to the status of purely 'functional élites'.[91] The élites overwhelmingly approved of the destruction in 1933 of the 'Party system' which they had so detested in the

---

[88] On Nazism's overlaps with and exploitation of 'mainstream' nationalism, see William Sheridan Allen, 'The Collapse of Nationalism in Nazi Germany', in John Breuilly, ed., *The State of Germany* (London, 1992), pp. 141–53.

[89] See, on this last point, the contributions to Otto Dov Kulka and Paul R. Mendes-Flohr, eds., *Judaism and Christianity under the Impact of National Socialism* (Jerusalem, 1987).

[90] A term coined by Hans Mommsen. See his article, 'Der Nationalsozialismus: Kumulative Radikalisierung und Selbstzerstörung des Regimes', in *Meyers Enzyklopädisches Lexikon*, 16 (1976), pp. 785–90.

[91] See Klaus-Jürgen Müller, *Armee, Politik und Gesellschaft in Deutschland 1933–1945* (Paderborn, 1979), p. 44.

Weimar Republic; they publicly lauded Hitler for the mass murder of the SA leaders in 1934; they rejoiced in the dismantling of Versailles and Locarno in 1935 and 1936; they accepted the lurch into the economics of autarky and war preparation in 1936; they voiced no criticism of the mounting wave of 'aryanization' of Jewish property and its accompanying wave of terror in 1937-8; they acquiesced in the removal of conservatives from all influential positions in military and foreign policy in February 1938; and they bathed in the reflected glory of their Leader following the Anschluß the following month. Only thereafter, in the wake of the growing Sudeten crisis, did the growing alarm of a minority within the élites come to form the embryo of the later conspiracy against Hitler.[92] Even then, further foreign policy coups by Hitler – including, of course, that achieved through the capitulation of the western powers at Munich – and later the astounding run of military successes, which only ground to a halt before the gates of Moscow, continued to disarm opposition.[93] Meanwhile, the bureaucratic, military, and economic élites were becoming more and more deeply implicated in the increasing barbarism of the regime in Poland then in the Soviet Union. The longer the war went on, the more they had burnt their boats along with the Nazi regime.

Though the Stauffenberg conspiracy gained an impressive circle of recruits in the officer corps of the army and the higher ranks of the civil service, as well as incorporating a number of churchmen and figures from the socialist and trade union movement, it is hard to sustain the claim that those involved were necessarily representative even of those groups from which they were drawn. And other social groups, such as the lower middle class, were hardly represented at all. The view articulated in, especially, the older literature on the 20 July plot, that the resistance was representative of all sections of the German people, can scarcely be upheld.[94]

Pater Delp's enquiries about the likely popular reaction to a putsch were disheartening in the extreme, while, in a memorandum written in 1944, Adam von Trott indicated widespread passivity among workers and little expectation of mass backing.[95] There was, therefore, much justified scepticism among the conspirators about the possibility of a 'revolution from below' to succeed an assassination of Hitler, and whether such an upheaval alone would give the 'revolution from above' its legitimation.[96] Hans Mommsen's assessment that in contrast to the resistance movements in the occupied territories the political thinking and planning of the German resistance was 'much influenced by the uncertainty of how the population, most of whom for most of the time supported Hitler, would react', seems apposite.[97]

[92] See Hoffmann, *Widerstand, Staatsstreich, Attentat*, chs. 3-4; Harold C. Deutsch, *The Conspiracy against Hitler in the Twilight War* (Minneapolis, 1968), remains valuable for the early stages of the conspiracy; on the key figure of Beck, the authoritative study is Klaus-Jürgen Müller, *General Ludwig Beck* (Boppard am Rhein, 1980).

[93] On the stance of the different élite groups in the early phase of the war, see Martin Broszat and Klaus Schwabe, eds., *Die deutschen Eliten und der Weg in den Zweiten Weltkrieg* (Munich, 1989).

[94] Mommsen in Schmädeke, pp. 8-9; and in *The German Resistance to Hitler*, p. 59, where he commented that the conspiracy of 20 July 1944 was, in its social structure, 'comparatively homogeneous' in its 'predominantly upper-class membership' and 'wholly unrelated to the mass of working people'.

[95] Cited by Mommsen, in *The German Resistance to Hitler*, p. 59.

[96] *The German Resistance to Hitler*, p. 63.

[97] *The German Resistance to Hitler*, p. 59.

Can 'resistance', then, be said to have been 'without the people'? Hundreds of thousands of ordinary citizens from all walks of life were persecuted in the Third Reich for political 'offences' against the Nazi state. These victims of Nazi repression include the many who willingly entered into the perilous terrain of the illegal worker underground organizations. But the worker organizations were of necessity small, as time went on increasingly secretive, and therefore isolated even within their own social milieu. It would, moreover, be hard to see many of the 'offences' (for which often draconian punishment was meted out) as falling within an understanding of resistance which did not widen that concept in a way which ultimately renders it more or less meaningless. One might conclude that while political dissent and opposition to specific measures of the Nazi regime were indeed widespread, 'resistance' in its fundamental sense lacked a popular base of support.

Of course, in conditions of terroristic dictatorship resistance capable of posing a successful challenge has, as we have noted, in the first instance to arise from within disaffected élites, that is from within the system itself.[98] Atomization and fear – fragmentation of opposition and all-pervasive anxiety among ordinary citizens – are normally sufficient to rule out mass risings from below against dictatorship. They were certainly sufficient in the case of the Nazi dictatorship. Keeping one's head below the parapet was a natural and normal reaction to terror – more than ever when the days of the regime were patently numbered. We have argued, however, that not only was resistance to Hitler carried out – inevitably, one might say – without *active* support from the mass of the people, but that even *passive* support was largely lacking for those risking everything to overthrow the system.

The reasons for this are implicit in what has already been said about the character of the regime and about the consensus which underpinned it. The very 'total claim' which the regime made on German society, and the apparent omnipresence of its representatives in all walks of life politicized (and criminalized), as we have seen, many 'normal' forms of behaviour, but at the same time restricted rather than enhanced the likelihood of grievances amalgamating into comprehensive and fundamental opposition among the mass of the population. If there was much to complain about in Nazi Germany, there was also much which appealed. Local Party functionaries could be lambasted in the same breath that lavished praise on Hitler; economic injustice could be bemoaned while ringing the cheers for the recovery of the Rhineland; attacks on the Churches could be decried while mouthing anti-Bolshevik slogans; law and order could be eulogized while Jewish synagogues were being vandalized. The young, unpolitical typist who shocked her socialist travelling companion by jumping out of a tram to offer an enthusiastic and unsolicited Hitler-salute to a passing SS column, pointing out afterwards that she had not done this for the SS or the Nazis, but from 'patriotic duty . . . , because I'm a German', epitomized such schizophrenia.[99] By the time the bomb exploded in Hitler's headquarters at

[98] It appears that the recent collapse of the GDR is no exception. Even though events were prompted by external circumstances – Gorbachev's support for a more liberal policy and the consequent pressure on the GDR as its borders became more porous – and were soon rapidly swept along by a popular uprising, the decisive undermining of the GDR regime took place, as the crisis deepened, initially at the élite level, with Gorbachev supporting a fronde against Honecker. See Günter Schabowski, *Das Politbüro* (Reinbek bei Hamburg, 1990).

[99] Archiv der Sozialen Demokratie, Bonn, Emigration Sopade, M65, report for Feb. 1937.

Rastenburg in July 1944, the mass of the German people longed more than anything else for the end of the war, and many recognized that it would only be possible through the end of the Hitler regime. But the war itself, the lack of alternative posed by 'unconditional surrender', and the fear of a victorious Soviet Union provided continuing negative bonds between regime and society. A successful overthrow of Hitler would probably have polarized opinion.[100] The putschists certainly had the potential to gain support through the ending of the war on terms which Germans could have seen as favourable. But in the context of Allied 'unconditional surrender' policy and attitudes towards the German resistance, that would have been unlikely.[101] On the other hand, they would have had to live with a new 'stab-in-the-back' taint.[102] And inability to reach a favourable settlement with the Allies would have been fatal to the chances of consolidating whatever popularity they enjoyed at the outset.

Compared with numerous other authoritarian systems which have eventually collapsed, and fully acknowledging the many spheres where there was considerable and serious dissent from Nazi policies and ideology, the Nazi regime could, then, rely upon a substantial level of popularity and an underlying consensus (if only of a negative kind) which it retained in all essentials until deep in the war. This provided a climate in which resistance to Hitler, from the beginning, lacked broad support from the mass of the people.

This is arguably, as we hinted earlier, where the 'fundamentalist' and 'societal' approaches become complementary, not contradictory. Instead of the 'fundamentalist' rejection of the 'societal' approach as one which distorts the perspective on 'resistance' proper, and detracts from its moral stature, it could be claimed that *only* through the societal approach can the position of the groups (élite and worker) engaged in all-out resistance to the regime gain full recognition.[103] The contradictions in 'everyday' attitudes and behaviour, the partial character of dissent and opposition, and the wide spheres of assent and collaboration, are far from irrelevant to considerations of fundamental resistance: they are, in fact, essential components of an explanation of its isolation, socially and ideologically, and in turn, therefore, of the reasons for its failure.

The societal approach has, furthermore, also been able to locate the historic weaknesses of élite resistance in ways rendered impossible by concentration on the ethical motivation and the moral lessons of the actions of élite groups – lessons which are already fading, as they are bound to do, like all monuments to heroism, in the course of time. This has enabled greater understanding of the conditions within which their fundamental opposition took shape – an understanding which, on the moral plane, ought to magnify rather than undermine admiration for their actions.

[100] Some of the varied reactions to news of the assassination attempt are presented in my study, *The 'Hitler Myth'. Image and Reality in the Third Reich* (Oxford, 1987) pp. 215–29.
[101] For a hightly critical view – if embedded in a dubious presumption reflected in the book's title – of the British government's dealings with the German resistance, see Patricia Meehan, *The Unnecessary War. Whitehall and the German Resistance to Hitler* (London, 1992).
[102] According to opinion surveys, the Stauffenberg attempt encountered disapproval even in the 1950s among a substantial minority of the population. See, for example, E. Noelle and E.P. Neumann, *Jahrbuch der öffentlichen Meinung 1947–1955* (Allensbach, 1956), p. 138.
[103] Some of the contributions, particularly those by Martin Broszat and Hans Mommsen, to the most recent anthology on essays on different aspects of resistance – David Clay Large, ed., *Contending with Hitler. Varieties of German Resistance in the Third Reich* (Cambridge, 1991) – point in this direction.

The isolation of resistance, made fully comprehensible through the societal approach, places the spotlight in resistance historiography above all on the inexorable radicalization of the regime – parasitically sucking in client groups and deepening complicity, if not euphoric support, at the same time that it created by its very radicalization the resistance among its own 'servants'. This resistance came close to destroying it, but those undertaking it did so in the full recognition that it was 'resistance without the people'.

The ineffectiveness and failure of German resistance to Nazism had its roots in the strife-torn political climate of the Weimar Republic. The internecine conflict on the Left, the enthusiasm of the conservative Right to act as grave-diggers to the Republic, and the massive popular readiness to embrace authoritarianism and reject the only form of democracy then known to Germany explain divisions within, slowness to act of, and lack of popular support for, resistance during the dictatorship. The moral courage of those who stood up to Nazi tyranny remains, and will remain, an example to all subsequent ages. But a historical understanding of the weaknesses and failure of resistance is crucial. Apart from illustrating the self-evident truth that it is easier to prevent a would-be dictator from taking power than to remove such a dictator once he has the might of the state at his disposal, the historiographical and conceptual debates about resistance, surveyed in this chapter, have increasingly demonstrated the very complexity of the problem of resistance under Nazism. Perhaps more than all else, it has been their merit to emphasise more and more as time has passed one cardinal aspect of the problem: that the story of dissent, opposition, and resistance in the Third Reich is indistinguishable from the story of consent, approval, and collaboration.

9

'Normality' and Genocide: The Problem of 'Historicization'

The problem of the so-called 'historicization' (*'Historisierung'*) of National Socialism, a term which first entered serious discussion when advanced by Martin Broszat in an important and programmatic essay published in 1985,[1] revolves around the question of whether, nearly half a century after the collapse of the Third Reich, it is possible to treat the Nazi era in the ways that others eras of the past are treated – as 'history' – and what new perspectives such a shift in conceptualization and method would demand. In intellectual terms, the controversy which Broszat's article provoked raises distinctive theoretical and methodological problems, involving consideration of the contribution and potential of what has, in many respects, proved a most fruitful approach in research on the Third Reich in recent years, that of *'Alltagsgeschichte'* ('the history of everyday life').

During the past two decades, new and exciting avenues of research have been explored in a massive outpouring of studies ranging over most of the important aspects of the impact of Nazism on German society. Yet just as the time seems ripe – almost quarter of a century since the appearance of Schoenbaum's wide-ranging social history of the Third Reich, seen as 'Hitler's social revolution', and Dahrendorf's equally influential interpretation of Nazism as 'the German revolution'[2] – for a new full-scale study which would synthesize and incorporate much of this work and offer a revised interpretation of German society under Nazism, the 'historicization' controversy casts doubt upon even the theoretical possibility of constructing such a social history without losing sight of the central aspects of Nazism which provide it with its lasting world-historical significance and its moral legacy. The first part of this chapter offers an outline of this important controversy, while the second part seeks to evaluate its implications for a potential history of German society in the Third Reich.

[1] Martin Broszat, 'Plädoyer für eine Historisierung des Nationalsozialismus', *Merkur* 39 (1985), pp. 373–85, reprinted in Martin Broszat, *Nach Hitler. Der schwierige Umgang mit unserer Geschichte* (Munich, 1986) pp. 159–73. All references which follow are to the latter version. An English translation is available in Peter Baldwin, ed., *Reworking the Past. Hitler, the Holocaust, and the Historians Debate* (Boston, Mass., 1990) pp. 77–87.
[2] David Schoenbaum, *Hitler's Social Revolution*, New York/London, 1966; Ralf Dahrendorf, *Society and Democracy in Germany*, London, 1968, ch. 25.

The 'Historicization' Approach

A major breakthrough in deepening awareness of the complexity of German society in the Third Reich, it is universally recognized, was the research undertaken and published between the mid 1970s and early 1980s within the framework of the 'Bavaria Project', which helped to offer an entirely new dimension to the understanding of relations between state and society in Nazi Germany.[3] The project, it seems clear, was an important impulse, among others, in the rapid development of the 'everyday life' approach to the Third Reich. The very concept of '*Alltagsgeschichte*' ('the history of everyday life'), and the methods deployed by its exponents, have provoked much stringent criticism – some of it well justified – particularly from the leading protagonists of the 'critical history' and 'history as social science' ('*historische Sozialwissenschaften*') approach.[4] Such criticism has, however, not been able to stem the continued spread of '*Alltagsgeschichte*', and some, even of the sharpest critics, have accepted that, properly conceptualized, '*Alltagsgeschichte*' can have much to offer in deepening understanding.[5] The remarkable resonance of the 'everyday life' approach, exploring subjective experiences and mentalities at the grass roots of society, presumably reflects in part, not least through the opening up of previously taboo areas of consideration, a need, particularly strong among the younger generation, to come to grips with the Third Reich not just as a political phenomenon – as a horrific regime providing a resort for political and moral lessons in a post-fascist democracy – but also as a social experience, in order to understand better the behaviour of ordinary people – like their own relatives – under Nazism. By making past behaviour and mentalities more explicable, more understandable, more 'normal' – even if to be condemned – it is arguable that '*Alltagsgeschichte*' has contributed to deepened awareness of the problems of historical identity in the Federal Republic, and of the relationship of the Third Reich not just to political continuities and discontinuities, but now also to social strands of continuity pre-dating Nazism and extending well into the post-war era. This further prompts the need to locate the Third Reich as an integral component of German history, not one which can be bracketed out and detached as if it did not really belong to it. These are some of the considerations behind Martin Broszat's well known 'plea for the historicization of National Socialism', premised upon the assertion that the history of the Nazi *era*, as opposed to that of the political system of the dictatorship, still remains to be written.[6]

Broszat's use of the term '*Historisierung*' ('historicization') relates to the problems of historians, and specifically West German historians, in dealing

---

[3] Martin Broszat *et. al*, eds., *Bayern in der NS-Zeit*, 6 vols., Munich 1977–83.
[4] See e.g. Hans-Ulrich Wehler, 'Königsweg zu neuen Ufern oder Irrgarten der Illusionen? Die 'westdeutsche Alltagsgeschichte: Geschichte "von innen" und "von unten" ', in F.J. Brüggemeier and J. Kocka, eds., '*Geschichte von unten – Geschichte von innen*'. *Kontroversen um die Alltagsgeschichte*, Fernuniversität Hagen, 1985, pp. 17–47. And for a lively debate about the merits and disadvantages of '*Alltagsgeschichte*', see *Alltagsgeschichte der NS-Zeit. Neue Perspektive oder Trivialisierung?*, Kolloquien des Instituts für Zeitgeschichte, Munich, 1984.
[5] See e.g. the thoughtful assessment of the limitations but also possibilities of '*Alltagsgeschichte*' by Jürgen Kocka in reviews in *Die Zeit* Nr.42 vom 14.Okt.1983 ('Drittes Reich: Die Reihen fast geschlossen') and *taz* vom 26. Jan. 1988 ('Geschichtswerkstätten und Historikerstreit').
[6] Broszat, *Nach Hitler*, S. 167.

with the Nazi past. Even approximately half a century after the end of the Third Reich, the distance which the historian puts between himself and the subject matter of Nazism provides, in Broszat's view, a major obstacle to the possibility of approaching the scholarly study and analysis of Nazism in the same way that other periods of history are tackled – with the degree of intuitive insight which 'normal' historical writing demands. Yet, without the proper integration of Nazism into 'normal' historical writing, he saw the Third Reich remaining an 'island' in modern German history,[7] a resort for lessons of political morality in which routine moral condemnation excludes historical understanding, reducing Nazism to an 'abnormality' and serving as a compensatory alibi for a restored historicism (*'Historismus'*) with regard to the more 'healthy' epochs before and after Hitler.[8] The position is summed up in the following way:

> A normalization of our historical consciousness and the communication of national identity through history can not be achieved by avoiding the Nazi era through its exclusion. Yet it seems to me that the greater the historical distance becomes, the more urgent it is to realize that bracketing the Hitler era out of history and historical thinking also occurs in a way when it is only dealt with from a political-moral perspective and not with the same differentiated applied historical method as other historical epochs, when treated with less carefully considered judgement and in a cruder, more general language, or when, for well-intentioned didactic reasons, we grant it a sort of methodological special treatment.[9]

A 'normalization' of methodological treatment would mean the application of the normal rigours of historical enquiry in a meticulous scholarship deploying 'mid-range' concepts subjectable to empirical investigation in place of bland moralization, whether from a liberal-conservative perspective or from sterile economistic determinant theories of a marxist-leninist or 'new Left' variety.[10] This in itself would refine moral sensitivity through the increased understanding derivable from greater differentiation, as in the relativization of 'resistance' through its 'de-heroization' and recognition of the chequered grey nature of the boundaries of opposition and conformity between the 'Other Germany' and the Nazi regime.[11] It would allow, too, Nazism's function as the exponent of modernizing change comparable with that in other contemporary societies to be properly incorporated in an understanding of the era, and hence a deeper awareness of the social forces and motivation which the Nazi

---

[7] See Broszat, *Nach Hitler*, pp. 114–20 ('Eine Insel in der Geschichte? Der Historiker in der Spannung zwischen Verstehen und Bewerten der Hitler-Zeit').

[8] Broszat, *Nach Hitler*, p. 173. See chapter 1, pp. 5–7, for brief comments on the philosophy of traditional historicism in Germany.

[9] Broszat, *Nach Hitler*, p. 153 (and back cover). As one who wrote extensively and with great sensitivity about Nazi concentration camps, in which the term 'special treatment' (*'Sonderbehandlung'*) was a euphemism for murder, Broszat's use of the same term in the present context was a remarkable and unfortunate linguistic lapse.

[10] Broszat, *Nach Hitler*, pp. 104 ff, cf. also pp. 36–41. In his exchange of letters with Saul Friedländer, Broszat spoke of a 'plea for the normalization of method, not of evaluation'. – Martin Broszat, Saul Friedländer, 'Um die "Historisierung des Nationalsozialismus"'. Ein Briefwechsel', *VfZ*, 36 (1988), pp. 339–72, here p. 365, (henceforth cited as 'Briefwechsel'). This letter exchange is now published in translatioin in Peter Baldwin, ed., *Reworking the Past. Hitler, the Holocaust, and the Historians' Debate*. (Boston, Mass., 1990), pp. 102–34. All references in this chapter are, however, to the German version.

[11] Broszat, *Nach Hitler*, pp. 170–1.

Movement could mobilize and exploit.[12]

The relevance of the 'Bavaria Project' and the emphasis upon '*Alltags-geschichte*' to this line of thought is self-evident. The underlying notion behind the whole concept of 'historicization' is that below the barbarism and the horror of the regime were patterns of social 'normality' which were, of course, affected by Nazism in various ways but which pre-dated and survived it. The role of Nazi ideology hence becomes 'relativized' in the context of a 'normality' of everyday life shaped for much of the time by non-ideological factors. Nazism can be seen to accelerate some and put the brake on other trends of social change and development which form a continuum from pre-Nazi times into the Federal Republic.[13] Beneath the barbarity, society in Nazi Germany can thus be more easily related to other eras in German history, and more easily compared with other contemporary societies. The long-term structural change and modernization of German society becomes thereby more explicable, as does the role of Nazism – deliberate or unwitting – in relationship to that change. This perspective challenges – and in some ways displaces – the traditional emphasis upon the ideological, political, and criminal terroristic aspects of Nazism. One of Broszat's critics has, for example, suggested that the approach which he is advocating looks to a comparison with the modernizing tendencies of other advanced western societies at the expense of neglecting the crucial differences in the essence of their development. From such a perspective, therefore, 'the racialist aspect . . . and particularly the "Final Solution of the Jewish Question" seem to be regarded as somehow irrelevant' since the 'unique duality' of the German modernizing experience is ignored.[14]

The suggested 'historicization' can, therefore, be summarized in the following claims: that Nazism should be subjected to the same methods of scholarly enquiry as any other era of history; that social continuities need to be much more fully incorporated in a far more complex picture of Nazism and the emphasis shifted away from heavy concentration upon the political-ideological sphere as a resort for moral lessons (since moral sensitivity can only arise from a deeper understanding, which 'historicization' offers, of the chequered complexities of the era); and that the Nazi era, at present almost a dislocated unit of German history, – no longer suppressed but reduced to no more than 'required reading' ('*Pflichtlektion*')[15] – needs to be relocated in wider evolutionary development.[16]

Criticism of 'Historicization'

The main critics of Broszat's 'historicization' plea are the Israeli historians Otto Dov Kulka, Dan Diner, and, especially, Saul Friedländer. They recognize the

[12] Broszat, *Nach Hitler*, pp. 171–2.
[13] For an excellent collection of essays summarizing much valuable research and locating Nazism within a context of long-term social change, see W. Conze and M.R. Lepsius, *Sozialgeschichte der Bundesrepublik Deutschland* (Stuttgart, 1983).
[14] Otto Dov Kulka, 'Singularity and its Relativization. Changing Views in German Historiography on National Socialism and the "Final Solution" ', *Yad Vashem Studies*, 19 (1988) pp. 151–86, here p. 170.
[15] Broszat, *Nach Hitler*, p. 161.
[16] See Saul Friedländer, 'Some Reflections on the Historization of National Socialism', *Tel Aviver Jahrbuch für deutsche Geschichte*, 16 (1987), pp. 310–24, here p. 313.

problem of 'historicization' as expounded by Broszat as an important methodological and theoretical issue, as representing in some respects a legitimate
perspective, and as raising a problem which 'belongs within the realm of a fundamental scholarly-scientific dialogue' between historians who 'share some
basic concerns as far as the attitudes towards Nazism and its crimes are concerned'. As such, they are anxious to distinguish it from the apologetics
advanced by Ernst Nolte in the '*Historikerstreit*'.[17] Even so, it is noted in passing that the exhortation, almost 50 years on, to treat the Nazi era like any other
period of history is also Nolte's starting point.[18] Leaving Nolte completely to
one side, there are still the implications of Andreas Hillgruber's approach to
the historical treatment of the German army on the Eastern Front for the concept of 'historicization', to which we will return.[19]

The most direct and structured critique of Broszat's 'historicization' plea has
been advanced by Saul Friedländer.[20] He sees three dilemmas in the 'historicization' notion, and a further three problems which the approach raises.

The first dilemma he points to is that of periodization and the specificity of
the dictatorship years themselves, the period 1933–1945.[21] The 'historicization' approach seeks to incorporate the Third Reich into a picture of long-term
social change. Broszat himself uses the example of the wartime social planning
of the German Labour Front both as an episode in the development of social
welfare schemes which pre-dated Nazism and extended into the modern system
of the Federal Republic, and as a parallel to what was taking place under
entirely different political systems, as in the British Beveridge Plan.[22] These
various long-term processes of social change, in this instance in social policy,
can be seen, therefore, as taking place in detachment from the specifics of Nazi
ideology and the particular circumstances of the Third Reich. The emphasis
shifts away from the singular characteristics of the Nazi period to a consideration of the relative and objective function of Nazism as an agent forcing (or
retarding) modernization.

The question of the intended or unintended 'modernization push' of Nazism
has, of course, been at issue ever since Dahrendorf and Schoenbaum wrote, as
we saw in the previous chapter. Friedländer accepts that recent studies have
extended knowledge on numerous aspects of this 'modernization'. However, in
his view, when taken as a whole such studies reveal a shift in interest from the
specificity of Nazism to the general problems of modernization, within which
Nazism plays a part. The issue is, therefore, one of 'the relative relevance'

[17] Friedländer, 'Reflections', pp. 310–11, 318; Kulka, 'Singularity and its Relativization' pp. 152,
167. The two contributions by Ernst Nolte, which were at the forefront of the '*Historikerstreit*' are
reproduced in '*Historikerstreit*'. *Die Dokumentation der Kontroverse um die Einzigartigkeit der
nationalsozialistischen Judenvernichtung* (Munich, 1987), pp. 13–35, 39–47.
[18] Friedländer, 'Reflections', pp. 317–18; Kulka, 'Singularity and its Relativization', pp. 167 ff.
[19] Friedländer, 'Reflections', p. 320; Dan Diner, 'Zwischen Aporie und Apologie', in Dan Diner,
ed., *Ist der Nationalsozialismus Geschichte? Zu Historisierung und Historikerstreit* (Frankfurt am
Main, 1987), pp. 62–73, here p. 66. An English translation is available in Baldwin, *Reworking the
Past*, pp. 135–44. The work referred to is the first essay ('Der Zusammenbruch im Osten 1944/45
als Problem der deutschen Nationalgeschichte und der europäischen Geschichte') of Andreas
Hillgruber's, *Zweierlei Untergang. Die Zerschlagung des Deutschen Reiches und das Ende des
europäischen Judentums*, Berlin, 1986.
[20] Friedländer, 'Reflections'.
[21] Friedländer, 'Reflections', pp. 314–16.
[22] Broszat, *Nach Hitler*, pp. 171–2.

of such developments in an overall history of the Nazi era.[23] And, in Friedländer's judgement, the danger – in fact, the almost inevitable result – is the relativization of the political–ideological–moral framework peculiar to the period 1933–45.[24]

The second dilemma arises from the recommended removal of the distance, founded on moral condemnation, which the historian of Nazism places between himself and the object of his research, and which prevents him from treating it as a 'normal' period of history. This raises, says, Friedländer, inextricable problems in the construction of a global picture of the Nazi era, since if few spheres of life were themselves criminal, few were completely untouched by the regime's criminality. Separation of criminality from normalcy is, therefore, scarcely an easy task. No objective criteria can be established for distinguishing which areas might be susceptible to empathetic treatment, and which still cannot be handled without the historian's distance from his subject of enquiry.[25]

The third dilemma arises from the vagueness and open-endedness of the concept of 'historicization', which implies a method and a philosophy but gives no clear notion of what the results might be. The implications of 'historicization' are, however, by no means straightforward, but might be interpreted in radically different ways – as indeed Nolte and Hillgruber demonstrate in their controversial interpretations of the Nazi era which provoked the 'Historikerstreit'.[26]

Friedländer is prepared to discount Nolte's writings in this context. But he uses the illustration of Hillgruber's essay on the Eastern Front to demonstrate the potential dangers of 'historicization', and links this squarely with the problems of the 'everyday history' approach itself, and with the open-ended nature of the 'Resistenz' concept used in the 'Bavaria Project'.[27] Not only the relativization of distance from the Nazi era, he argues, but also the emphasis in 'Alltagsgeschichte' on the ordinariness of many aspects of the Third Reich, on the non-ideological and non-criminal spheres of activity, and on ever more nuanced attitudes and behavioural patterns, creates significant problems. Friedländer accepts that 'criminality' is not necessarily excluded, and that a continuum can be constructed involving 'criminality' in everyday life and normality in the regime's 'criminal' system. However, he suggests that in an overall perspective of the Third Reich premised upon the relativization and normalization of the Nazi era advocated in the 'historicization' approach, the tendency

---

[23] Friedländer, 'Reflections', p. 315.
[24] Friedländer, 'Reflections', p. 314. Kulka's criticism in 'Singularity and its Relativization', pp. 168–73, runs along similar lines. Diner ('Zwischen Aporie und Apologie', p. 67) also criticizes the inevitable loss of the specifics of the period 1933–45 when, as in the 'Alltagsgeschichte' approach, the emphasis is placed on 'normality'. With reference to the oral history project directed by Lutz Niethammer on the experiences of Ruhr workers, he points out that 'the good and bad times' in subjective memory by no means accord with the significant developments of the period 1933–45. A 'considerable trivialization of the Nazi era' is allegedly the consequence. The reference is to Ulrich Herbert, 'Die guten und die schlechten Zeiten', in Lutz Niethammer, ed., 'Die Jahre weiß man nicht, wo man die heute hinsetzen soll.' Faschismuserfahrungen im Ruhrgebiet (Bonn, 1986), S.67–96.
[25] Friedländer, 'Reflections', pp. 316–17.
[26] Friedländer, 'Reflections', pp. 317.
[27] Friedländer, 'Reflections', pp. 317–21.

to overweight the 'normality' end of the continuum can scarcely be avoided. Despite Broszat's disclaimers, fears Friedländer, the passage from 'historicization' to 'historicism' ('*Historisierung*' to '*Historismus*') in regard to the Third Reich is a real danger.[28] Hillgruber defended his controversial empathizing and identification with the German troops in the east by comparing his approach with that of 'everyday history', as applied to other areas of research.[29] Accepting that there is some force in this defence, Friedländer suggests that one might justifiably apply the concept of '*Resistenz*' to the behaviour of the German soldiers defending the Eastern Front in the final phase of the war. Hence, many units were relatively immune to Nazi ideology and were only doing their job like soldiers in any army in defending the Front. On the other hand, of course, the *Wehrmacht* was system-supporting more than almost any other institution. This reveals to Friedländer not only that '*Resistenz*' is 'much too amorphous a concept to be of any great use',[30] but also the vacuous nature of 'historicization', which 'implies many different things' so that 'within the present context it may encourage some interpretations rather than others'.[31]

From the dilemmas arise, in Friedländer's view, three general problems. The first is that the Nazi past is still too overwhelmingly present to deal with it in the 'normal' way that one might, for example, tackle the history of sixteenth-century France. The self-reflection of the historian necessary to any good historical writing is decisive in approaching the Nazi era. The Third Reich simply cannot be regarded in the same way or approached with the same methods as 'normal' history.[32]

The second general problem is what Friedländer calls 'differential relevance'.[33] The history of Nazism, he says, belongs to everyone. The study of everyday life in the Third Reich may indeed be relevant to Germans in terms of self-perception and national identity, and thereby be a perspective which commends itself to German historians. But for historians outside Germany, this perspective might be less relevant in comparison with the political and ideological aspects of the Third Reich, and in particular the relationship of ideology to politics.

The same point is made in slightly different fashion by other critics of 'historicization'. Otto Dov Kulka sees the emphasis upon the 'normal' aspects of the Third Reich as a reflection of the present-day situation and self-image of the Federal Republic as an affluent, modern society – an image into which Nazi ideology and the 'criminality' of the regime can scarcely be accommodated. From this present-day West German perspective, he accepts the examination of, for example, long-term trends in the development of social policy as both justified and important. But the world-historical uniqueness of Nazism, he emphasizes, resides specifically in the duality of a society where 'normal' trends of modernization were accompanied by the slave labour and extermination 'in industrially rational fashion' of those ideologically excluded from the 'national

[28] Friedländer, 'Reflections', p. 318.
[29] Friedländer, 'Reflections', pp. 319–21; and see Diner, 'Zwischen Aporie und Apologie', pp. 66, 69.
[30] Friedländer, 'Reflections', p. 319.
[31] Friedländer, 'Reflections', p. 321.
[32] Friedländer, 'Reflections', pp. 321–2.
[33] Friedländer, 'Reflections', p. 322.

community'. And in the event of a victorious Third Reich, modern German society would have looked very different from the present democratic welfare state of the Federal Republic and of the socialist German Democratic Republic.[34]

The third – and most crucial – problem is, therefore, how to integrate Nazi crimes into the 'historicization' of the Third Reich. In Friedländer's view – and he accepts that this is a value-judgement – the specificity, or uniqueness, of Nazism resides in the fact that it 'tried "to determine who should and who should not inhabit the world" '.[35] The problem – and the limits – of 'historicization' lie consequently in its inability to integrate into its picture of 'normal' development 'the specificity and the historical place of the annihilation policies of the Third Reich'.[36]

Evaluation

The objections to the 'historicization of National Socialism' raised by Friedländer, Kulka, and Diner cannot lightly be dismissed. They touch upon important philosophical and methodological considerations which have a direct bearing on any attempt at writing the history of German society under Nazism.

Friedländer's concern about the omission or down-playing of the political, ideological, and moral aspects of Nazism permeates his critique. But it could at the outset be queried whether the traditional concentration on the political-ideological–moral framework could lead to further major advances in the depth of that understanding which provides the basis of enhanced moral awareness. This 'traditional' emphasis, epitomized perhaps most clearly in the work of Karl-Dietrich Bracher, produced many lasting gains.[37] A 'historicized' treatment would not need to discard them. But rigidly to confine scholarship to the traditional framework would be sterile and perhaps ultimately even counterproductive, since it would put a block on precisely the approaches which have led to much of the most original – and most morally sensitive – research in recent years. Moreover, the implications of 'historicization' might be less serious both in theory and in practice than Friedländer fears.

[34] Kulka, 'Singularity and its Relativization', pp. 169, 171–2, and as cited by Herbert Freeden, 'Um die Singularität von Auschwitz', *Tribüne*, 26, Heft 102 (1987), pp. 123–4.
[35] Friedländer, 'Reflections', p. 323. The phrase is taken from the closing lines of Hannah Arendt, *Eichmann in Jerusalem* (London, 1963), p. 256.
[36] Friedländer 'Reflections', p. 323. Diner ('Zwischen Aporie und Apologie', pp. 67–8, 71–3) is even more unyielding in his criticism, emphasizing the centrality of Auschwitz as a 'universalist point of departure from which to measure the world-historical significance of National Socialism', the impossibility of 'historicizing' Auschwitz, the diametrically opposed experiences of 'perpetrators' and 'victims', and the theoretical impossibility of combining the 'normality' experiences of the former and the experiences of the later of an 'absolutely exceptional situation' in one narrative history. He adds (p. 68) that any notion of 'daily routine' ('*Alltag*') has of necessity to begin from its conceptual opposite of the 'specifically exceptional'. Apparently accepting (p. 71) that some synthesis might after all be possible, he draws the conclusion with reference to the Holocaust that 'only proceeding from this extreme case allows one to make that divided simultaneity – divided, that is, by the close-up perspective into everyday history and mass murders – of the banality of the unreally shaped real normal situation on the one hand and its monstrous upshot on the other hand even approximately comprehensible'.
[37] Most classically in Karl-Dietrich Bracher, *The German Dictatorship*, New York, 1970.

It seems questionable whether the first dilemma posed by Friedländer –
the incompatibility of doing justice to the specific character of the Nazi era in
a treatment which concentrates upon the unfolding of long-term social
change – is a necessary one. It might, in fact, be countered that the specific
features of the period 1933–45 can only be highlighted by a 'longitudinal'
analysis crossing those chronological barriers and placing the era in a development context of elements of social change which long preceded Nazism and continued after its demise. Friedländer's fear is that there would be an inevitable
shift in focus to the problem of modernization, and that a 'relativization' of
the dictatorship era by its new location in a long-term context of 'neutral' social
change would be bound to lose sight of, or reduce in emphasis, crucial events
or policy decisions in the period of Nazi rule itself.

The fear does not appear to be borne out by recent works on social change,
some of which have adopted a long-term perspective and have deliberately
addressed the issue of modernization and the 'social revolution' argument.
Obviously, the 'criminal' side of the Third Reich is not the dominant focus in
such works. But in the stress of Nazi social policy, the significance of ideology
is by no means underplayed, and the relationship of this ideology to the core
racial–imperialist essence of Nazism is made abundantly plain. For instance,
the wartime social programme of Robert Ley – to take the example, from
Marie-Louise Recker's study of wartime social policy, which Broszat cites and
Friedländer sees as an example of the dangers implicit in 'historicization' –
indeed reveals a number of superficial similarities to Beveridge's social
insurance provisions in Britain. But what is most striking in Recker's
analysis – though, admittedly, not in Broszat's reference to her findings – is
the specific and unmistakable Nazi character of the programme.[38] Not only is
it legitimate (and necessary) to deploy a 'longitudinal' and also a comparative
perspective in analysis of Ley's programme, but such a perspective contributes
directly to a clearer definition of the peculiarly Nazi essence of social policy in
the years 1933–45. The same can be said of Michael Prinz's admirable analysis
of Nazi attempts to eradicate the status barrier between white- and blue-collar
workers, in which the long-term perspective serves to depict particularly clearly
both the specific features of Nazi social policy towards white-collar workers,
and the anchorage of this policy in Nazi ideological precepts.[39]

Applied to other subject areas, the 'longitudinal' approach highlights precisely the political–ideological–moral framework which Friedländer suspects
will be ignored or downplayed – if in ways different to, and often more
challenging than, the traditional approach. An instance would be Ulrich
Herbert's excellent analysis of the treatment of foreign labour in Germany since
the nineteenth century, which allows both the continuities which cross the Nazi
era, but also the specific barbarities of that era itself, to come more clearly into
view.[40] Herbert was, of course, a leading participant in the Ruhr oral-history
project which was so closely linked to perceived experiences of the 'normality'
of 'everyday life'. It is all the more significant, therefore, that he was the

---

[38] Marie-Louise Recker, *Nationalsozialistische Sozialpolitik im Zweiten Weltkrieg* (Munich
1985). See Broszat, *Nach Hitler*, p. 171.
[39] Michael Prinz, *Vom neuen Mittelstand zum Volksgenossen* (Munich, 1986).
[40] Ulrich Herbert, *Geschichte der Ausländerbeschäftigung in Deutschland 1880 bis 1980* (Bonn,
1986).

historian to contribute an outstanding monograph on foreign workers which offers the first major analysis of one of the most barbarous aspects of the Third Reich, and that he not only brings out fully the ideologically rooted nature of the regime's policy towards foreign workers, but also the extent to which 'racism was not just a phenomenon to be found among the Party leadership and the SS, . . . but a practical reality to be experienced as an everyday occurrence in Germany during the war'.[41]

The moral dimension is also more than evident in recent research on professional and social groups – such as the medical, legal, and teaching professions, technicians and students.[42] And there has been little difficulty in such studies in blending together long-term patterns of development and change (into which the Nazi era has to be fitted) and specific facets in such processes peculiar to Nazism. The same is abundantly true of research on the position of women. Continuities in anti-feminism have not prevented an elaboration of the specific contours of the 1933–45 era, as in Gisela Bock's work, for example, in which a direct association is made of Nazi anti-feminism and racial policy by way of an analysis of compulsory sterilization.[43] As in this instance, most other recent publications, many of them excellent in quality, on women in the Third Reich have placed particular emphasis upon the central issue of race – precisely the issue which Friedländer fears will lose significance through a social rather than political history perspective.[44]

It is difficult to see how any scholarly attempt to construct an overall picture of society under Nazism could ignore the findings of such important research. We still face, however, Friedländer's second dilemma: the inability of the historian, having removed the previously automatic 'distance' from Nazism, taken the epoch out of its 'quarantine', and abolished the 'syndrome of "required-reading" ',[45] to apply objective criteria to separate 'criminality' from 'normalcy' in the construction of a 'global' picture of the Nazi era.

Friedländer's worry is evidently that spheres of empathetic understanding might now be found in the 'normality' of everyday life under Nazism. The previous general consensus resting upon a total and complete rejection of this era would thereby be broken. But the historian, now faced with a choice other than rejection,[46] would have no objective criteria for drawing distinctions. In

[41] Ulrich Herbert, *Fremdarbeiter. Politik and Praxis des 'Ausländer-Einsatzes' in der.Kriegswirtschaft des Dritten Reiches* (Berlin/Bonn, 1985), back cover. See also Herbert's essay, 'Arbeit und Vernichtung. Okonomisches Interesse und Primat der "Weltanschauung" ', in Diner, ed., *Ist der Nationalsozialismus Geschichte?*, pp. 198–236.
[42] Not surprisingly, moral issues are particularly close to the surface in research, which has made considerable strides forward in recent years, on the place of the Third Reich in the professionalization of medical practice. For surveys of the literature see: Michael H. Kater, 'Medizin and Mediziner im Dritten Reich. Eine Bestandsaufnahme', *Historische Zeitschrift*, 244 (1987), pp. 299–352; and Michael H. Kater, 'The Burden of the Past: Problems of a Modern Historiography of Physicians and Medicine in Nazi Germany', *German Studies Review*, 10 (1987). See also the monographs by Robert Jay Lifton, *The Nazi Doctors* (New York, 1986) and Michael H. Kater, *Doctors under Hitler* (Chapel Hill/London, 1989).
[43] Gisela Bock, *Zwangssterilisation im Nationalsozialismus* (Opladen, 1986).
[44] See particularly Renate Bridenthal, Atina Grossman, and Marion Kaplan, eds., *When Biology became Destiny. Women in Weimar and Nazi Germany*, New York, 1984; and Claudia Koonz, *Mothers in the Fatherland. Women, the Family, and Nazi Politics* (New York, 1986).
[45] Friedländer, 'Reflections', p. 316.
[46] Some 25 years ago, Wolfgang Sauer pointed out that a characteristic feature of writing on Nazism was that the historian faced no other choice than rejection. (Reference in chapter 1 note 39.)

the context of the philosophy of 'historicism' (*'Historismus'*), and in the realm of pure theory, the problem of 'distance' or 'empathy', which Friedländer poses, does indeed appear insoluble. But even at the theoretical level, the problem is hardly peculiar to the Third Reich, and poses itself implicitly in all historical writing. In many areas of contemporary history in particular, one might think, the problem seems hardly less acute than in the case of Nazism. Whether the historian writing on Soviet society under Stalin, on the society of Fascist Italy or Franco's Spain, on the Vietnam war, on South Africa, or on British imperialism faces a fundamentally different dilemma might be questioned. Objective criteria resting on the historian's 'neutrality' arguably play no part in any historical writing. Selection on the basis of subjectively determined choices and emphases is inescapable. A rigorous critical method and full recognition of subjective factors shaping the approach deployed and evaluation of the findings provide the only means of control. The historian of Nazism is in no different position to any other historian in this respect.

Broszat's writings are in places certainly less clear and unambiguous than they might be on the difference between the method he advocates and the traditional or 'restored' historicism which he contrasts with it.[47] He explicitly presents 'distance' and intuitive insight or 'empathy' (*'Einfühlen'*) as opposites, and speaks of the possibility of 'a degree of sympathetic identification' (*'ein Maß mitfühlender Identifikation'*) both with victims and with 'wrongly invested achievements and virtues' (*'fehlinvestierten Leistungen und Tugenden'*). At the same time, however, he makes sufficiently plain that the counter to an uncritical, positive identification with the subject matter lies precisely in the critical historical method, applied to Nazism as to other periods of history, and ultimately promoting enhanced moral sensibility precisely through meticulous scholarship which includes empathy but does not uncritically embrace it.[48] The result is the methodological tightrope which all historians have to walk, in which the choice between empathy or moral distance is reshaped by the critical method into the position which characterizes a great deal of good historical writing – that of rejection through 'understanding'. This, the premise that 'enlightenment' (*'Aufklärung'*) comes through 'explanation' (*'Erklärung'*),[49] seems the basis of Broszat's approach in his collected papers, and certainly in his own work on the 'Bavaria Project' and elsewhere.

The best work arising from *'Alltagsgeschichte'*, in fact, clearly demonstrates that a concern with everyday behaviour and mentalities by no means implies empathetic treatment. Detlev Peukert's work, in which 'normality' is rooted in a theory of the 'pathology of modernity', provides an outstanding example.[50]

---

[47] See Broszat, *Nach Hitler*, pp. 120, 161, for the phrases cited in the following sentence, and pp. 100–1, 173 for comments on 'historicism' (*'Historismus'*).

[48] See, in particular, the essay: 'Grenzen der Wertneutralität in der Zeitgeschichtsforschung: Der Historiker und der Nationalsozialismus', in *Nach Hitler*, pp. 92–113.

[49] Broszat, *Nach Hitler*, p. 100. See also 'Briefwechsel', p. 340, where Broszat re-emphasizes his dependence upon a 'principle of critical, enlightening (*aufklärerischen*) historical understanding which . . . is to be clearly distinguished from the concept of understanding (*Verstehens-Begriff*) of German historicism in the nineteenth century . . .'

[50] See Detlev Peukert, *Volksgenossen und Gemeinschaftsfremde. Anpassung, Ausmerze und Aufbegehren unter dem Nationalsozialismus* (Cologne, 1982) (Engl. trans., *Inside Nazi Germany. Conformity and Opposition in Everyday Life*, London, 1987). Friedländer offers a qualified acceptance of the merits of *Alltagsgeschichte* in 'Briefwechsel' pp. 354–5, though this is far from satisfying Broszat (see 'Briefwechsel', pp. 362–3).

The dilemma posed by Friedländer is scarcely visible here. 'Everyday normality' is not presented as a positive counterpoint to the 'negative' aspects of Nazism, but as a framework within which 'criminality', arising from a 'pathological' side of 'normality', becomes more readily explicable. Nor is the concern that a continuum from 'normality' to 'criminality' inevitably means in practice that the dominant emphasis falls upon the former upheld in Peukert's work, which is all the more impressive in that he has offered so far practically the only wide-ranging attempt to synthesize research emanating from a wide variety of monographs falling within the 'history of everyday life' approach to German society in the Third Reich. And, though Peukert deliberately excluded it from his consideration in his book, there is no reason why the 'road to Auschwitz' could not be fully incorporated into an analysis premised on such an approach to 'normality'. By expressly linking 'daily life and barbarism', through association with the destructive potential built into modern society's emphasis upon advances in production and and efficiency, he himself indicated how an 'everyday history of racism', which is still in its beginnings, could contribute to a deeper understanding of the behaviour and mentalities which made the Holocaust possible.[51] Here, too, the dilemma of empathy or distance would be premised upon a false dichotomy and would not in practice present itself.

Friedländer's third dilemma arises from the vagueness and open-endedness of the term 'historicization', which is subject to different – some unattractive – interpretations. It can be readily conceded that 'historicization' is indeed an imprecise and unclear concept.[52] In some respects it is ambiguous if not outrightly misleading. The proximity of the term to 'historicism', which is the opposite of what it denotes, does not help clarity. And it seems related to 'normal' in at least three different ways: to the proposed 'normalization' of 'historical consciousness'; to the application of 'normal' historical method in approaching the Third Reich; and to the 'normality' of 'everyday life'. As an ordering or analytical concept, it has no obvious value, and is purely suggestive of a method of approach. The discarding of the term would arguably be no great loss. It confuses more than it clarifies. But the approach and method signified by 'historicization' could not be dispensed with. Even so, it would be necessary to distinguish the three different uses of 'normal'. The application of 'normal' historical method, and the extension of the sphere of analysis to the 'normality' of 'everyday life' can be more easily defended than can the inclusion of the Nazi era in a supposed 'normalization of historical consciousness'. This last usage, as the '*Historikerstreit*' has demonstrated, and as Friedländer and others fear, indeed appears either to elide the Nazi era altogether, or to erase or dilute the moral dimension by shifting the spotlight to parallel (and allegedly 'more original') barbarities of other 'totalitarian' states, particularly those of Bolshevik Russia. It is in the context of such distortions that Friedländer poses his third dilemma, by pointing to the use of the same term,

[51] Detlev Peukert, 'Alltag und Barbarei. Zur Normalität des Dritten Reiches', in Diner, ed., *Ist der Nationalsozialismus Geschichte?*, pp. 51–61, esp. pp. 53, 56, 59–61.
[52] See the comments of Adelheid von Saldern, which offer some support to Friedländer's objections, in her critique, 'Hillgrubers "Zweierlei Untergang" – der Untergang historischer Erfahrungsanalyse', in Heide Gerstenberger und Dorothea Schmidt, ed., *Normalität oder Normalisierung? Geschichtswerkstätten und Faschismusanalyse* (Münster, 1987), esp. pp. 164, 167–8. Broszat himself came to accept ('Briefwechsel', pp. 340, 361–2) that the 'historicization' concept is 'ambiguous and misleading'.

'historicization', in the context of an intended 'normalization' of historical consciousness in the face of a 'past which will not not pass away', by Nolte and, implicitly, by Hillgruber.[53]

The argument that the notion of 'historicization' advanced by Broszat, with its connotations of heightened moral sensitivity towards the Nazi past, might be misused 'in the present ideological context'[54] to result in the diametrically opposed 'relativization' of the regime's criminality, as in Nolte's essays which prompted the '*Historikerstreit*',[55] is certainly a serious criticism of the vagueness of the concept, but not convincing in itself as a rejection of the approach – largely based on an 'everyday history of the Nazi era' – which Broszat's concept is meant to denote.

If, however, as Friedländer himself suggests, Nolte's eccentric argumentation is left on one side, there still remains the question of Hillgruber's declared adaptation of the approach of '*Alltagsgeschichte*' to the problem of the troops on the Eastern Front, with the dubious conclusions he drew.[56] Friedländer astutely points out that the empathetic approach can produce startling results, and suggests that Hillgruber's essay demonstrates how Broszat's supposed 'historicization', aimed precisely at avoiding traditional 'historicism', can lead to a return of 'historicism', now dangerously applied to the Third Reich itself.[57] But the point about Hillgruber's essay is that it is squarely rooted in a crude form of the 'historicist' tradition which presumes that 'understanding' ('*Verstehen*') can only come about through empathetic identification. It is precisely the claim that the historian's only valid position is one of identification with the German troops fighting on the Eastern Front which invoked such widespread and vehement criticism of Hillgruber's essay.[58] The critical method, which in his other work – not excluding his essay on 'the historical place of the extermination of the Jews' in the same volume as the controversial treatment of the Eastern Front – made him a formidable historian whose strength lay in the careful and measured treatment of empirical data, entirely deserted him here and is wholly lacking in this one-sided, uncritical empathizing

---

[53] Friedländer, 'Reflections', pp. 317–21. Nolte's article, 'Vergangenheit, die nicht vergehen will', is in '*Historikerstreit*', pp. 39–47. Hillgruber's work referred to is the first essay in *Zweierlei Untergang*.

[54] Friedländer, 'Reflections', p. 324.

[55] See '*Historikerstreit*' pp. 13–35, 39–47. Klaus Hildebrand, for example, praised Nolte in a review for the way in which he undertook 'to incorporate in historicizing fashion (*historisierend einzuordnen*) that central element for the history of National Socialism and of the "Third Reich" of the annihilatory capacity of the ideology and of the regime, and to comprehend this totalitarian reality in the interrelated context of Russian and German history'. – *Historische Zeitschrift*, 242 (1986), p. 465.

[56] See Hillgruber's remarks in '*Historikerstreit*', pp. 234–5.

[57] Friedländer, 'Reflections', pp. 320–1. See the further debate between Broszat and Friedländer on this point in 'Briefwechsel', pp. 346, 355–6, 360–1.

[58] See Diner, 'Zwischen Aporie und Apologie', pp. 69–70, and von Saldern, 'Hillgrubers "Zweierlei Untergang" ', pp. 161–2, 168 for comments on Hillgruber's argument in the context of the 'historicization' problem. The most devastating critique of Hillgruber's position can be found in Hans-Ulrich Wehler, *Entsorgung der deutschen Geschichte?* (Munich, 1988), pp. 46 ff, 154 ff. See also the excellent review article by Omer Bartov (whose own book, *The Eastern Front 1941–45. German Troops and the Barbarisation of Warfare* (London, 1985), offers a necessary and an important counter-interpretation to that of Hillgruber): 'Historians on the Eastern Front. Andreas Hillgruber and Germany's Tragedy', *Tel Aviver Jahrbuch für deutsche Geschichte*, 16 (1987), pp. 325–45.

with the German troops. Though Hillgruber claimed to be applying the technique of 'Alltagsgeschichte' and the approach advocated by Broszat and others to experience events from the point of view of those at the base of society directly affected by them, it is precisely the absence of critical reflection which provides the gulf between his depiction and the work of Broszat, Peukert and others, who indeed looked to 'grassroots' experiences, but did not detach these from a critical framework of analysis.

The example of Hillgruber appears, therefore, misplaced. What, apart from the dubious value of the actual term 'historicization', it illustrates, is that, in his zeal to emphasize the need for greater empathetic understanding of 'experience', Broszat appeared to have posed a false dichotomy with the 'distance' which is an important control mechanism of the historian of any period, not just of the Nazi era. In reality, Broszat's own historical writing – even his last short book in a series founded on the necessity to 'historicize' German history – plainly did not abrogate 'distance' in the interests of uncritical empathy. Neither here, nor in Broszat's other recent writing, could it be claimed that the narrative approach ('Erzählen') which he missed in historical treatment of the Third Reich[59] had come to dominate or to replace critical, structured analysis and reflection. 'Distance' as well as empathetic understanding might be said to be vital to the historian of any period.

The preservation of a critical distance in the case of National Socialism is, in fact, far from being dispensable, a crucial component of the new social history of the Third Reich. But it is precisely the virtue of this new social history located in description and structured analysis of 'everyday' experience, that it breaks down the unreflected distance which has traditionally been provided by abstractions such as 'totalitarian rule' and compels a deeper comprehension through greater awareness of the complexity of social reality.[60] If I understand it correctly, this was the essence of Broszat's plea for 'historicization', and for a structured 'Alltagsgeschichte' as the most fruitful method of approach. And the findings of the 'Bavaria Project' alone demonstrate how enriching such an approach can prove.

It seems plain that Friedländer is correct to stress that the Nazi era, from whichever perspective it is approached, can not be regarded as a 'normal' part of history in the way that even the most barbarous episodes of the more distant past can be viewed. The emotions which rightly still colour attitudes to Nazism obviously rule out the detachment with which not only sixteenth-century France (Friedländer's example) but also many more recent events and periods in German history and in the history of other nations can be analysed. In this sense, Wolfgang Benz is quite right when he claims: 'Detached concern with Nazism as an era of German history among others and work on it devoted to purely scholarly interest seems then not so easily possible. The mere distance of 40 or 50 years does not yet make the Nazi era historical'.[61] But of course

---

[59] See Martin Broszat, Die Machtergreifung (Munich, 1984) (Engl. trans., Hitler and the Collapse of Weimar Germany, Leamington Spa, 1987). For Broszat's remarks on the concept behind the series Deutsche Geschichte der neuesten Zeit, see Nach Hitler, p. 152. And for his advocation of narrative (Erzählen) as historical method, see Broszat, Nach Hitler, pp. 137, 161.

[60] See Broszat, Nach Hitler, pp. 131–9, 'Alltagsgeschichte der NS-Zeit'.

[61] Wolfgang Benz, 'Die Abwehr der Vergangenheit. Ein Problem nur für Historiker und Moralisten?' in Diner, ed., Ist der Nationalsozialismus Geschichte?, p. 33.

this does not rule out the application of 'normal' historical method to the social, as well as to the political, history of Germany in the Nazi era. Even if a wide-ranging interpretative analysis of the Nazi era based on such methods will, as Benz adds, naturally be unable 'to do justice to the longing of the citizens of the post-war society to be released from the shadow of the past', this does not mean that it cannot be written.[62] And, while the historian's relationship to his subject of study is different in the case of Nazism than, say, in that of the French Revolution, it could be argued that, even accepting the uniqueness of the Holocaust, the problems posed by 'historicization' are little different in theory to those facing the historian of, say, Soviet society under Stalin.

Like the French and Russian Revolutions, the Third Reich embraces events of world-historical importance. Its history can certainly be approached as part of the pre-history of the Federal Republic (and of the German Democratic Republic), but, as Friedländer rightly says, 'the history of Nazism belongs to everybody'.[63] Perspectives inevitably vary. The polarization of German and Jewish collective memory of the Nazi era – epitomized in the films *Heimat* and *Shoah* – is plausibly advanced by Friedländer as an important element in the current debates about approaches to the Third Reich.[64] The differences in emphasis are unavoidable, and each has its own legitimacy. It is difficult to see how they can satisfactorily be blended together in any history which, purely or largely based upon the notion of 'experience' and constructed upon a narrative method (*'Erzählen'*), attempts a 'global' description of the Nazi era. Even if one suggests that in some ways the historian who shares neither collective memory possibly has an advantageous perspective, the attempt seems in any case bound to founder on the assumption that it is theoretically possible to write the 'total' history of an entire 'era' based upon collective 'experience'.[65] Equally impossible is the construction of a history built solely around the actions or 'experiences' of the 'historical actors' themselves and detached from the often impersonally structured conditions which in good measure shape or predetermine those 'experiences'.[66] Only the application of constructs, concepts, and even theories which reside outside the sphere of historical experience can provide order and make sense of experience in a historical analysis which is bound to be less than 'total' or 'global'.[67] If this appears to stand in con-

---

[62] Benz, p. 19. Norbert Frei's recent short book, *Der Führerstaat* (Munich, 1987), (Engl. trans., *Nazi Germany. A Social History* (Oxford, 1992)) offers some pointers towards the potential of such an approach.

[63] Friedländer, 'Reflections', p. 322.

[64] Saul Friedländer, 'West Germany and the Burden of the Past: The Ongoing Debate', *Jerusalem Quarterly*, 42 (1987), pp. 16–17. And see also 'Briefwechsel', pp. 346, 366–7, on the 'dissonance between memories'.

[65] See here the pertinent remarks of Wehler, 'Königsweg', p. 35. On the potential, but also substantial problems, of 'experience analysis' (*Erfahrungsanalyse*) with reference to the Third Reich, see von Saldern, 'Hillgrubers "Zweierlei Untergang" '. Friedländer emphasizes the limits of narrative as a method in 'Briefwechsel', pp. 370–1, while Diner ('Zwischen Aporie und Apologie', p. 67) is adamant that 'experienced everyday routine and existential exception can theoretically no longer be narrated as one history'.

[66] See Wehler, *Entsorgung*, p. 54, referring to the problems involved in Hillgruber's identification with the German troops on the Eastern Front.

[67] See the comments by Klaus Tenfelde and Jürgen Kocka in *Alltagsgeschichte der NS-Zeit*, pp. 36, 50–4, 63–4, and by Kocka – on the need for theory in '*Alltagsgeschichte*' – in a review in *taz*, 26 Jan. 1988 (see note 5, this chapter).

tradiction to Broszat's 'historicization' plea, it is scarcely out of synchroniza-
tion with his practice in his own writing on the history of the Nazi era.

If the assumption is abandoned that the history of the Nazi era (or any
other 'era'), in the sense of any 'total' grasp of the complexity of all the contradictory and often unrelated experiences which occur in a given period of
time, is theoretically and practically possible, then it becomes feasible to conceive of a history of German society under Nazism which could incorporate in
a structured analysis the findings of recent social historical research, in particular that of '*Alltagsgeschichte*', but which at the same time would embed this
in the political–ideological–moral framework which Friedländer is anxious not
to lose. Such an approach would have to jettison notions of the 'historicization'
of Nazism in terms of regarding it as any other period of history or 'relativizing'
its significance. But it would find indispensable the normal methodological
rigour of historical enquiry, deployed as a matter of course in dealing with other
eras (and already, one might add, deployed in countless scholarly works
on Nazism). Applied to the social sphere of 'daily life' as well as to the
political–ideological domain, conventional critical historical method would be
sufficient to eliminate the modern antiquarianism which has rightly been
criticized as a feature of the poorer strains of '*Alltagsgeschichte*'. Finally, it
would not only be legitimate, but essential, to proceed in such an approach by
way of a critical exploration of the continuum which stretches from 'normality'
to barbarism and genocide, in order better to comprehend the social as well as
political context in which inhumane ideologies become implemented as practical policies of almost inconceivable inhumanity. 'Auschwitz' would, therefore, inevitably form the point of departure from which the thin ice of modern
civilization and its veneer of 'normality' could be critically examined.[68]

The last, and ultimately fundamental, issue preoccupying Friedländer seems
resolvable in such an approach. The integration of Nazi crimes against
humanity into a 'global' interpretation of society in the Third Reich ought to
become, in fact, more rather than less possible in the light of the developments
made in the empirical social history of Nazism in the past decade. Peukert's
synthesis has, in many respects, pointed the way towards an integration of 'normality' and 'barbarism'.[69] I have attempted in my own work explicitly to
relate lack of humanitarian concern with regard to the 'Jewish Question' to
spheres of dissent and protest in 'everyday' matters.[70] My working hypothesis
in such research was the notion that, especially under 'extreme' conditions,
'normal' daily and private concerns consume such energy and attention that
indifference to inhumanity, and thereby indirect support of an inhumane
political system, is significantly furthered. Robert Gellately, building upon the
work of the late Reinhard Mann, has extended such suggestions to the areas of
social consensus and active support for 'policing' measures in racial issues.[71]

[68] See Peukert, 'Alltag und Barbarei', p. 61; and Diner, 'Zwischen Aporie und Apologie',
pp. 71–2.
[69] Peukert, *Volksgenossen und Gemeinschaftsfremde*. Engl. trans., *Inside Nazi Germany*, (see
note 50, this chapter); see also his 'Alltag und Barbarei' (see note 51, this chapter).
[70] Ian Kershaw, *Popular Opinion and Political Dissent in the Third Reich* (Oxford, 1983).
[71] Reinhard Mann, *Protest und Kontrolle im Dritten Reich* (Frankfurt am Main/New York,
1987); Robert Gellately, 'The Gestapo and German Society: Political Denunciation in the Gestapo
Case Files', *Journal of Modern History* 60 (1988), pp. 654–94; and, especially, *The Gestapo and
German Society. Enforcing Racial Policy* (Oxford, 1990).

To posit a clear divorce between the concerns of '*Alltagsgeschichte*' and the political–ideological–moral framework which focuses upon the genocidal criminality of the Nazi regime is to adopt a misleading perspective. Out of recent work on the social history of the Third Reich, which Broszat did more than most to promote, emerges the realization that there can be a social context in 'civilized society' in which genocide becomes acceptable. Research on the 'grassroots' history of the Nazi era has significantly deepened awareness of the troublesome reflection that 'many features of contemporary "civilized" society encourage the easy resort to genocidal holocausts'.[72]

[72] Leo Kuper, *Genocide* Harmondsworth, 1981, p. 137.

10

Shifting Perspectives: Historiographical Trends in the Aftermath of Unification

By the mid 1980s, the self-image of the Federal Republic of Germany had become increasingly schizoprenic: on the one hand, the material success-story of the post-war era – prosperous, stable, highly-developed; on the other hand, doomed it seemed for ever to live under the shadow of the crimes committed in Germany's name during the Third Reich. It was little wonder that conservative politicians and publicists came to feel it more and more necessary to draw a line under the Hitler era, to emerge – as one leading politician put it – from 'the shadows of the Third Reich', and be proud to be Germans again.[1]

Times have changed more rapidly than anyone in the mid 1980s could have imagined. Now, in a changed Germany within a transformed Europe, drawing the line under the Nazi past appears less easily possible. The reawakened problems of fascism, racism, and nationalism straddle the decades and ensure a continuing preoccupation with the Hitler era. Nazism truly remains 'a past which will not pass away'.[2]

Historical perspectives are, however, never static. They quite properly and naturally change over time. Those on the Nazi era, as was pointed out in chapter 1, are more than most affected by a variety of influences outside the strict bounds of historical scholarship. One substantial impact on historiography of the political changes in eastern Europe can be noted at the outset: the effective demise of marxist analyses of Nazism. These have lost, at best, a good deal of their former appeal; at worst they have lost credibility. The orthodox, strait-

---

[1] James M. Markham, 'Whither Strauß – Bavaria or Bonn? Premier Campaigns for "Emergence From Third Reich" ', *International Herald Tribune*, 15 Jan. 1987.

[2] The title of Ernst Nolte's essay which prompted the *Historikerstreit* ('historians' dispute') of 1986. See '*Historikerstreit*' (ref. in ch. 9, note 17), pp. 39–47. Extensive analyses of the *Historikerstreit* are provided by Hans-Ulrich Wehler, *Entsorgung der deutschen Vergangenheit? Ein polemischer Essay zum 'Historikerstreit'* (Munich, 1988), Charles Maier, *The Unmasterable Past: History, Holocaust, and German National Identity* (Cambridge, Mass., 1988), and Richard J. Evans, *In Hitler's Shadow. West German Historians and the Attempt to Escape from the Nazi Past* (New York, 1989), while important contributions and commentaries were made available in English in *Yad Vashem Studies*, 19 (1988), pp. 1–186 and Peter Baldwin, ed., *Reworking the Past. Hitler, the Holocaust, and the Historians' Debate* (Boston, Mass., 1990). The number of articles, commentaries, reports etc. on the *Historikerstreit*, within and outside Germany, runs into the hundreds. Bibliographical references to much of the important literature can be found in the notes to Wehler, pp. 212 ff., the commented guide to further reading in Evans, pp. 186–9, the bibliography in Baldwin, pp. 295–304, and in Geoff Eley, 'Nazism, Politics, and the Image of the Past. Thought on the West German *Historikerstreit* 1986–87', *Past and Present*, 121 (1988), pp. 171–208, here pp. 177–8 notes 12–13.

jacketed marxism-leninism which underpinned the official state ideologies in the GDR and other former Soviet bloc states falls into the latter category: it now finds few willing to defend its manifold and fundamental flaws as a theoretical framework of interpretation. It is rare to find historians of the Third Reich (or other historians) who regret its demise. More regrettable (though not every scholar would agree) is the current dismissiveness shown towards *all* variants of marxist analysis, even those (such as Bonapartist and Gramscian approaches) which, at the very least, have been intellectually fertile and heuristically stimulating approaches.

What this means is that the marxist contributions to the debates on Nazism which have been outlined in the chapters of this book no longer enjoy major currency, and that the continuing debates are all conducted within the framework of liberal historiography, now facing – for the first time since political and scholarly analysis of fascism began in the early 1920s – no serious challenge from a fundamentally opposed, alternative philosophy.

Of course, many of the genuine advances in scholarly understanding of the Third Reich which have been explored in the preceding chapters will stand the test of time, whatever the political climate. But this book began by commenting that past and present cannot be clinically separated, that conflicting interpretations of Nazism are inextricably bound up with the continuing reappraisal of the political identity of the Federal Republic and the changing ways in which it attempts to cope with the moral burden of the past. It would have been remarkable, therefore, had the unification of the divided Germany in 1990 left no influence on historiographical trends. In the midst of a vast transformation whose outcome we cannot foretell, with Germany under great internal strain and Europe in turmoil, it is possible only to hint at some changes which are becoming visible. But it appears that the issues which surfaced in the debates of the mid 1980s still, if in a different context, provide the framework for the current historiographical agenda.

The trends highlighted in what follows focus, firstly, on the ways in which perceptions of the location of Nazism in German history might be refashioned in the light of a changing sense of German identity since the recent unification; secondly, on the place of Nazism in the long-term modernization of Germany and how this could alter the perspective on Nazi barbarism; and, thirdly, on how the end of Soviet Communism could influence attitudes towards the horrors of the Third Reich. The link between the three trends selected for discussion is the continuing preoccupation – reflected in different ways – with 'historicization', which we encountered and explored in the previous chapter.

I

Nazism and National Identity

A key theme in the *Historikerstreit*, particularly in the contributions of Michael Stürmer, was the role of history in creating a positive sense of national identity, and the blockage imposed by the Third Reich on that identity.[3] The long span

---

[3] The writings of Michael Stürmer most relevant to the *Historikerstreit* include: 'Kein Eigentum der Deutschen: die deutsche Frage', in Werner Weidenfeld, ed., *Die Identität der Deutschen* (Munich/Vienna, 1983), pp. 83–101; *Dissonanzen des Fortschritts. Essays über Geschichte und*

of German history, rather than the negative concentration on the Third Reich, was in his view the key to finding that identity – an identity capable of uniting, not dividing and morally repelling. He spoke of the Germans in a divided Germany needing to find their identity, which had to be a national identity – though a national identity which, as it seemed at the time, had no prospect of deriving from a German nation-state.[4]

The diametrical opposite to this view was the 'critical history' approach, which we encountered in chapter 1, associated above all with Hans-Ulrich Wehler and Jürgen Kocka and intellectually dominant in the decade before the *Tendenzwende* – the political and intellectual challenge to the social-liberal values which had prevailed in the previous two decades or so – set in around the end of the 1970s and early 1980s. It was this approach – emphatic in its self-conscious, politically and morally informed critical approach to the national past, vehemently upholding a sense of post-nationalist identity formed through 'constitutional patriotism' and bonds with western liberal values, and methodologically wedded to applied social science techniques affording a comparative history of society – which was represented in the dispute by Jürgen Habermas.[5]

Application of the 'critical history' approach has not been confined to the Third Reich. In fact, more of the scholarly analyses and monographs influenced by this approach – and many of the most important works of its chief exponents – have focused on nineteenth- rather than on twentieth-century history.[6] However, implicitly if not explicitly, the search for an explanation of how Hitler's triumph in 1933 was possible has been central to the the work of the 'critical' historians. And the legacy of the Nazi abomination was pivotal to the socio-political as well as historical philosophy which underpinned the 'critical approach'. As Habermas expressed it, 'a commitment to universalist constitutional principles rooted in conviction has only been feasible in the cultural nation of the Germans after – and through – Auschwitz. . . . Anyone wishing to recall the Germans to a conventional form of their national identity destroys the only reliable basis of our bonds with the West'.[7]

The contrast between the two approaches to the German past could scarcely be plainer. An attempt to create a sense of national identity through an approach to the national past which does not attempt to conceal the crimes of

---

Politik in Deutschland (Munich, 1986); 'Geschichte in geschichtslosem Land', in '*Historikerstreit*', pp. 36–9; 'Was Geschichte wiegt', in '*Historikerstreit*', pp. 293–5; 'Weder verdrängen noch bewältigen: Geschichte und Gegenwartsbewußtsein der Deutschen', *Schweizer Monatshefte*, 66 (1986), pp. 689–94; 'Suche nach der verlorenen Erinnerung', in *Das Parlament*, 36 (1986), Nr. 20–21, 17–24 May 1986. On Stürmer's metamorphosis from former adherent of the 'critical history' school to publicist of German conservatism, see Volker R. Berghahn, 'Geschichtswissenschaft und Große Politik', *APZ*, B11/87, 14 Mar. 1987, pp. 25–37; Hans-Jürgen Puhle, 'Die neue Ruhelosigkeit: Michael Stürmers nationalpolitischer Revisionismus', *Geschichte und Gesellschaft*, 13 (1987), pp. 382–99; and Hans-Ulrich Wehler, *Entsorgung der deutschen Vergangenheit? Ein polemischer Essay zum 'Historikerstreit'* (Munich, 1988), pp. 28–36.
[4] Stürmer, 'Kein Eigentum der Deutschen', p. 98.
[5] Jürgen Habermas, 'Eine Art Schadensabwicklung. Die apologetischen Tendenzen in der deutschen Zeitgeschichtsschreibung', in '*Historikerstreit*', pp. 62–76 (Eng. trans., 'A Kind of Indemnification', *Yad Vashem Studies*, 19 (1988), pp. 75–92.
[6] See, for an overview, Roger Fletcher, 'Recent Developments in West German Historiography: The Bielefeld School and its Critics', *German Studies Review*, 7 (1984), pp. 451–80.
[7] Habermas, 'Eine Art Schadensabwicklung', in '*Historikerstreit*', pp. 75–6.

Nazism but to transcend them by 'historicizing' them into a longer-term and wider perspective embracing a multiplicity of facets of national history is confronted by an approach which sees Auschwitz as the essential starting-point of all that is positive in a post-national form of identity.

At the time of the *Historikerstreit*, many thought that the critics of the 'revisionist' positions of Stürmer, Hillgruber, and Nolte had enjoyed the better of the rancorous exchanges. German unification has, however, inevitably altered perspectives on national identity and, therefore, on how the national past could be viewed. The effect has been, it appears, to place the 'critical history' approach even further on the defensive than it had already become during the *Tendenzwende* of the 1980s and to offer greater encouragement than was ever possible during the 1960s and 1970s to an emphasis upon German nationality as the cornerstone of historical analysis – with the dangers of a neo-historicism which that entails.[8]

The events of 1989–90 gave German conservatives the opportunity to marry what for four decades had seemed irreconcilable alternatives: the Adenauer heritage of the bonds with the West and the accomplishment of national unity. Since then, the deep divisions, problems of integration, and 'identity crisis' within the new German national state formed by the incorporation of the GDR in the Federal Republic have seemed to lend new urgency – and far more obviously than had been the case in the mid 1980s – to attempts to place the emphasis upon national history as a foundation of national identity and, eventually, cultural and political unity. That an attempt to create national identity through history is as philosophically flawed as it is ideologically tendentious will presumably provide no barrier to this shift in perspective.[9]

Different ways of regarding the German past opened up with the fall of the Berlin Wall and the subsequent unification of the two Germanies. Since the War, it had only been possible to see national history in terms of the temporary, imperfect, and ill-fated unity of the German Reich, leading within little over 70 years to catastrophe and seemingly permanent division. The events of 1989–90 brought not only the Cold War and, with it, the post-war era, to a close. They also seemingly restored to Germany the 'normality' of the existence of a nation-state; as Saul Friedländer put it, they have 'given back national continuity to German history'.[10] The presumed German '*Sonderweg*' ('special path') could

---

[8] In a percipient article published in 1984 (see ch 1 note 23 for the full reference), Irmline Veit-Brause already saw the wind of change blowing in the replacement of the 'critical history' approach by the new 'paradigm' of 'national identity'. Hans-Ulrich Wehler's deep antagonism to the shift in focus is shown in his attack on Stürmer's approach in *Entsorgung*, pp. 69–78, 138–45, 171–89, and in his ferocious response to an article by Harold James ('Die Nemesis der Einfallslosigkeit', *Frankfurter Allgemeine Zeitung*, 17 Sept. 1990) suggesting that national myths were needed to compensate for material dissatisfaction and could help to create stability. See Hans-Ulrich Wehler, 'Aufforderung zum Irrweg: Wiederbelebung des deutschen Nationalismus und seiner Mythen', *Der Spiegel*, 24 Sept. 1990, and 'Welche Probleme kann ein deutscher Nationalismus heute überhaupt noch lösen? Wider die Apostel der nationalen "Normalität": Der Verfassungs- und Sozialstaat schafft Loyalität und Staatsbürgerstolz', *Die Zeit* (24 Sept. 1990); and also his review of Harold James's book, *A German Identity* (London, 1989), 'Im Irrgarten des ökonomischen Determinismus', *Die Zeit*, 11 Oct. 1991.

[9] For powerful objections to the presumption that identity can be exclusively drawn from history, see Maier, *The Unmasterable Past*, pp. 149–56.

[10] Saul Friedländer, 'Martin Broszat und die Historisierung des Nationalsozialismus', in Klaus-Dietmar Henke and Claudio Natoli, eds., *Mit dem Pathos der Nüchternheit. Martin Broszat, das*

now be seen to be at an end.[11] The development of the nation-state had no longer been abruptly truncated and irrevocably terminated by the split of the nation into two states. The future might look bleak; but there *was* a future for the nation, and that future was open. History had not been closed off by war and genocide. Hitler's legacy had not, after all, been the end of the German nation, merely a lengthy interruption to the 'normality' of national unity. Within this changed perspective, it is plain, Auschwitz could no longer continue to serve as the reference point for post-war identity, as Habermas had wanted it to remain.

An obvious problem, nevertheless, is the kind of 'normality' which *could* serve as a reference point for national identity. The Reich that existed between 1871 and 1945, can scarcely offer a model; yet it is the only previous experience of a German nation-state. In his latest book, Michael Stürmer poses alternative vantage points, looking to long-term developments since the Thirty Years' War of the seventeenth century for 'national and trans-national traditions and patterns worth cherishing'. These include, in his view, federalism, religious tolerance, civic institutions, and balance between centre and periphery. He regards an emphasis upon such traditions as offering potential for the creation of a German sense of historical identity within an increasingly closely-knit (western) Europe.[12]

Whether such cultural currents are sufficient to outweigh the ideological baggage of the historical nation-state in shaping new forms of German identity might be doubted; and whether Stürmer's vision of the new Europe will come to pass is at present no more than a matter for speculation. But his 'bid to prevent Hitler remaining the final, unavoidable object of German history, or indeed its one and only starting point'[13] is both understandable and seems likely, with the passage of time in post-unification Germany, to be increasingly successful.

Whatever aspirations to national identity exist today, they contain only in the strident tones of the radical Right discordant echoes of the nationalist-chauvinist clamour of the 1871–1945 era. In a world of nation-states – whatever the post-national wishes of those, in Germany as elsewhere, who strive for closer European unity – aspiring to a stronger sense of German national identity is in itself certainly not unnatural and must not necessarily prove unhealthy. But German history provides few models for constructing a national identity. Such a presumed German 'national identity' has not been a historical constant, but is a product of only the last two centuries, has fluctuated greatly over time, and been based on shifting sets of 'German' territorial borders since the Napoleonic conquests.[14] Nor, arguably, can any common identity – the prospect of which at the time of writing seems to be receding

---

Institut für Zeitgeschichte und die Erforschung des Nationalsozialismus (Frankfurt am Main/New York, 1991), pp. 155–71, here p. 159.

[11] The debate on the *Sonderweg* can be followed from the references in ch. 2 note 2.

[12] Michael Stürmer, *Die Grenzen der Macht. Begegnung der Deutschen mit der Geschichte* (Berlin, 1992).

[13] Michael Stürmer in an interview with David Walker, *The Times Higher*, 24 July 1992.

[14] See now on this the excellent introduction by John Breuilly, 'The National Idea in Modern German History' in John Breuilly, ed., *The State of Germany. The National Idea in the Making, Unmaking, and Remaking of a Modern Nation-State* (London, 1992).

rather than advancing, given the internal divisions and deep-rooted cultural dissonances of the new Germany – be an artificial product resting solely, or even mainly, upon historical perspectives but will grow, if at all, organically over time out of common cultural experience and common political and social institutions.

Historiographically, however, the 'paradigm shift' in perspective means not only 'historicizing' the Third Reich in the long span of German history and ceasing to treat it as the central point or even end-point of that history. It also means that some elements within the history of the Third Reich become emphasized more than others, as they fit better into the changed perspective. Some implications of this were already evident in the debate over Broszat's 'historicization plea'.[15] The changing perspective comes out particularly clearly in the way 'historicization' is used to treat the question of Nazism and modernization.

II

Nazism and Modernization

When we explored the question of Nazism and modernization in chapter 7, we saw that the central areas of debate – leaving aside the marxist rejection of the modernization question in its entirety – were whether the Nazi regime had, despite an anti-modern ideology, unwittingly brought about Germany's 'revolution of modernity', or whether it amounted to social reaction. The modernization theme was then taken up in a different way by Broszat in his 'historicization plea' through the suggestion that the social planning of the German Labour Front could be viewed, in a sense, in detachment from the specifics of Nazi ideology and the particular circumstances of the Third Reich, as an episode in the development of social welfare schemes which pre-dated Nazism and extended into the modern system of the Federal Republic, and as a parallel to what was taking place under entirely different political systems, as in the British Beveridge Plan.[16]

The suggestion that the phenomenon of Nazism could be better understood by locating it in the continuities of German development which extended over and beyond the Third Reich, by looking away from the barbarities long established as its hallmark to underlying social 'normality', has been avidly taken up by a number of, predominantly younger, German scholars, and 'historicization' has provided the cue for examining Nazism and the links with modernization after 1945 in a variety of new ways.[17] But underlying them now is an assumption which differs sharply from earlier treatment of the 'modernization' theme: the claim that the Nazi leadership not only brought about a

---

[15] Broszat, *Nach Hitler*, pp. 159–73 (ref. in ch. 9 note 1).
[16] Broszat, *Nach Hitler*, pp. 171–2. The comparison with the Beveridge Plan, which Broszat suggested, was apparently first made by Hans-Günther Hockerts, 'Sicherung im Alter. Kontinuität and Wandel der gesetzlichen Rentenversicherung 1889–1979', in Conze and Lepsius, *Sozialgeschichte der Bundesrepublik*, p. 309. It has been more recently repeated in Ronald Smelser, *Robert Ley. Hitler's Labour Front Leader* (Oxford/New York/Hamburg, 1988), p. 307.
[17] See Michael Prinz and Rainer Zitelmann, eds., *Nationalsozialismus und Modernisierung* (Darmstadt, 1991).

modernizing revolution in Germany, but, in fact, *intended* to do so.[18] Claims that Mussolini's regime (among other fascist-style dictatorships) was a modernizing dictatorship are, as we noted in chapter 2, not new, though they have frequently encountered the criticism that they ignore the essence of fascism by concentrating on the by-product of modernization. In the case of Nazism, that criticism can be reinforced. Placing the emphasis upon 'modernization' unavoidably leads to a shift in perspective on the Third Reich.[19] In turn, this change in perspective can rapidly engender a trivialization of Nazism, whose crimes are not ignored, but nonetheless largely taken for granted and displaced by the image of the Third Reich as an important era of modernization in the long-term historical treatment of German national development.

This new approach to Nazism and modernization, and its implications for the way the Third Reich is viewed, have been strongly influenced by the work of the young Berlin political scientist Rainer Zitelmann and come out plainly in his study of Hitler's social ideas, published in 1987.[20] Zitelmann explicitly regarded his book as a contribution to that 'historicization' of National Socialism for which Broszat had pleaded. Young Germans, Zitelmann argued, had hitherto been faced only with stark alternatives, both unacceptable, and both rendering Hitler and the generation which supported him utterly incomprehensible: either total moral condemnation – a demonization of Hitler which turned him into the incarnation of evil – or apologetics and distortion of historical reality. His study of Hitler's social aims and philosophy was an attempt to overcome such incomprehensibility and to break down the sense of distant unreality about the Nazi regime and its leader.[21]

Concentrating not, in the conventional fashion, on Hitler's anti-semitic and *Lebensraum* obsessions (which he took as read), Zitelmann saw a logical cohesion to the German dictator's views on social and economic matters and was not dismissive, as most historians have been, of such views. Not only were Hitler's ideas (within the context of his racist-darwinist philosophy) coherent; they were, Zitelmann argues, in many senses distinctly 'modern'. Hitler was not looking backwards (as were Darré and Himmler) to the recreation of an agrarian wonderland, but forwards to a highly developed, advanced industrial and technological society – resting of course on supplies of raw materials and forced labour extracted from conquered territories, but modern for all that. The decadent bourgeoisie would be replaced by upwardly-mobile workers, with abundant chances of enhanced status and opportunities for social advancement. Industry would fall into line or be taken over by the state. A planned economy would in any case follow after the war. The model, as the brutal agent

[18] See Uwe Backes, Eckhard Jesse and Rainer Zitelmann, eds., *Die Schatten der Vergangenheit. Impulse zur Historisierung des Nationalsozialismus* (Frankfurt am Main/Berlin, 1990), pp. 42–3.
[19] For powerful rejections of the interpretation of the Third Reich as a form of 'modernizing dictatorship, see Jens Albers, 'Nationalsozialismus und Modernisierung', *Kölner Zeitschrift für Soziologie und Sozialpsychologie*, 41 (1989), pp. 346–65; and Hans Mommsen, 'Nationalsozialismus als vorgetäuschte Modernisierung', in Walter H. Pehle, ed., *Der historische Ort des Nationalsozialismus. Annäherungen* (Frankfurt am Main, 1990), pp. 11–46; see also Manfred Rauh, 'Anti-Modernismus im nationalsozialistischen Staat', *Historisches Jahrbuch*, 108 (1987), pp. 94–121.
[20] Rainer Zitelmann, *Hitler. Selbstverständnis eines Revolutionärs* (Hamburg/Leamington Spa/New York, 1987).
[21] Zitelmann, *Hitler*, p. 20.

of a modernizing dictatorship, was Stalin, whom Hitler admired. Instead of regarding Hitler's social ideas, whatever coherence one attaches to them, as a means to the end of racial purification and conquest, Zitelmann came close to inverting the order by seeing the racial programme as the means to bring about revolutionary plans for a transformation of German society by a dictator who saw himself – and deserves to be treated by historians – as a social revolutionary.[22]

Since the publication of the book which made his name, Zitelmann, an enormously productive writer, has argued in much the same vein in a plethora of essays, reviews, and newspaper articles. He has also been extremely active in stimulating other young scholars to collaborate on essay collections framed around the themes of modernization and 'historicization'. The core of Zitelmann's argument is that 'modernization' ought to be decoupled from any normative links with 'progress', humanitarian values, pluralist participatory political systems, and democratization, and seen as 'value-free' – simply a pragmatic tool of empirical investigation and scholarly analysis. It then becomes perfectly possible, he suggests, to speak of modernization taking place (intentionally, not just 'accidentally') in totalitarian states, as well as in liberal systems.[23] This was self-evidently the case under Stalinism (a point which most historians would readily accept), and, on analysis of the thought (and practice) of Hitler and other Nazi leaders such as the Labour Front boss Robert Ley or Albert Speer, the conclusion should to be reached, he concludes, that Nazism not only unwittingly contributed towards modernization in Germany, but intended to bring it about and strove to do so.[24]

Powerfully argued though the case is, it seems to contain both methodological and conceptual flaws. As regards method, a kind of coherence in Hitler's social views can be found by taking his comments on 'social matters' from speeches in the early 1920s through to monologues in his field headquarters in the war, and piecing them together. But this approach pays little attention to the precise context or intended function of Hitler's comments and is in danger, therefore, both of exaggerating the coherence and, especially, of elevating the importance of such ideas within Hitler's *Weltanschauung*. In the comprehensive collections of Hitler's speeches and writings between 1919 and 1928 which are now available,[25] for instance, the vision of a modern society is largely confined to the monotonous repetition of the aim to wipe out the distinction between 'workers of the brain' and 'the fist' by the creation of a 'national community' based on racial purity, principles of struggle, and strength to ensure survival through territorial conquest attained by the sword.

[22] The interpretation summarized in the above paragraph is advanced consistently throughout Zitelmann's study with the main points brought together in the conclusion, pp. 453–66.
[23] See Rainer Zitelmann's essays, 'Nationalsozialismus und Moderne. Eine Zwischenbilanz', in Werner Süß, ed., *Übergänge. Zeitgeschichte zwischen Utopie und Machbarkeit* (Berlin, 1990), pp. 195–223; and especially 'Die totalitäre Seite der Moderne', in Prinz and Zitelmann, *Nationalsozialismus und Modernisierung*, pp. 1–20.
[24] Zitelmann, *Hitler*, p. 7; 'Nationalsozialismus und Moderne', pp. 221, 223; 'Die totalitäre Seite der Moderne', pp. 12–20. Smelser, *Robert Ley*, p. 305 and note 3, saw his study of the Labour Front leader 'dovetailing' into Zitelmann's interpretation of Hitler as envisaging 'the Nazi revolution as a major tool to push Germany forward into a more modern society'.
[25] Eberhard Jäckel and Axel Kuhn, eds., *Hitler. Sämtliche Aufzeichnungen 1905–1924* (Stuttgart, 1980); and Clemens Vollnhals and Bärbel Dusik, eds., *Hitler. Reden, Schriften, Anordnungen Februar 1925 bis Januar 1933*, 3 vols. (Munich, 1992).

The 'social vision' – essentially an offshoot of the preoccupation with the 'living space question', obsessive anti-semitism, and pervading racial philosophy which dominate the speeches and writings – is a primitive derivative of nineteenth-century racialist and social-darwinist ideologies, not a blueprint for 'modernization'. It is difficult, in particular, to accept Zitelmann's inversion of means and ends, reducing the obsessions with destroying the Jews and the acquisition of 'living space' at the expense of the Soviet Union to the functional purpose of the revolutionary modernization of German society.

Conceptually, the attempt to remove all normative connotations from modernization and to treat the term as value-free makes it analytically as good as unusable.[26] Of course, it is possible to *describe* 'modernizing' elements in Nazism – which, in fact, have seldom been denied in the literature on the Third Reich. But it cannot suffice *analytically*, in a thesis of Nazism as an intentionally modernizing dictatorship, to evade a definition of 'modernization' or 'modernity' on the grounds that 'modernity' is still too recent to distinguish what is 'normal' and 'exceptional' about it, or that, 'in the present state of discussion', a general definition applicable to the time in question would be 'extraordinarily difficult'.[27] In the absence of any attempt at definition, what the modernizing elements present in the Nazi era ought to signify is difficult to see. In any event, as Charles Maier has pointed out, 'a modern labour market imposes structural demands even on governments that have murderous agendas', so there are scant grounds for surprise at many of the 'modernizing' elements in Nazism. 'What are morally significant', Maier adds, 'are the few institutions that were murderous, not the many normal aspects of running a society'.[28]

The main problem with Zitelmann's approach to the alleged 'modernizing intentions, of the Hitler regime is that it comes close to substituting the accidental for the essential in Nazism as a *historical* phenomenon – that is, for Nazism as it actually developed.[29] Also the worry is – as Saul Friedländer has remarked – that, in the new Germany and with the passage of time gradually eliminating those with living experience of the Third Reich, the perspective could become indelibly shifted from the unique characteristics of the 1933–45 era to the more 'comprehensible' – because more 'normal' – elements which can be regarded as the part of the 'pre-history' of the Federal Republic.[30]

[26] For arguments, which I support, that an explicitly stated normative usage, clear definition, and precise instrumentalisation are prerequisites for the deployment of the 'modernization' concept as an analytical device, see the contributions by Wolfgang J. Mommsen, Jürgen Kocka, and Hans-Ulrich Wehler to the discussion following the presentation by Matzerath and Volkmann (reference in ch. 2 note 53), pp. 107, 111–16. I pointed out some reservatons about its applicability to the Third Reich in ch. 7. However, the central argument of Matzerath and Volkmann (cited in ch. 7), that, as regards the 'modernization issue', National Socialism was an attempted special route out of a modernization crisis which led into a disastrous blind alley, is one that I share. See, for a similar argument, the penetrating assessment by Gerald Feldman, 'The Weimar Republic: A Problem of Modernization?', *Archiv für Sozialgeschichte*, 26 (1986), pp. 1–26.

[27] See the introduction to Prinz and Zitelmann, *Nationalsozialismus und Moderne*, p. x, and Zitelmann's own contribution, 'Die totalitäre Seite der Moderne', p. 11.

[28] Maier, *The Unmasterable Past*, p. 96.

[29] A good corrective, in concentrating on the racial essence of Nazism, is provided by Michael Burleigh and Wolfgang Wippermann, *The Racial State. Germany 1933–1945* (Cambridge, 1991).

[30] See Friedländer, 'Martin Broszat und die Historisierung des Nationalsozialismus', pp. 161–2, 168–72. For some perceptive and sensible remarks on how the continuities across 1945 might be approached, see the interesting review article by Harold James, 'The Prehistory of the Federal Republic', *JMH*, 63 (1991), pp. 98–115.

Zitelmann has remarked that Broszat's 'historicization plea' had a 'liberating effect' on himself, and on other younger colleagues.[31] His approach reveals, however, that the vague and intepretationally open-ended concept of 'historicization' can lead in directions never intended by Broszat.[32] There is, of course, no suggestion that Zitelmann's motives in advocating a radical break with the way Nazism has conventionally been treated in scholarship, teaching, and public discourse are anything other than scholarly and honourable. And he has much which is correct and important to say about the attractiveness of Nazism to the German population, and the reasons why it could have such drawing power – inexplicable simply on grounds of the persecution of the Jews, the apparatus of repression, or other aspects of the Third Reich which have rightly been central themes of historical research.[33] But when he rhetorically asks: 'How heavily, as opposed to the atrocities perpetrated against the Jews and other minorities, are social-political progress and increased upward mobility chances for "people's comrades" to be weighted?', and 'can one, given the sufferings of the victims, speak at all of those sides of reality which many people have experienced as positive?';[34] and when we take into account that even before the Berlin Wall fell a third of the population of the Federal Republic thought the Third Reich was too negatively depicted in school lessons, while more than two-thirds of those questioned were in favour of drawing a line under the Nazi past;[35] then the implicit tendency is plain to see.

III

Nazism and Stalinism

A third way in which approaches to the Third Reich is being affected by the changes in Europe since 1989 has different links with the notion of 'historicization'. In sharp distinction to Broszat's use of this concept – and demonstrating once again how unsatisfactory is the woolliness of the term – Ernst Nolte, in the *Historikerstreit*, had sought a rethinking of Nazism's place in German history by regarding it as the reaction and counterpoint to Soviet communism in the 'European civil war' between 1917 and 1945.[36] Nolte's line of argument

---

[31] Rainer Zitelmann, 'Vom Umgang mit der NS-Vergangenheit', in Rolf Italiaander *et al.*, eds., *Bewußtseins-Notstand. Ein optimistisches Lesebuch* (Düsseldorf, 1990), pp. 69–79, here p. 76.

[32] See the critical review of the book co-edited by Zitelmann, *Die Schatten der Vergangenheit* (reference in note 18, this chapter) by Norbert Frei, 'Die neue Unbefangenheit. Oder: Von den Tücken einer "Historisierung" des Nationalsozialismus', *Frankfurter Rundschau*, 5 Jan. 1991.

[33] See here, for example, his comments in 'Vom Umgang mit der NS-Vergangenheit', p. 70.

[34] Zitelmann, 'Vom Umgang mit der NS-Vergangenheit', p. 72.

[35] Zitelmann, 'Vom Umgang mit der NS-Vergangenheit', p. 70.

[36] Ernst Nolte, 'Vergangenheit, die nicht vergehen will', in '*Historikerstreit*', pp. 39–47, originally published in *Frankfurter Allgemeine Zeitung*, 6 June 1986. Some of the more contentious assertions, and a more extended version of the basic argument, were contained in Nolte's earlier essay, 'Zwischen Geschichtslegende und Revisionismus?', in '*Historikerstreit*', pp. 13–35. This latter essay was, in fact, the basis of a lecture delivered by Nolte as far back as 1980. It was published in abbreviated form the same year in the *Frankfurter Allgemeine Zeitung* under the title 'Die negative Lebendigkeit des Dritten Reiches. Eine Frage aus dem Blickwinkel des Jahres 1980' (see '*Historikerstreit*', p. 35), and was subsequently included in revised form (and with editorial interpolations) in English translation, entitled 'Between Myth and Revisionism? The Third Reich in the Perspective of the 1980s', in H.W. Koch, ed., *Aspects of the Third Reich* (London, 1985), pp. 17–38. The full statement of his thesis appeared in Ernst Nolte, *Der europäische Bürgerkrieg*

prompted bitter debate about the singularity of the Nazi genocide against the Jews, and the extent to which it could be seen as comparable to other twentieth-century genocides or even as a response to the Bolshevik 'class genocide' in the Russian civil war. For Nolte, the fates of the Soviet Union and Germany were, therefore, historically intertwined through the 'European civil war': the fight to the last of opposite, but related, ideologies. In his assertion that Nazism was *reactive*, a preventive attempt to stave off destruction by the equal, if not greater, evil of Soviet Bolshevism, Nolte was coming close to turning the Nazis' own justification of the war they launched into scholarly interpretation. But the furore he instigated certainly succeeded in turning the spotlight in a new way on to the intertwined history of Bolshevism and Nazism and the ideological origins of the genocidal war in the Soviet Union.

This theme, again against the backcloth of the notion of 'historicization', but from a wholly different perspective, was soon to be at the centre of another controversial study. In his book on the 'Final Solution' published just before the events of 1989, in which he claimed he was attempting to 'historicize' the 'Judeocide' (more usually called 'the Holocaust'), the American historian Arno Mayer – whose leftist inclinations set him poles apart from Nolte's stance – also took the German-Soviet relationship as the intrinsic element in what he described as a second 'Thirty Years War'.[37] But his approach was diametrically opposed to Nolte's. There was no suggestion of a 'preventive' attempt to stave off destruction by the Bolsheviks. Rather, Mayer saw the German invasion of the Soviet Union and the war of annihilation which followed as an ideological crusade inspired by a widespread and deep-rooted morbid fear of Bolshevism long prevalent in Germany's bourgeoisie and ruling classes and easily able to marry with and subsume paranoid Nazi images of 'Judeo-Bolshevism'. Far from presuming Nazism to be a reaction to prior Bolshevik barbarity, he laid the stress on unprovoked extreme and lethal anti-Bolshevism (extending way beyond Nazi hard-core support) as the prime, more extensive motivator, and interpreted the genocide against the Jews as arising from the war rather than being planned long before it.

For Mayer as for Nolte, therefore, the war with the Soviet Union and the ideological clash between Nazism and Bolshevism formed the core of any attempt at historical understanding of the Nazi phenomenon. In both cases, but from entirely different starting-points and on the basis of opposing interpretations, the emphasis, as Peter Baldwin summed up, had 'shifted away from the Jews to the Soviets'. But whereas 'for Nolte, the Bolsheviks were the main aggressors', for Mayer, they were 'the primary victims'.[38]

1917-1945. Nationalsozialismus und Bolschewismus (Berlin, 1987), published about a year after the '*Historikerstreit*'. Nolte's defence of his position in *Das Vergehen der Vergangenheit. Antwort an meine Kritiker im sogenannten Historikerstreit* (Berlin, 1987), further embittered the controversy, producing allegations that he had deliberately distorted (by brief and misleading paraphrasing) the content of lengthy and highly critical letters received from the Israeli historian Otto Dov Kulka, with the effect of making them appear relatively favourable to his own position. See Otto Dov Kulka, 'Der Umgang des Historikers Ernst Nolte mit Briefen aus Israel', *Frankfurter Rundschau*, 5 Nov. 1987, and the letters to the *Frankfurter Rundschau* which followed on 17 Dec. 1987 from Wolfgang Schieder, 15 Jan. 1988 from Ernst Nolte, and 19 Feb. 1988 from Otto Dov Kulka.

[37] Arno Mayer, *Why did the Heavens not Darken? The 'Final Solution' in History* (New York, 1988).

[38] Baldwin, *Reworking the Past*, p. 26.

Since the fall of Communism and the possibilities of attaining deeper insights into the inner workings of the Soviet system, the relationship between Nazi Germany and the Soviet Union – especially the comparison and contrast of the Hitler and Stalin regimes – has naturally prompted intensified interest. One offshoot has been a revival of the concept of 'totalitarianism'. The scholarly limitations of this concept were discussed earlier, in chapter 2. Although after its hey-day in the 1950s the concept fell into some disuse and discredit, particularly by historians and political scientists who inclined to the Left, it by no means lost all its appeal as long as the Cold War lasted and, in fact, enjoyed something of a revival in the later 1970s and 1980s. Since the collapse of the Soviet system, the rehabilitation of the concept has been well-nigh complete.

This is not a surprising development. The sharpened focus on the scale and nature of repression – especially under Stalin but more generally in the Soviet system and, not least, in the GDR – together with stories that can now be told of deeply moving personal experiences of repression by the police state have given new vitality to the 'totalitarianism' concept.[39]

The danger is that this will provide sustenance to a simplistic popular image which implicitly posits an identification with Nazism not only of the Stalinist regime in the Soviet Union, but of 'Stalinism' – a concept extended to cover the political system of the GDR.[40] Within Germany, this could easily mean that the perfectly understandable preoccupation with the inhumanity of the GDR system, which has only recently disappeared and is therefore much more vivid in memory, increasingly displaces the fading memory of Nazism, trivializing the horrors perpetrated under Hitler by naive and shallow comparison with the crimes of the Honecker regime.[41] The distance to a relativization of German inhumanity under Nazism, at the centre of the storm provoked by Nolte's contributions to the *Historikerstreit*, is then a short one. Hitler could be viewed as a wicked tyrant – though less wicked than Stalin; and the Holocaust could be seen as no worse than the Stalinist mass-murders, and as no more than a horrible by-product of the life-and-death struggle of totalitarian systems, in which major atrocities were committed by both sides.

The renewed interest in 'totalitarian' systems does not, however, have only a negative side to it. Given the new research into the functioning of Soviet rule and the sophistication of research since the 1960s into the power structures and repressive apparatus of the Third Reich, the comparative analysis of 'Stalinism' and 'Hitlerism' need not be a retrograde step, and holds out the prospect of a deeper understanding of both systems and the societies upholding them. It ought, for instance, to be possible now, as never before, to instrumentalize the notion of 'total claim' (referred to earlier in chapter 2) in order not only to analyse comparatively the repressive structures of the respective police states, but also to explore and evaluate the patterns of consent, approval, and collaboration providing the levels of social control which enabled the repressive

[39] The evocation of images similar to those advanced by Hannah Arendt in her path-breaking work of the 1950s (discussed in ch. 2) is unmistakable. See Hannah Arendt, *The Origins of Totalitarianism* (New York, 1951).

[40] For a thoughtful and balanced comparison – legitimate and necessary, however repulsive the task – of the scale and character of the mass murder perpetrated by the regimes of Stalin and Hitler, see Maier, *Unmasterable Past*, 'Preserving Distinctions', pp. 71–84.

[41] Eberhard Jäckel, 'Die doppelte Vergangenheit', *Der Spiegel*, 23 Dec. 1991, pp. 29–43, offers some pertinent comments on this point.

apparatuses to function so effectively. The pioneering work of Peter Hüttenberger, Martin Broszat, Reinhard Mann, and Robert Gellately in examining denunciations and other ways in which the population willingly co-operated in self-policing techniques in Nazi Germany, is a starting-point from which more extensive and sophisticated research and analysis than ever before into the structure and functioning of the Gestapo can be undertaken.[42] This in turn would fit into a major interdisciplinary research project currently being established in Germany on the comparative structures of differing kinds of twentieth-century dictatorships in Europe.[43] The new insights to be expected from such comparative research into the Third Reich, as into the operation of other dictatorial systems of rule, are considerable.

Reflections

The advances made in the historiography of the Third Reich over the past 30 or so years have been truly impressive. A genuinely international scholarship - if more often than not, working to an agenda strongly influenced by inner German developments and preoccupations - has explored practically all aspects of Nazi rule and laid bare in detailed empirical examination almost every significant facet of the Third Reich. A great body of specialized knowledge has been accumulated in a vast literature, much of it of very high calibre. With the opening up of East German and Soviet archives, there is every prospect of significant advances in empirical research on themes and topics which have, until recently, been impossible to pursue.[44]

There appears, nevertheless, to be an inverse ratio between the massive extension of empirical research on the Third Reich and the embodiment of such research in a full-scale synthesis. Of course, as in all major areas of historical analysis, there are many issues on which interpretations quite legitimately differ. Open questions naturally remain. But it is still, above all, the moral dimension mentioned in chapter 1 which means that any attempt to provide a general interpretation of the Third Reich is guaranteed to give rise to further controversy quite different from, and extending far beyond that of, conventional historical debate.

Though, as we have seen, leading historians of the Third Reich have over several decades engaged vigorously in intense debate about central areas of interpretation, some key issues of understanding and explanation have seemed,

---

[42] Peter Hüttenberger, 'Heimtückefälle vor dem Sondergericht München 1933-1939', in: *Bayern in der NS-Zeit*, iv. 435-526; Martin Broszat, 'Politische Denunziation in der NS-Zeit. Aus Forschungserfahrungen im Staatsarchiv München', *Archivalische Zeitschrift*, 73 (1977), pp. 221-38; Reinhard Mann, *Protest und Kontrolle im Dritten Reich. Nationalsozialische Herrschaft im Alltag einer rheinischen Großstadt* (Frankfurt am Main/New York, 1987); Robert Gellately, 'The Gestapo and German Society: Political Denunciation in the Gestapo Case Files', *Journal of Modern History*, 60 (1988), pp. 654-94; and Robert Gellately, *The Gestapo and German Society. Enforcing Racial Policy 1933-1945* (Oxford, 1990).

[43] The project, funded by the Volkswagen Foundation, has the title: 'Diktaturen im Europa des 20. Jahrhunderts: Strukturen, Erfahrungen, Überwindung und Vergleich'.

[44] Regional studies of the eastern parts of Germany, desirable in order to counter the earlier inevitable overreliance in regional and local studies on areas within the former borders of the Federal Republic, represent merely one genre of research which is likely to develop in highly interesting fashion in the new circumstances.

if anything, less rather than more resolvable as time has passed and perspectives have changed. If, as the earlier chapters of this book have on occasion suggested, levels of synthesis are often possible, and the scholarly divides are not always the gaping chasms that historians themselves often claim them to be, the 'historicization' debate, and the outright polemics which ensued as it merged into the *Historikerstreit*, have shown that the distance between historians on fundamental, overarching perspectives of interpretation is vast. Further empirical research and new archival work alone cannot bridge the gulf.

I have suggested, in fact, that the historiographical impulses stimulated or enhanced by the unification of Germany have probably widened the gap still further. The end of the Cold War and German unification may lead to a drawing of the line under the Nazi era and its aftermath – the division of Germany – and, as Saul Friedländer has claimed, must almost certainly induce over the next few years 'a transformation of historical consciousness'.[45]

This transformation will be reinforced in at least three ways. Firstly, as the generation which experienced the Third Reich at first hand gradually dies out, 'collective memory'[46] will be replaced by an indirectly conveyed rather than directly experienced 'collective image' of the Hitler era – only partially shaped by the scholarly works of professional historians of the Third Reich (almost all of whom now belong to the post-war generation).[47] Secondly, it seems likely – in fact the trend has been unmistakably visible for some years – that the prevailing popular image will show increasing impatience in the new Germany with an image of the Third Reich which places heavy – at times near exclusive – emphasis upon German atrocities, war crimes, racial persecution, and genocide against the Jews, all symbolized by the name 'Auschwitz'. The fanatical quest for racial purification – the essence of Nazism which subsumed every aspect of policy in the Third Reich – is a negative feature of the recent past (and scarcely an obvious area of identification or comprehensibility) that many young Germans, who feel no personal responsibility for what took place, not unnaturally want to shake off. The 'historicization' approaches accord in good measure with this shifting popular image. The presumption is that the Third Reich can be most easily 'integrated' into historical consciousness by focusing upon those elements which fit into, and can be understood as part of, the development of a post-war modernized, technocratic, economically advanced, social welfare state. As regards actual popular memory of what life was like under Hitler, oral history techniques have revealed the extent to which the Third Reich – particularly the peace-time years between 1933 and 1939 – were seen as 'normal years' sandwiched between economic misery and war, and years which had many positive sides to them.[48] 'Strength through Joy' works outings, Hitler Youth rambles, the building of the motorways, the clearing away of unemployment, and the promise of the 'people's car' outweigh in

---

[45] Friedländer, 'Martin Broszat und die Historisierung des Nationalsozialismus', p. 159.

[46] See Friedländer, 'Martin Broszat und die Historisierung des Nationalsozialismus', p. 166.

[47] The changing popular images of the Third Reich, and the tension between depicting 'everyday normality' and unprecedented 'political criminality' in a single analysis informs the conceptualization of the recent impressive general work by Jost Dülffer, *Deutsche Geschichte 1933–1945. Führerglaube und Vernichtungskrieg* (Stuttgart, 1992). Engl. trans. in preparation.

[48] See Niethammer, *'Die Jahre weiß man nicht . . .'*, particularly the contribution by Ulrich Herbert which drew sharp criticism from Dan Diner (reference in ch. 9 note 24).

such memory the 'seamier' side of the Third Reich – concentration camps, pogroms, deportations, and the mass murder of designated 'racial inferiors'. And thirdly, the popular image of the Soviet Union as an 'evil Empire' which committed crimes at least equal to, and probably worse than, those of the Third Reich, will inevitably help to shape the refashioned view of the Hitler era.

The different facets of the 'historicizing' image are not in themselves false, or fabrications. They are legitimate components of the experience and memory of the Nazi past, a genuine element of remembered reality. But for the victims of Nazism – most of whom were not German – the experience and memory are different, utterly contrasting, but certainly no less genuine or legitimate. From their perspective, a portrayal of the Hitler era which does not have at its core Auschwitz as the epitome of the evil and horror of the Third Reich, can only be a flawed, misguided, or deliberately tendentious one. The unique features of the 1933–45 era must not be diluted or erased through a concentration of the underlying 'normality' of those years for most 'ordinary' Germans.

The divergence in memory, experience, and popular image poses the dilemma for historians which the 'historicization' debate between Broszat and Friedländer highlighted.[49] The 'historicization' approaches do not in any sense excuse or deny the barbarities of Nazism. But the change in emphasis, matching the changing popular image of the Third Reich, contains the implicit danger of substituting the *accidentals* (such as enhanced social mobility, technological improvements, and 'modernization') for the *substance* (the drive to create a racially and biologically pure society through conquest, exploitation, and extermination of those deemed racially impure and inferior). From the contrasting vantage-point, as the generation of the victims of Nazism dies out, the emphasis upon Auschwitz as the symbol of this inhumanity is becoming if anything magnified rather than diminished among their descendants – possibly producing its own 'monumentalization' with dubious consequences for furthering the understanding of how the Nazi 'Final Solution' could come about.[50]

Two general trends are, therefore, at present discernible in the historiography of the Third Reich. First, research has increasingly fragmented into specialist monographs on every conceivable aspect of Nazi rule. Not just the sheer volume of published research, but the proliferation of different approaches – methodological, thematic, theoretical – means that a comprehensive study of the Third Reich is far more difficult to write today than it was when Karl Dietrich Bracher produced his standard work on *The German Dictatorship*, resting on a combination of the *Sonderweg* and 'totalitarianism' concepts for its interpretation, and amounting to a classical political history 'from above' with as good as no attention paid to social, economic, and cultural aspects.[51] Bracher's study has been described as a 'monumental summation' which, nevertheless, today seems 'clinical, remote from the lived experience of the Germans' and 'is to what actually happened as macroeconomic theory is to the worker who gets laid off'. In other words: it does not 'historicize'.[52]

---

[49] See ch. 9 note 10 for the reference.

[50] See the comments of Friedländer, 'Martin Broszat und die Historisierung des Nationalsozialismus', p. 167.

[51] Karl Dietrich Bracher, *Die deutsche Diktatur* (Cologne, 1969); Engl. trans., *The German Dictatorship* (Harmondsworth, 1973).

[52] Maier, *The Unmasterable Past*, p. 101.

The second major trend follows from what has been said above. The 'historicizing' approach to Nazism within Germany, which has continued to gain ground following unification, is diverging, and will probably diverge still further, from those approaches which continue to see the world-historical significance of Nazism in its genocidal race philosophy and policy, and therefore place Auschwitz at the centre of all contemplation of the Nazi phenomenon. This divergence means that, as we move further from the Third Reich, the problem of constructing a comprehensive history of the Third Reich which satisfies the demands of both conflicting approaches is probably becoming harder.

Given these trends – the fragmentation of research and divergence of interpretation – a comprehensive study, a full-scale work of synthesis, aiming at embracing, but not artifically harmonizing, the disparate strands of research findings in a broad framework of enquiry, is more necessary than ever.[53] As Broszat pointed out in his 'historicization plea', the history of the Nazi *era* – blending in fully societal analysis with political development – remains to be written.[54] But the problems in conceptualizing and writing such a study are daunting. Can it be written? Can the opposing poles of interpretation be in any way drawn together? Are new approaches possible which might overcome the apparent dilemma? Can 'normality' and genocide be linked rather than presented as irreconcilable opposites? This is the challenge which faces any new general interpretative study of the Third Reich.

The analysis of the 'historicization' debate in the previous chapter ended by suggesting that recent research indeed signalled some ways forward in posing connections between 'everyday normality' and 'barbarity'. It is important to search for and develop such connections if the latent polarization of fundamentally opposed interpretations is not to harden into trivialization on the one side and monumentalization on the other.

As Charles Maier has commented, attempts at 'historicization' 'must risk apology but need not lead to it'.[55] Whether they do lead in an apologetic direction rests heavily upon perspective and emphasis. Questions about modernizing trends under Nazism, about social welfare systems, 'everyday normality' in conditions of repressive dictatorship, or Nazism as part of a longer-term question of national identity are in themselves legitimate and proper. So are questions which seek to explore 'everyday life' in the Third Reich in the context of continuities beyond 1945 – as the 'prehistory' of the Federal Republic.[56] There

---

[53] The massive general study by Hans-Ulrich Thamer, *Verführung und Gewalt. Deutschland 1933–1945* (Berlin, 1986) is stronger on political and ideological than social developments. Norbert Frei's good, succinct volume, *Der Führerstaat. Nationalsozialistische Herrschaft 1933 bis 1945* (Munich, 1987) focuses less on governmental structure and central policy-making than on German society under Hitler. For the recent English version under the title *National Socialist Rule in Germany. The Führer State 1933–1945* (Oxford, 1993), Frei has, however, added a brief treatment of anti-Jewish policy, bracketed out of the German original, along with foreign policy, because these topics were treated in accompanying volumes in the same series.

[54] Broszat, *Nach Hitler*, p. 167.

[55] Maier, *Unmasterable Past*, p. 93.

[56] Important works on this issue are Hans Woller, *Gesellschaft und Politik in der amerikanischen Besatzungszone. Die Region Ansbach und Fürth* (Munich, 1986), and, particularly, Martin Broszat, Klaus-Dietmar Henke and Hans Woller, eds., *Von Stalingrad zur Währungsreform* (Munich, 1988). See also the review article by Harold James, mentioned in note 30, this chapter.

is no prescription in historical writing on any period to determine precise weighting.

In the case of the Third Reich, however, any attempt to shift the overall emphasis in a general interpretation from what was singular about the regime to what was 'normal' about it would be fundamentally misguided, even where it did not become apologetic. The uniqueness of Nazism, its distinctiveness as a regime – distinguishing it even from other forms of brutal dictatorship – was its implementation of systematic genocide. This is not simply an arbitrary choice of emphasis by the historian. The emphasis reflects the essence of the Nazi phenomenon. Auschwitz need not be at the centre of every perspective of German history. But there could be no satisfactory or relevant way of displacing Auschwitz – as shorthand for Nazi barbarism – as the centre of analysis of any general interpretation of the Third Reich. And where Auschwitz fits in to German history, how German history 'produced' Auschwitz, and how German history has been moulded by Auschwitz are questions whose importance can not fade with time, and will continue to transcend the bounds of 'normal' historical enquiry. It is certainly difficult to imagine them being supplanted by questions about the place of Nazism in the development of modern German social insurance policy or social welfare programmes.

Historical scholarship cannot be determined by popular images of the past. Nor can it concentrate on preserving experience or memory – which will inevitably vary from individual to individual as well as collectively. Its task is to explain the past, to make it more understandable – perhaps contributing indirectly, therefore, to the refinement of popular images. Many aspects of 'everyday life' in the Nazi era are relevant to this task, and can certainly be called upon to show Nazism's popular appeal and how ordinary people as well as élite groups adjusted to the demands of the regime and, in numerous ways, became implicated in its policies. In this way, exploration of 'normality' and 'everyday life' can, in ways indicated in the previous chapter, contribute to a deepened understanding of the growing barbarism of the regime.

The mooted 'historicization' of National Socialism, however, far from offering fertile new approaches to the Third Reich, is a historiographical red-herring. Quite apart from the evident emptiness of the term itself, 'historicization' represents an approach to the Nazi past which, in contrast to the other approaches analysed in this book, is quite *peculiarly* German. That is to say, it can have little meaning to those who do not view the Third Reich from some form of German perspective. For those who see the catastrophe of the Third Reich as a part not only of German history, but also of European and world history, whatever meaning might be attached to 'historicization' is either an irrelevance or a distraction.[57] Current German concerns about Nazism's

---

[57] This point was made by Friedländer, 'Reflections', p. 322. It may be, however, that in his comments there, and in subsequent criticisms of the concept of 'historicization' in 'Briefwechsel' (see ch. 9 note 10 for reference) and 'Martin Broszat und die Historisierung des Nationalsozialismus' (see note 10, this chapter, for reference), Friedländer has been rather too pessimistic about changing perspectives on the Third Reich. If the change in historical consciousness which exercises him so much does indeed come about, it will be largely confined to specifically German attitudes towards the past. These will in turn, if the perspective is distorted, be open to criticism, as indeed was the case in the *Historikerstreit*, both from outside Germany and from the many critics of the new revisionism inside Germany.

place in a sense of national identity, or its role in the development of a modern technologically-based society, have as little claim to affect a historical understanding of the phenomenon of Nazism as current concerns of Christians have claim to shape our understanding of the Reformation. From the point of view of what Nazism *was*, rather than how it might be approached in different ways in accordance with a present-day sense of German identity, there is no cause to attempt the 'historicization' of the Third Reich. In fact, there is every reason to retain the focus on what was the genuine historical significance of Nazism. Auschwitz is unique *so far* in history; but there are, unfortunately, no grounds to assume that a collapse of civilization with similar horrifying results could never occur elsewhere, with different victims and different perpetrators. That is all the more reason for Auschwitz to serve as a central reference-point not only for German history, but for modern history in general.

The key question, which will surely be the focus of any genuine attempt to locate Nazism in the unfolding of modern German history, inevitably remains, therefore, how such an unprecedented and rapid 'collapse of civilization' could have come about in the highly developed state system of a modern industrial society. This question cannot, of course, be approached by a concentration on continuities in 'everyday normality' which implicitly or explicitly excise the intrinsic barbarism of the regime. However, the 'mystification' of Auschwitz, which seems inherent in the statement of Dan Diner, one of the most vehement critics of 'historicization', that 'Auschwitz is a no-man's-land of understanding, a black box of explanation, a vacuum of extrahistorical significance which sucks in attempts at historiographic interpretation', leaves only the option of despairing incomprehensibility.[58]

To accept that there are no ways of explaining some of the most momentous events in world history – the collapse of civilization which produced Auschwitz – would indeed be to capitulate to mystification. Such counsels of despair must be resisted. Ways have to be found of tackling the polarized, but actually interlinked, 'normality' and genocide. And to do this, the methods of the new social history and a political structural history need to be reconciled, not kept apart. A two-fold approach offers some possibilities.

A construct drawn from sociological writing – 'the pathology of modern

---

[58] Diner, 'Zwischen Aporie und Apologie' (ref. in ch. 9 note 19), p. 73 (translation taken from Baldwin, *Reworking the Past*, p. 144). Diner's essay, which we encountered in the previous chapter, is a complex and thought-provoking attack on 'historicization' and the distortions of the 'revisionist' historians in the *Historikerstreit*. The quotation cited occurs in the conclusion, after he has dismissed alternative 'intentionalist' and 'functionalist' approaches to 'understanding Auschwitz'. The former approach he regards as producing a 'mysterious mechanism of actualization' defying intentionalist understanding – a 'methodological black box'. The latter approach, he argues, can bring some illumination but reduces the horror of what took place to the banality of its constituent parts. In this, he believes the perspective of the victims, whose experience of the anti-rationality (not irrationality) of the Nazis he sees as the only valid point of departure in contemplating mass extermination, to be obscured. Hence, attempts to understand Auschwitz are, for Diner, bound to lead to relativization, banalization, and rationalization. – Diner, 'Zwischen Aporie und Apologie', pp. 70-3 (in Baldwin, p. 141-4). It has to be said that Diner's use of language mystifies more than it clarifies. A 'black box' is, in today's parlance (as in aeroplane disasters) something which assists rather than defies understanding. Perhaps he means 'black hole', which might make more sense. What he means by 'a vacuum of extrahistorical significance' (ein . . . *außerhistorische* Bedeutung annehmendes Vakuum) is anyone's guess.

civilization' – forms one conceptual starting-point in the attempt to grasp how the social and political conditions in which anti-humanitarian and anti-emancipatory impulses, present in many forms and processes of modern industrial society, can gain wide – and murderous – popularity.[59] The phenomenon of Nazism throws into sharpest relief the Janus-face of modernity and the disasters to which the crises of modern societies and state systems can lead.[60] The Third Reich might perhaps be seen as a sort of Chernobyl in the history of modern society: a disaster which did not *have* to happen, but one which arose from a potential present in the very character of modern society.[61] The 'normal' state of a nuclear reactor is not a blow-out. But it *did* and *can* happen. The 'normal' state of a modern, advanced, industrial society is not the sort of 'societal blow-out' of the Third Reich. But it *did* and *can* happen. The disturbing trends of our own times, not least since the collapse of the Soviet empire, would tend to indicate that the potential is still there.[62]

---

[59] See Peukert, *Volksgenossen* (reference in ch. 2 note 45), esp. pp. 13–17, 289–96, for stimulating ideas pointing in this direction (drawing on Foucault and Habermas).

[60] On the 'Janus-face' of modernity, see the remarks by Detlev Peukert, *Max Webers Diagnose der Moderne* (Göttingen, 1989), esp. pp. 55 ff.

[61] Detlev Peukert deployed a slightly different metaphor of the nuclear power plant as modern society in 'The Weimar Republic – Old and New Perspectives', *German History*, 6 (1988), pp. 133–44, here p. 143.

[62] An argument for seeing the Holocaust as a product, not a failure, of modernity – a 'test of the hidden possibilities of modern society', which retains the potential for a future variant of Auschwitz – is powerfully advanced by Zygmunt Baumann, *Modernity and the Holocaust* (Oxford, 1989), here pp. 5, 12, 84–5. Baumann stresses that 'the "Final Solution" did not clash at any stage with the rational pursuit of efficient, optional goal-implementation', arising, on the contrary, 'out of a genuinely rational concern, and . . . generated by bureaucracy true to its form and purpose' (p. 17). Modern genocide, according to this view, is 'an element of social engineering, meant to bring about a social order conforming to the design of the perfect society' (p. 91). This challenging argument (with which I largely concur) contains some points of overlap with the claims of two young German historians, Susanne Heim and Götz Aly. In a number of publications which have stirred heated controversy in Germany, Heim and Aly have posited an 'economy of the "Final Solution" '. They attribute extermination policy to the conceptions of technocrats based in planning agencies in Poland, whose vision of a 'new order' presupposed the extermination of east European Jews, whose presence – as a poverty-stricken urban proletariat, monopolizing pre-modern sectors of the economy, and occupying housing potentially available for an influx of the rural population to be deployed in industry – posed an obstacle to the reduction of overpopulation, chances of social mobility, and the rationalization of industrial production. Of their numerous publications, see, especially, Götz Aly *et al.*, *Sozialpolitik und Judenvernichtung. Gibt es eine Ökonomie der Endlösung?* (Berlin, 1987). Heim and Aly deserve credit in my view for their detailed empirical research in Polish archives. This has uncovered new sources demonstrating, as they claim, genocidal visions related to conceptions of population and social policy among the planners of the *Generalgouvernement*. The interpretation they place on their evidence, however, goes too far. Effectively, they are arguing that the planners were the initiators of genocide, that without their schemes Nazi race policy would not have gone beyond pogroms. Compelling objections to this line of argument have, to my mind, been lodged especially by Christopher Browning and Ulrich Herbert, who return the emphasis on the *cause* of genocide to the combination of an internally coherent interpretation of modern society as racial-biologically determined (a world-view which, though not specifically Nazi, became encapsulated in Nazi ideology and encountered new possibilities of being translated into ever more radical policy in the 'enabling' conditions of the Third Reich); and the new availability of technological and bureaucratic killing-capacity (tried out in the 'euthanasia action') allowing the regime to embark on all-out genocide following the invasion of the Soviet Union. Given the perceived centrality of the task of 'solving the Jewish problem', all sorts of modern technocratic and bureaucratic planning-agencies, including those highlighted by Aly and Heim, set to work, using a variety of modes of rationalisation (overcrowding, disease, poverty, hunger, etc.) to legitimize the policies of genocide. The controversy over the theses of

The Chernobyl blow-out, to stay with the metaphor for a moment longer, was not a 'works accident' arising 'out of the blue', without structural, systemic causes, and undetached from human errors and miscalculations. A different kind of reactor, or different management of the reactor, could well have prevented, or substantially reduced the risk of, disaster – even if a potential danger remained inherent in the deployment of such a reactor. In the 'societal blow-out' of Nazism, this returns us to the need – which earlier chapters of this book have emphasized – to integrate structural and personal factors in attempting to explain the collapse of civilization in the Third Reich.

The single most crucial role in the 'societal blow-out' of the Third Reich was played by the leader of the Nazi regime. Any attempt at a satisfactory synthesis, or general interpretation, will have to do justice to the 'Hitler factor'. This cannot satisfactorily be achieved by the 'intentionalist' approaches discussed above, whose deficiencies have been examined. On the other hand, in their anxiety to combat over-personalized interpretations, 'structuralist' approaches have at times appeared almost to write Hitler out of the script. Hitler's aims and actions *were* of critical importance. But what has to be done is to explore the ways in which society and regime – different social groups and the various components of an increasingly fragmented political system – enabled personalized power to gather increasing momentum in a crisis-ridden system driven by millenarian goals, sustained Hitler's extraordinary form of leadership even when it was taking them over the abyss, made his arbitrary form of decision-making operable, and turned his ideological vision into horrifying practical reality. The potential is there – a consideration embryonically visible in some of the chapters of this book – through the deployment of a theoretical conceptualization stemming from Max Weber of the growth, character, and function of charismatic leadership, the conditions of its emergence, and its pivotal role in the government and society of the Third Reich.[63]

The Nazi past raises passionate feelings of moral denunciation in those who have to confront it. It is right that it does so. Yet, as justified and even necessary as such feelings are, moral denunciation in the long run will not suffice and can easily become the stuff of legend, not understanding.[64] Moral outrage and revulsion need constantly to be reinforced by genuine historical scholarship and understanding. The past does shape the present – in very obvious ways in

---

Heim and Aly has, as usual in the German context, been overheated and unduly polemicized, with high moral claims and denunciations on both sides. The contributions to the debate are conveniently brought together in Wolfgang Schneider, ed., *Vernichtungspolitik. Eine Debatte über den Zusammenhang von Sozialpolitik und Genozid im nationalsozialistischen Deutschland* (Hamburg, 1991). In my view, the work of Heim and Aly offers an interesting contribution to the approach to modernity offered by Baumann without satisfying the overstated claims the authors make for it as a novel explanation of the Nazi genocide against the Jews.

[63] This is emplicitly encouraged by Hans-Ulrich Wehler, '30. January 1933 – Ein halbes Jahrhundert danach', *APZ*, 29 Jan. 1983, pp. 43–54, here p. 50. My recent brief study, *Hitler* (London, 1991), attempts to develop the approach.

[64] See Broszat's pertinent comment ('Briefwechsel', p. 365): 'The danger of the suppression of this era does not in my opinion consist only of the normal forgetting, but in this case, almost paradoxically, because for didactic reasons people are too much "at pains" about this chapter of history. From the original, authentic continuum of this history, an arsenal of teaching sessions and statuesque images are pieced together, which more and more develop an existence of their own, especially then in the second and third generation coming to take the place of the original history, before being finally naively misunderstood as the actual history'. See also Broszat, *Nach Hitler*, pp. 114–20.

Germany, and by no means always or only in negative fashion.

Never since the war – with new forms of fascism and racism more menacing than thought imaginable only a few short years ago – has it been more important to understand the disaster which Nazism wrought on Germany and Europe. Doubtless, the contribution of the specialist historian of Nazism to countering the worrying and depressing reawakening of fascism can be only a small one. But it is nevertheless vitally important that the contribution, however modest, is made. Knowledge is better than ignorance; history better than myth. These truisms are more than ever worth bearing in mind where ignorance and myth spawn racial intolerance and a revival of the illusions and idiocies of fascism.

Suggestions for Further Reading

Most of the works consulted in the writing of this book are in German, and full references are provided in the relevant footnotes. I have confined suggestions for further reading to a selection of available works in English which have a particular bearing on the debates explored above, and have attempted where possible to include recent publications.

General Historiographical Surveys

Pierre Ayçoberry, *The Nazi Question* (London, 1981).
John Hiden and John Farquharson, *Explaining Hitler's Germany. Historians and the Third Reich* (London, 2nd edn., 1989).
Klaus Hildebrand, *The Third Reich* (London, 1984).
(A selection of documents relating to all the aspects of Nazism dealt with above, and with an admirable commentary by Jeremy Noakes, is available in the greatly revised, three-volume edition of Jeremy Noakes and Geoffrey Pridham, eds., *Nazism, 1919–1945. A Documentary Reader* (Exeter Studies in History, Exeter, 1983–8). A fourth and final volume is in preparation).

1 Historians and the Problem of Explaining Nazism

George Iggers, *The German Conception of History* (Middletown, Conn., 1968).
Tim Mason, 'Intention and Explanation: A Current Controversy about the Interpretation of National Socialism', in Gerhard Hirschfeld and Lothar Kettenacker, eds., *Der 'Führerstaat': Mythos und Realität* (Stuttgart, 1981), pp. 23–40.
Tim Mason, 'Open Questions on Nazism', in Raphael Samuel, ed., *People's History and Socialist Theory* (London, 1981), pp. 205–10.
Jörn Rüsen, 'Theory of History in the Development of West German Historical Studies: A Reconstruction and Outlook', *German Studies Review* 7 (1984), pp. 14–18.
Hans-Ulrich Wehler, 'Historiography in Germany Today', in Jürgen Habermas, ed., *Observations on 'The Spiritual Situation of the Age'. Contemporary German Perspectives* (Cambridge, Mass./London, 1984), pp. 221–59.

2. The Essence of Nazism: Form of Fascism, Brand of Totalitarianism, or Unique Phenomenon?

Hannah Arendt, *The Origins of Totalitarianism* (New York, 1951).
David Beetham, *Marxists in Face of Fascism* (Manchester, 1983).
Karl Dietrich Bracher, *The German Dictatorship* (Harmondsworth, 1973).
Roger Eatwell, 'Towards a New Model of Generic Fascism', *Journal of Theoretical Politics* 4 (1992), pp. 161-94.
Geoff Eley, 'What produces Fascism: Preindustrial Traditions or a Crisis of the Capitalist State?', *Politics and Society* 12 (1983), pp. 53-82.
Roger Griffin, *The Nature of Fascism* (London, 1990).
Martin Kitchen, *Fascism* (London, 1976).
Jürgen Kocka, 'German History before Hitler: The Debate about the German *Sonderweg*', *JCH* 23 (1988), pp. 3-16.
Stein Ugelvik Larsen *et al.*, *Who were the Fascists? Social Roots of European Fascism* (Bergen, 1980).
Juan J. Linz, 'Some Notes towards a Comparative Study of Fascism in Sociological Historical Perspective', in Walter Laqueur, ed., *Fascism. A Reader's Guide* (Harmondsworth, 1979), pp. 13-78.
Nicos Poulantzas, *Fascism and Dictatorship* (London, 1974).
Leonard Schapiro, *Totalitarianism* (London, 1973).

3. Politics and Economics in the Nazi State

Avraham Barkai, *Nazi Economics* (London, 1990).
Berenice Carroll, *Design for Total War: Arms and Economics in the Third Reich* (The Hague, 1968).
Gustavo Corni, *Hitler and the Peasants. Agrarian Policy of the Third Reich 1930-1939* (New York/Oxford/Munich, 1990).
John R. Gillingham, *Industry and Politics in the Third Reich* (London, 1985).
Peter Hayes, *Industry and Ideology. IG Farben in the Nazi Era* (Cambridge, 1987).
Harold James, *The German Slump, Politics and Economics 1924-1936* (Oxford, 1986).
Tim Mason, 'The Primacy of Politics – Politics and Economics in National Socialist Germany', in Henry A. Turner, ed., *Nazism and the Third Reich* (New York, 1972), pp. 175-200.
Alan Milward, *The German Economy at War* (London, 1965).
Alan Milward, 'Fascism and the Economy', in Walter Laqueur, ed., *Fascism. A Reader's Guide* (Harmondsworth, 1979), pp. 409-53.
Franz Neumann, *Behemoth. The Structure and Practice of National Socialism* (London, 1942).
Richard J. Overy, 'Hitler's War and the German Economy: A Reinterpretation', *Economic History Review* 35 (1982), pp. 272-91.
Richard J. Overy, *The Nazi Economic Recovery 1932-1938* (London, 1982).
Richard J. Overy, 'Germany, "Domestic Crisis", and War in 1939', *Past and Present* 116 (1987), pp. 138-68.
Alfred Sohn-Rethel, *The Economy and Class Structure of German Fascism* (London, 1987).

4. Hitler: 'Master in the Third Reich', or 'Weak Dictator'?

Karl Dietrich Bracher, 'The Role of Hitler: Perspectives of Interpretation', in Walter Laqueur, ed., *Fascism. A Reader's Guide* (Harmondsworth, 1979), pp. 193–212.
Martin Broszat, *The Hitler State* (London, 1981).
Alan Bullock, *Hitler. A Study in Tyranny* (London, rev. edn., 1964).
Alan Bullock, *Hitler and Stalin. Parallel Lives* (London, 1991).
Jane Caplan, 'Bureaucracy, Politics, and the National Socialist State', in Peter D. Stachura, ed., *The Shaping of the Nazi State* (London, 1978), pp. 234–56.
Jane Caplan, *Government without Administration. State and Civil Service in Weimar and Nazi Germany* (Oxford, 1988).
William Carr, *Hitler. A Study in Personality and Politics* (London, 1978).
Joachim C. Fest, *Hitler* (London, 1974).
Robert Koehl, 'Feudal Aspects of National Socialism', in Henry A. Turner, ed., *Nazism and the Third Reich* (New York, 1972), pp. 151–74.
Eberhard Jäckel, *Hitler's Weltanschauung. A Blueprint for Power* (Middletown, Conn., 1972).
Eberhard Jäckel, *Hitler in History* (Hanover/London, 1984).
Michael Kater, 'Hitler in a Social Context', *CEH* 14 (1981), pp. 243–72.
Ian Kershaw, *The 'Hitler Myth'. Image and Reality in the Third Reich* (Oxford, 1987).
Ian Kershaw, *Hitler* (London, 1991).
Hans Mommsen, 'National Socialism: Continuity and Change', in Walter Laqueur, ed., *Fascism. A Reader's Guide* (Harmondsworth, 1979), pp. 151–92.
Hans Mommsen, 'Hitler's Position in the Nazi System' in Hans Mommsen, *From Weimar to Auschwitz* (Oxford, 1991), pp. 163–88.
Edward N. Peterson, *The Limits of Hitler's Power* (Princeton, 1969).
J.P. Stern, *Hitler, The Führer and the People* (London, 1975).

5. Hitler and the Holocaust

David Bankier, 'Hitler and the Policy-Making Process on the Jewish Question', *Holocaust and Genocide Studies* 3 (1988), pp. 1–20.
David Bankier, *The Germans and the Final Solution. Public Opinion under Nazism* (Oxford, 1992).
Yehuda Bauer, *The Holocaust in Historical Perspective* (London, 1978).
Zygmunt Bauman, *Modernity and the Holocaust* (Oxford, 1989).
Richard Breitman, *The Architect of Genocide. Himmler and the Final Solution* (London, 1991).
Martin Broszat, 'Hitler and the Genesis of the "Final Solution" ', *Yad Vashem Studies* 13 (1979), repr. in H. W. Koch, ed., *Aspects of the Third Reich* (London, 1985), pp. 390–429.
Christopher Browning, *Fateful Months* (New York, 1985).
Christopher Browning, 'A Reply to Martin Broszat regarding the Origins of the Final Solution', *Simon Wiesenthal Center Annual* 1 (1984), pp. 113–32.

Christopher Browning, *The Path to Genocide. Essays on Launching the Final Solution* (Cambridge, 1992).

Christopher Browning, *Ordinary Men. Reserve Police Battalion 101 and the Final Solution in Poland* (New York, 1992).

Philippe Burrin, *Hitler and the Jews. The Genesis of the Holocaust* (Eng. trans., forthcoming, London, 1993).

Gerald Fleming, *Hitler and the Final Solution* (Oxford, 1986).

Robert Gellately, *The Gestapo and German Society. Enforcing Racial Policy, 1933-1945* (Oxford, 1990).

Hermann Graml, *Antisemitism and its Origins in the Third Reich* (Oxford, 1992).

Raul Hilberg, *The Destruction of the European Jews* (New York, 1961, 2nd rev. edn., 1983).

Gerhard Hirschfeld, ed., *The Policies of Genocide* (London, 1986).

Ian Kershaw, 'Improvised Genocide? The Emergence of the 'Final Solution' in the 'Warthegau', *Transactions of the Royal Historical Society*, 6th Ser., 2 (1992), pp. 51-78.

Helmut Krausnick, 'The Persecution of the Jews', in Helmut Krausnick *et al., Anatomy of the SS State* (London, 1968), pp. 1-124.

Otto Dov Kulka, 'Major Trends and Tendencies of German Historiography on National Socialism and the "Jewish Question" (1924-1984)', *Yearbook of the Leo Baeck Institute* 30 (1985), pp. 215-42.

Michael Marrus, *The Holocaust in History* (London, 1988).

Arno J. Mayer, *Why did the Heavens not Darken? The 'Final Solution' in History* (New York, 1988).

Hans Mommsen, 'The Realisation of the Unthinkable. The "Final Solution of the Jewish Question" in the Third Reich', in Hans Mommsen, *From Weimar to Auschwitz* (Oxford, 1991), pp. 224-53.

Walter H. Pehle, ed., *November 1938. From 'Kristallnacht' to Genocide* (New York/Oxford/Munich, 1991).

Karl A. Schleunes, *The Twisted Road to Auschwitz. Nazi Policy toward German Jews, 1933-1939* (Urbana/Chicago/London, 1970).

6. Nazi Foreign Policy: Hitler's 'Programme' or 'Expansion without Object'?

William Carr, *Arms, Autarky, and Aggression. A Study in German Foreign Policy, 1933-1939* (London, 2nd. edn., 1979).

William Carr, *Poland to Pearl Harbor. The Making of the Second World War* (London, 1985).

Wilhelm Deist, *The Wehrmacht and German Rearmament* (London, 1981).

Fritz Fischer, *From Kaiserreich to Third Reich. Elements of Continuity in German History, 1871-1945* (London, 1986).

Milan Hauner, 'Did Hitler want a World Dominion?', *Journal of Contemporary History* 13 (1978), pp. 15-32.

Klaus Hildebrand, *The Foreign Policy of the Third Reich* (London, 1973).

Konrad Jarausch, 'From Second to Third Reich: the Problem of Continuity in German Foreign Policy', *CEH* 12 (1979), pp. 68-82.

Meir Michaelis, 'World Power Status or World Dominion?', *The Historical Journal* 15 (1972), pp. 331–60.

Wolfgang Michalka, 'From the Anti-Commintern Pact to the Euro-Asiatic Bloc: Ribbentrop's Alternative Concept of Hitler's Foreign Policy Programme', in H.W. Koch, ed., *Aspects of the the Third Reich* (London, 1985), pp. 267–84.

Wolfgang J. Mommsen and Lothar Kettenacker, eds., *The Fascist Challenge and the Policy of Appeasement* (London, 1983).

Klaus-Jürgen Müller, *Army, Politics, and Society in Germany 1933–1945* (Manchester, 1987).

Rich, Norman, *Hitler's War Aims*, 2 vols. (London, 1973–74).

Esmond M. Robertson, ed., *The Origins of the Second World War* (London, 1971).

Jochen Thies, 'Nazi Architecture – A Blueprint for World Domination: The Last Aims of Adolf Hitler', in David Welch, ed., *Nazi Propaganda. The Power and the Limitations* (London, 1983).

Gerhard Weinberg, *The Foreign Policy of Hitler's Germany. Diplomatic Revolution in Europe 1933–36* (Chicago/London, 1970).

Gerhard Weinberg, *The Foreign Policy of Hitler's Germany. Starting World War II* (Chicago/London, 1980).

Jonathan Wright and Paul Stafford, 'Hitler, Britain, and the Hoßbach Memorandum', *MGM* 42 (1987), pp. 77–123.

7. The Third Reich: 'Social Reaction' or 'Social Revolution'?

Peter Baldwin, 'Social Interpretations of Nazism: Renewing a Tradition', *JCH* 25 (1990), pp. 5–37.

Richard Bessel, 'Living with the Nazis: Some Recent Writing on the Social History of the Third Reich', *European History Quarterly* 14 (1984), pp. 211–20.

Richard Bessel, ed, *Life in the Third Reich* (Oxford, 1987).

Gunnar C. Boehnert, 'The Third Reich and the Problem of "Social Revolution": German Officers and the SS', in Volker R. Berghahn and Martin Kitchen, eds., *Germany in the Age of Total War* (London, 1981), pp. 203–17.

Renate Bridenthal *et al., When Biology became Destiny. Women in Weimar and Nazi Germany* (New York, 1984).

Michael Burleigh and Wolfgang Wippermann, *The Racial State. Germany 1933–1945* (Cambridge, Mass., 1991).

Ralf Dahrendorf, *Society and Democracy in Germany* (London, 1968).

Norbert Frei, *Nationalist Socialist Rule in Germany* (Oxford, 1993).

Richard Grunberger, *A Social History of the Third Reich* (London, 1971).

Michael Kater, *Doctors under Hitler* (Chapel Hill/London, 1989).

Michael Kater, *Different Drummers: Jazz in the Culture of Nazi Germany* (Oxford, 1992).

Claudia Koonz, *Mothers in the Fatherland. Women, the Family, and Nazi Politics* (New York, 1976).

Tim Mason, *Social Policy in the Third Reich* (New York/Oxford/Munich, 1993).
Tim Mason, 'Women in Nazi Germany', *HWJ* 1 (spring 1976), pp. 74-113, 2 (autumn 1976), pp. 5-32.
Jeremy Noakes, 'Nazism and Revolution', in Noel O'Sullivan, ed., *Revolutionary Theory and Political Reality* (London, 1983), pp. 73-100.
Detlev Peukert, *Inside Nazi Germany. Conformity and Opposition in Everyday Life* (London, 1987).
Stephen Salter, 'Structures of Consensus and Coercion. Workers' Morale and the Maintenance of Work Discipline, 1939-1945', in David Welch, ed., *Nazi Propaganda. The Power and the Limitations* (London, 1983), pp. 88-116.
Stephen Salter, 'National Socialism, the Nazi Regime, and German Society', *The Historical Journal* 35 (1992), pp. 487-99.
David Schoenbaum, *Hitler's Social Revolution* (London, 1966).
Ronald Smelser, *Robert Ley. Hitler's Labour Front Leader* (Oxford/New York/Hamburg, 1988).
Jill Stephenson, *Women in Nazi Society* (London, 1975).
David Welch, 'Propaganda and Indoctrination in the Third Reich: Success or Failure?', *European History Quarterly* 17 (1987), pp. 403-22.

8. 'Resistance without the People'

Michael Balfour, *Withstanding Hitler in Germany 1933-1945* (London, 1988).
Hedley Bull, ed., *The Challenge of the Third Reich* (Oxford, 1986).
Harold C. Deutsch, *The Conspiracy against Hitler in the Twilight War* (Minneapolis, 1968).
Harold C. Deutsch *et al.*, 'Symposium: New Perspectives on the German Resistance against National Socialism', *CEH*, 14 (1981), pp. 322-99.
Hermann Graml *et al.*, *The German Resistance to Hitler* (London, 1970).
Peter Hoffmann, *The History of the German Resistance 1933-1945* (rev. edn., Cambridge, Mass., 1977).
Peter Hoffmann, *German Resistance to Hitler* (Cambridge, Mass., 1988).
Ian Kershaw, *Popular Opinion and Political Dissent in the Third Reich. Bavaria 1933-1945* (Oxford, 1983).
Klemens von Klemperer, *A Nobel Combat. The Letters of Shiela Grant Duff and Adam von Trott zu Solz 1932-1939* (Oxford, 1988).
Klemens von Klemperer, *German Resistance against Hitler. The Search for Allies Abroad 1938-1945* (Oxford, 1992).
David Clay Large, ed., *Contending with Hitler. Varieties of German Resistance in the Third Reich* (Cambridge, Mass., 1991).
Tim Mason, 'The Workers' Opposition in Nazi Germany', *HWJ* 11 (1981), pp. 120-37.
Patricia Meehan, *The Unnecessary War. Whitehall and the German Resistance to Hitler* (London, 1992).
Alan Merson, *Communist Resistance in Nazi Germany* (London, 1985).
Hans Mommsen, '20 July 1944 and the German Labour Movement', in Hans Mommsen, *From Weimar to Auschwitz* (Oxford, 1991), pp. 189-207.

Hans Mommsen, 'German Society and Resistance to Hitler', in Hans Mommsen, *From Weimar to Auschwitz* (Oxford, 1991), pp. 208–24.

Klaus-Jürgen Müller, 'The Structure and Nature of the National Conservative Opposition in Germany up to 1940', in H.W. Koch, ed., *Aspects of the Third Reich* (London, 1985), pp. 133–78.

Francis R. Nicosia and Lawrence D. Stokes, eds., *Germans against Nazism* (Oxford, 1990).

Jeremy Noakes, 'The Oldenburg Crucifix Struggle of November 1936. A Case study of Opposition in the Third Reich', in Peter D. Stachura, *The Shaping of the Nazi State* (London, 1978), pp. 210–33.

Terence Prittie, *Germans against Hitler* (London, 1964).

Gerhard Ritter, *The German Resistance: Carl Goerdeler's Struggle against Tyranny* (London, 1958).

Hans Rothfels, *The German Opposition to Hitler* (London, 1961).

Inge Scholl, *Students against Tyranny. The Resistance of the White Rose, Munich 1942–1943* (Middletown, Conn., 1983).

Christopher Sykes, *Troubled Loyalty: A Biography of Adam von Trott zu Solz* (London, 1968).

9. 'Normality' and Genocide: the Problem of 'Historicisation'

Peter Baldwin, ed., *Reworking the Past: Hitler, the Holocaust and the Historians' Debate* (Boston, Mass., 1990).

Martin Broszat, 'A Plea for the Historicisation of National Socialism', in Baldwin, *Reworking the Past*, pp. 77–87.

Martin Broszat and Saul Friedländer, 'A Controversy about the Historicisation of National Socialism', in *Yad Vashem Studies* 19 (1988), pp. 1–47, repr. in Baldwin, *Reworking the Past*, pp. 102–34.

Dan Diner, 'Between Aporia and Apology: On the Limits of Historicising National Socialism', in Baldwin, *Reworking the Past*, pp. 135–45.

Saul Friedländer, 'Some Reflections on the Historicisation of National Socialism', *Tel Aviver Jahrbuch für deutsche Geschichte* 16 (1987), pp. 310–24, repr. in Baldwin, *Reworking the Past*, pp. 88–101.

Jürgen Kocka, 'The Weight of the Past in Germany's Future', *German Politics and Society* (Center for European Studies, Harvard University) 13 (1988), pp. 22–9.

Otto Dov Kulka, 'Singularity and its Relativisation. Changing Views in German Historiography on National Socialism and the "Final Solution" ', *Yad Vashem Studies* 19 (1988), pp. 151–86, repr. in Baldwin, *Reworking the Past*, pp. 146–70.

10. Shifting Perspectives: Historiographical Trends in the Aftermath of Unification

John Breuilly, 'The National Idea in Modern German History', in John Breuilly, ed., *The State of Germany. The National Idea in the Making, Unmaking, and Remaking of a Modern Nation-State* (London, 1992), pp. 1–28.

Geoff Eley, 'Nazism, Politics, and the Image of the Past', *Past and Present* 121 (1988), pp. 171–208.

Richard J. Evans, *In Hitler's Shadow. West German Historians and the Attempt to Escape from the Nazi Past* (New York, 1989).

Saul Friedländer, 'West Germany and the Burden of the Past: the Ongoing Debate', *Jerusalem Quarterly* 42 (1987), pp. 3–18.

Jürgen Habermas, 'A Kind of Indemnification: The Tendencies toward Apologia in German Research on Current History', *Yad Vashem Studies* 19 (1988), pp. 75–92.

Geoffrey Hartman, ed., *Bitburg in Moral and Political Perspective* (Bloomington, 1986).

Harold James, *A German Identity, 1770–1990* (London, 1989).

Jürgen Kocka, 'German Identity and Historical Comparison: After the *Historikerstreit*', in Baldwin, *Reworking the Past*, pp. 279–93.

Charles Maier, *The Unmasterable Past: History, Holocaust, and German National Identity* (Cambridge, Mass., 1988).

Christian Meier, 'To Condemn and to Understand: A Turning Point in German Historical Remembrance', *Yad Vashem Studies* 19 (1988), pp. 93–105.

Ernst Nolte, 'Between Myth and Revisionism? The Third Reich in the Perspective of the 1980s', in H.W. Koch, ed., *Aspects of the Third Reich* (London, 1985), pp. 17–38.

Ernst Nolte, 'A Past That Will Not Pass Away (A Speech It Was Possible to Write, But Not to Present)', *Yad Vashem Studies* 19 (1988), pp. 65–73.

Index

Abelshauser, Werner, 138
Abraham, David, 40 n1
Abwehr circle, 155
Abyssinia, 93, 119
Adam, Uwe Dietrich, 87, 90, 98
Adenauer, Konrad, 153, 200
Agriculture, 53
Air Ministry, 51
Aktionsrichtung, 125
Allies, 1, 67, 131, 180–3
Alltaggeschichte, 190, 192, 193
Annales school, 7
Anschluss (1938), 93, 176
Anti-bolshevism, 115
Anti-feminism, 144, 189
Anti-Jewish legislation, 55, 83, 84, 87
Anti-Jewish policy, 3, 50, 55–6, 67, 82,
  ch. 5 *passim*
Anti-Marxism, 140
Anti-semitism, 63, ch. 5 *passim*
Arendt, Hannah, 20, 82
Argentina, 34
Aristocracy, 56
Army, 49ff, 56, 89, 116, 128,
  see also Wehrmacht
  High Command (1st World War), 116
'Aryanization', 52, 56, 87, 94
'Aryans', 92
Auschwitz, 4, 55, 61, 80, 83, 84, 86, 89,
  103, 191, 195, 199, 200ff, 210ff
Auschwitz-Birkenau, 104
Auslandsorganisation (of NSDAP), 113
Austria, 25, 52, 53, 94, 102, 119, 120
Autarky, 50, 61
Azores, 128

Babi-Yar, 101
Baltic, 100
Basic Law (*Grundgesetz*) of Federal
  Republic of Germany, 11, 154
Bauer, Otto, 24
Bauer, Yehuda, 80, n2 86

'Bavarian Project', 157ff, 166, 168, 181,
  183
Bayreuth, 113
Baldwin, Peter, 207
Beck, Ludwig, 155
Behemoth (Franz Neumann), 20
Belgium, 102
Belzec, 104
Benz, Wolfgang, 193, 104
Berlin, 2, 62, 72, 92, 117, 129
  Treaty of, 118
  Wall, 200, 206
Berufsverbot, 12
Beveridge Plan, 188
Big Business *See Industry*
Binion, Rudolf, 83
Bismarckian era, 116, 140, 147
Bismarck, Otto von, 140, 147, 155
Blitzkrieg, 42, 79, 85, 97, 126
'Blomberg-Fritsch Affair', 52
Blomberg, Werner von, 51, 113, 114
Boberach, Heinz, 161
Bock, Gisela, 189
Bolshevik Revolution, 17
Bolshevism, 20, 31, 54, 99, 114, 124ff,
  129, 155, 168, 175, 191, 207
Bomb plot, 56, 150, 154, 170, 171, 177
Bonaparte, Charles Louis Napoleon, 24
Bonapartism, 24, 25, 44, 45, 46
Bonhoeffer, Dietrich, 174
Borkenau, Franz, 20
Bormann, Martin, 70, 73
Botz, Gerhard, 169, 170
'Boycott Movement', 55, 91
Bracher, Karl Dietrich, 11–12, 15, 21,
  22, 37, 44, 62, 64, 187, 211
Brack, Viktor, 97, 104
Britain, 2, 76, 78, 117–18, 122, 123, 129
  see also England
  Naval Treaty with Germany, 118, 122
Broszat, Martin, 31, 33, 64, 65, 85, 88,
  98, 112, 113, 123, 158ff, 180, 181,

182, 184, 186, 188, 190, 193, 195, 206, 212
Browning, Christopher, 98–9, 102, 105
Bulgaria, 34
Bullock, Alan, 60, 61
Burckhardt, Carl, 125
Burrin, Philippe, 88 n31, 99, 101, 102, 104, 105

Caetano, Marcello, 34
Canaris, Wilhelm, 113
Capitalism, 18, 27, 40, 44, 47, 48, 53, 141
Carr, William, 48
Chelmno, 83, 104
Chernobyl, 215
Chile, 34
Christian Democracy, 11, 12
Churches, 66, 71, 74, 142, 145, 163, 164, 169
  leaders of, 71
  resistance of, 159, 171, 173, 174
  'struggle', 75, 145, 155
Ciano, Galeazzo, 96
Civil Service, 64, 91, 106
Cold War, 17, 19, 31, 152, 153, 200, 208, 210
Collective memory, 210, 211
Comintern, 10, 17, 23–5, 29, 43,
Communism, 11–13, 20, 30ff, 172
Communist International *See Comintern*
'Concept-pluralist' approach, 115
'Critical history' approach, 199
Croce, Benedetto, 25
'Crystal Night', 93, 94, 174
Czechoslovakia, 52, 53, 102, 119
Czichon, Eberhard, 43

Dachau, 174
Dahrendorf, Ralf, 27 135ff, 138, 181, 184
Danzig, 109, 117
Darré, Richard Walther, 137, 142, 203
Dawidowicz, Lucy, 16, 83, 98
Decision making, 9, 47, 66, 73, 118–20
Delp, Peter Alfred, 174, 176
Department stores, 75
Deputy Führer, 70
Die deutsche Katastrophe (Friedrich Meinecke), 6
Diels, Rudolf, 91
Dietrich, Otto, 69
Dilthey, Wilhelm, 15, 26
Dimitroff, Georgi, 10, 24
Diner, Dan, 183, 187, 214

Diplomatic corps, 92
Divide-and-rule strategy, 64, 70, 74
Dollfuss, Engelbert, 119
Dortmund, 157
Dülffer, Jost, 110, 119, 121
Duisberg, 157

East Germany, 2, 10, 14 *See also* German Democratic Republic
Economic Inspectorate of the *Wehrmacht*, 53
Eichholtz, Dietrich, 41, 43
Eichmann, Adolf, 82, 96, 102, 103
Eighteenth Brumaire of Louis Bonaparte (Karl Marx), 24
Einsatzgruppen, 97ff, 100–1
'Elites', 48, 66, 78, 90, 111, 122, 176
Elser, Georg, 171
Enabling Act (1933), 70
Engels, Friedrich, 43
England, 20, 77, 115 (*see also* Britain)
Entailed Farm Law, 144
Erdmann, Karl Dietrich, 44
Erfahrungsgeschichte, 157
Essen, 157
Eternal Jew, The, 97
Europa und die deutsche Frage (Gerhard Ritter), 6
'Euthanasia Action', 97, 104, 174

Faschismus in seiner Epoche, Der (Ernst Nolte), 22
Fascism, 3, 9, 10–13, 16, ch. 2 *passim*, 41, 44ff, 63, 133
  in Italy, 20, 22, 33
  Marxist theories and definitions, 11, 24, 30, 41, 43
Faust, Anselm, 138
Federal Republic of Germany, 2, 10, 22, 36, 82, 134, 135, 155, 181, 182, 186, 197, 198, 202, 205, 206
Feder, Gottfried, 142
Fest, Joachim C., 60–1, 83
'Final solution', The, 55, 56, 61, 66, 68, 82, ch. 5 *passim*, 146, 183, 207, 211
Fischer, Fritz, 7, 116
'Fischer controversy', 7, 153
'Flag Law' (1935), 93
Fleming, Gerald, 84, 86, 87, 90, 98
Foreign Minister, 92, 117
Foreign Ministry Office, 53, 93, 96
Foreign policy, 3, 9, 63, 66, 79, ch. 6 *passim*
Four year Plan, 42, 44, 51, 52, 124
Fraenkel, Ernst, 63

France, 24, 102, 187
Franco, Francisco, 34
Frank, Hans, 71, 96, 104
French Revolution, 133, 194
French-Soviet Pact, 119
Frick, Wilhelm, 70, 72, 92
Friedländer, Saul, 183, 185, 186, 188, 189, 191, 193, 200, 205, 210
Friedrich, Carl Joachim, 11, 20, 21, 29
Friedrich-Ebert-Stiftung, 157
Fritsch, Werner Freiherr von, 52
Fromm, Erich, 25
Führer, 63, 65, 68ff, 75, 86, 93, 111, 123, 124, 130, 141, 169
   authority, 7ff
   orders of, 73, 82, 85, 88, 95, 99, 103
Führer Chancellory, 96, 104
'Functionalist' approach, 16
   see also 'Structuralist' approach
'Fundamentalist' approach, 163, 167-8, 170, 178

Gauleiter, 64, 70, 71, 72, 75, 86, 91, 96, 99
Gellately, Robert, 195, 209
Generalgouvernement, 95, 96
Geneva, Disarmament Conference at, 117
Genocide, 43, 56, 64, 80 n 2, 98, ch. 5
Gentile, Giovanni, 20
German Democratic Republic, 2, 5, 11, 13, 34, 35, 41, 43, 48-9, 77, 81, 86, 108, 133ff, 151, 187, 198, 208, 209
German Dictatorship, The (Karl Dietrich Bracher), 211
German Labour Front, 70, 74, 184
Gestapo, 90-3, 104, 157, 173
   see also SS-Police-SD Complex
Geschichtsdidaktik, 8
Geschichte und Gesellschaft, 8
Geschichte von unten, 131
Ghettos, 56, 95
Gleichschaltung, 62, 135
'Globalist' approach, 110
Goebbels, Joseph, 78, 94, 120, 124, 125
Goerdeler, Carl, 14, 153, 155, 161, 162
Göring, Hermann, 51, 70, 93, 96, 97, 102, 113, 114, 119, 120, 124, 175
Graml, Hermann, 154, 155
Gramsci, Antonio, 25, 44
Greece, 34
Gregor, A.J., 27
Greiser, Arthur, 96
Griff nach der Weltmacht (Fritz Fischer), 7

Gruber, Paster, 174

Habermas, Jürgen, 199
Haffner, Sebastian, 61, 83
Hamburg, 75
Hassell, Ulrich von, 155
Hauner, Milan, 110, 111, 114
Heimat, 194
Hennig, Eike, 45
Herbert, Ulrich, 188
Hess, Rudolf, 70, 74
Heydrich, Reinhard, 95-6, 97, 100, 101, 103, 115, 143
Hilberg, Raul, 98
Hildebrand, Klaus, 8, 9, 15, 27, 37, 61, 63, 65, 110, 115
Hilferding, Rudolf, 20
Hillgruber, Andreas, 8, 9, 15, 27, 37, 44, 61-2, 97, 110, 125, 128, 185, 192, 193, 200
Himmler, Heinrich, 70, 71, 95-7, 104, 105, 203
Hindenburg, Paul von, 91
Historical-social science approach, 7-8
Historicism, 4-6, 7, 9, 181ff, 188, 191
Historicization, ch. 9 *passim*
Historigraphy, ch. 10 *passim*
Historikerstreit, 1-3, ch. 9 passim, 184, 190, 191
Historische Zeitschrift, 6
Hitler, Adolf, 1, 5, 7, 9-10, 19, 22, 33-4, 37, 38, 44, 47, 49, 51, 53, 54, 58, 182 *see also* Führer
   Alleinherrschaft of, 63
   attack on life of, 56, 150, 154, 170, 171, 177
   biographies of, 60ff
   charismatic authority of, 69, 74, 121
   factor, 215
   'Grund-Plan' ('Basic Plan') of, 123
   programme of 62, 79, 82ff, 108ff, 110ff, 123ff, 126
   resistance to, ch. 8 *passim*
   Weltanschauung of, 63, 65, 68, 84, 107, 110
   World Domination, aims of, 110, 126 ff, 129
Hitler-centrism, 61, 64, 113
'Hitler-Fascism', 11, 43
'Hitlerism' approach, 19, 38, 60, 63, 65, 82, 86, 111
Hitler Speaks (Hermann Rauschning), 126, 127
Hitler State, The (Martin Broszat), 64

'Hitler Wave', 60
Hitler Youth, 91, 210
Holocaust, 4, ch. 5 *passim*, 191, 194, 207
Hofer, Walther, 84, 100, 162, 164
Höss, Rudolf, 103, 106
Honecker, Erich, 208
Hossbach, Friedrich, 120
Hungary, 34
Hüttenberger, Peter, 49, 158, 167, 209

IG Farben, 50, 53, 57
Imperial era, 146
Imperialism, 9, 11, 18
India, 128
Industry 3, ch. 3 *passim*, 108, 117, 136, 144
    profits of, 57
Industrial workers, 172
Institute of Contemporary History
    (*Institut für Zeitgeschichte*) Munich, 157
'Intentionalist' approach, 8, 60, 67, 68, 69, 89, 104, 115
Irving, David, 82
Israel, 2, 81, 86
Italy, 18, 19, 27, 34–7, 45, 180

Jäckel, Eberhard, 61, 63, 97, 110
Jacobinism, 26
Jacobsen, Hans-Adolf, 114
Japan, 124, 128
Jaspers, Kari, 16
'Jewish Bolshevism', 54, 93
'Jewish Question', 50, 56, 83, 87, 88, ch. 5 *passim*, 108, 112, 130, 183
Jews, 4, 31, 64, ch. 5 *passim*
    definition, 9
    departmental stores belonging to, 75
    deportation of, 85, 94, 103ff
    'eastern', 100
    emigration of, 85, 94, 96, 100
    extermination of, 55, 56, 63, 64, 192, ch. 5 *passim* 206
    Polish, 55, 85, 93, 97, 103
    Russian, 85, 99, 100ff, 103
Jünger, Ernst, 20

Kaufmann, Karl, 75
Kehr, Eckart, 12
Keppler, Wilhelm, 120
Kersten, Felix, 97, 104
'Keynesian revolution', 138

Kleist-Schmenzin, Ewald von, 164
Kocka, Jorgen, 18 n 1, 36, 199
Koehl, Robert, 71
Kommissarbefehl, 85, 100
Kommunistische Partei Deutschlands
    (KPD), 151, 152, 158, 173
Kowno, 84,
Krauch, Karl, 53
Krausnick, Helmut, 97
Kreisau Circle, 154, 155
Knochel, Wilhelm, 173
Kuhn, Axel, 110
Kühnl, Reinhard, 45
Kulka, Otto Dov, 183, 186, 187
Kuper, Leo, 177

Labour Front, 70, 74, 184
Lammers, Hans Heinrich, 73
Länder governments, 70
Latin America, Hitler's views on, 127
'Law for the Protection of German
    Blood and German Honour' (1935), 92
League of Nations, 117, 124
Lebensraum, 33, 55, 65, 67, 86, 108, 110, 112, 113, 116, 123, 124, 125, 126, 129, 137, 203, 205
Leistungsgesellschaft, 148
Lenin, Vladimir Ilyich, theory of
    imperialism of, 43
Ley, Robert, 70, 74, 188
Library of Resistance, 156
Limits of Hitler's Power, The (Edward
    N Peterson), 64, 75
Linz, 84, 120
Linz, Juan, 36
Lichtenberg, Father, 174
Lipset, Seymour, 28
Lithuania, 101
Locarno, treaty of, 119
Łódź (Litzmannstadt), 96
Lohse, Hinrich, 104
Löwenthal, Richard, 20
Luftwaffe, 128
Luzk, 100

'Madagascar Plan', 96
Maier, Charles, 205, 212
Mann, Golo, 25
Mann, Reinhard, 195, 209
Marx, Karl, 25, 46, 58, 79
Marxism, 22, 141
    -Leninism, 10, 24–6, 34, 43, 81, 198
    New Left and, 12, 15, 23–4, 35, 37

and theories of social change, 133ff
'Marxist' historians, 16, 36, 42ff 46, 137, 140
  see also German Democratic Republic
Mason, Tim, 10, 15, 42, 44, 52, 56, 67, 75–9, 113, 165
Matzerath, Horst, 138,
Mayer, Arno, 98, 207
Meinecke, Friedrich, 6–7, 14, 25
Mein Kampf (Adolf Hitler), 83, 85, 108, 126
Messerschmidt, Manfred, 161
Mexico, Hitler's views on, 127
Michalka, Wolfgang, 115
Michaelis, Meir, 20 n.4
Middle east, 110
Milward, Alan, 42, 52, 57
Ministry for the Occupied Eastern Territories, 55, 104
Ministry of War Production, 57
Mittelstand, 142, 144
'Modernization' approaches, 133ff, 148
Moltke, Graf Helmut von, 155
Moltmann, Günther, 110, 125
Mommsen, Hans, 9, 32, 60, 65, 66, 84–5, 98, 100, 105, 111, 112, 113,
Monocratic rule, 65, 69
Moore, Barrington, 27
Müller, Klaus-Jürgen, 160
Munich Conference, 120, 176
Mussolini, Benito, 17, 19, 35, 96, 119, 203
  'March on Rome' of, 17

National, Community, 64, 141
National Identity, 198ff, 214
Navy, 118, 122, 128
Nazi Lawyers' Association, 71
Nazi Party (NSDAP), 32, 44, 64, 69, 88, 90ff, 130, 131, 135, 141ff, 158, 175
Nazi Seizure of Power, 2, 12, 141
Nazi-Soviet Pact, 17, 54, 115
Neumann, Franz, 20, 33, 49, 63, 136
Neurath, Constantin Freiherr von, 92, 117–18
New Deal, 136
'New Order' 137
'New Social History', 7ff
Niemöller, Martin, 174
'Night of the Long Knives', 141, 171
Nitti, Francesco, 20
Noakes, Jeremy, 139
Nolte, Ernst, 13, 22, 26, 27, 37, 44,
184, 185, 192, 200, 206, 207, 208,
Nuremberg Laws, 92, 164

Occupied territories, 56, 86
Operation Barbarossa, 83, 99, 101
Organski, A.F.K., 27
Origins of *Totalitarianism* (Hannah Arendt), 20
Oster, Hans, 155
Ostland, 104
Overy, Richard, 44

Palestine, 96
Papen, Franz von, 75, 117
Parsons, Talcott, 25, 138
Pasewalk, 83, 84
Pätzold, Kurt, 81, 86
Peasantry, 144
Peterson, Edward N., 75
Petzina, Dieter, 42, 169, 170
Peukert, Detlev, 190, 191
'Pluralist' approaches, 113ff
Pogrom (November 1938), 94
Poland, 34, 53, 55, 56, 85, 94–6, 97, 112, 117, 122, 176
  Germany's non-aggression treaty with, 117, 122
Police, 94, 100
  see also Gestapo *and* SS-Police-SD Complex
'Polycracy' of Nazi government, 113ff
Popitz, Johannes, 155
Portugal, 34, 97
Poulantzas, Nicos, 25, 45, 46, 47
Power-Cartel in Nazi State, 50ff, 57ff
President of Danzig Senate, 109
Price Commissioner, 161
'Primacy of domestic politics', 13, 122
'Primacy of economics', 58
'Primacy of politics', 48, 58, 108, 138
Prinz, Michael, 188
'Programmatist' approach, 61, 84, 115
Prussia, 6, 17, 19, 35, 149
'Psycho-historical' approach, 61
Puhle, Hans-Jürgen, 38

Race policy, 107, 115 *see also* Anti-Jewish policy
'Racial defilement, 91
Racial Political Office, 93
Rademacher, Franz, 96
Radkau, Joachim, 53, 54
Ranke, Leopold von, 5
Rapallo, Treaty of, 118

Rauschning, Hermann, 72, 109, 126, 127, 129, 139
Rearmament, 50, 78, 120, 124
Recker, Marie-Louise, 188
Red Army 134
Reich Cabinet, 64, 71, 75, 124
    Chancellor, 70, 116, 155
    Chancellory, 73, 97
    Citizenship Law (1935), 93
    Commissar for Price Surveillance, 76
    Commissar for the Ostland (Baltic), 104
    Constitution, 70
    Doctors' Leader, 92
    Governors, 70, 72
    Ministry of Economics, 50, 75, 93
    Ministry of Food and Agriculture, 77
    Ministry of the Interior, 70, 72, 92, 93
    Ministry of Labour, 75
    Organization Leader (of Nazi Party), 70
    Reform, 70
    Security Head office, 105, 106
    Vice Chancellor, 75
Reich, Wilhelm, 25
Reichsbank, 50
Reichstag, 92, 99
Reichswehr, 49
Reichswerke-Hermann-Göring, 52
Reitlinger, Gerald, 98
'Relativist' historians, 15
Resistance, 14, 33, 57, ch. 8 *passim*, 181, 184
'Resistenz', 181, 184, ch. 8 *passim*
'Revisionist' historians, 2, 14, 84, 200
Revolution of 1918, 76, 84, 138
Rhineland, reoccupation of, 119, 122
Ribbentrop, Joachim von, 96, 115, 118
Rich, Norman, 60, 111
Riga, 104
Ritter, Gerhard, 6, 7, 14, 25, 153
Röhm, Ernst, 70, 141
Rosenberg Agency, 64, 93
Rosenberg, Alfred, 64, 142
Rumania, 54
Rüsen, Jörn, 5 n.7
Russia *See* USSR
Russian Revolution, 194

SA, 27 n 29, 50, 70, 90, 141, 171
Salazar, Antonio de Oliveira, 34
Sauckel, Fritz, 56
Sauer, Wolfgang, 14
Schacht, Hjalmar, 45, 50, 51, 55, 74

Schieder, Wolfgang, 9, 113, 114
Schleicher, Kurt von, 117
Schieunes, Karl A., 87, 90
Schmidt, Helmut, 2
Schmitt, Carl, 2, 65
Schoenbaum, David, 27, 135ff, 138, 142, 145, 154, 180
Schumpeter, Joseph, 112
Schweitzer, Arthur, 41
Second Book (Adolf Hitler), 108, 126, 127
Seldte, Franz, 75
Shoah, 194
Social Democracy, 11, 153, 172
Socialist Unity Party (SED) of German Democratic Republic, 134
Social reaction, ch. 7 *passim*
Societal approach, 163, 178
Sohn-Rethel, Alfred, 45, 47, 124
Sonderweg, 18 n2
SOPADE, 165
SPD, 156, 165
South Africa, 190
South Vietnam, 2
Soviet Army, 97, 134
Soviet Bloc, 198
Soviet Union *See* USSR
Sozialgeschichte von unten 157
Spain, 34, 97, 180
Spanish Civil War, 113, 124
Speer, Albert, 42, 51, 76, 143
SS, 86, 90, 93, 97, 98, 143, 146, 177
    see also SS-Police-SD Complex
SS-Police-SD Complex, 47, 49, 52, 56, 85, 90, 106
Stalin, Josef, 21, 32, 39, 55, 125, 204
Stalinism, 19, 21, 32, 39
    Nazism and, 206
State Monopoly Capitalism, 9, 43, 46, 47, 48
Stauffenberg, Claus Graf Schenk, 152, 153, 175, 176
Steinbach, Peter, 161
Steinhart, Marlis, 160
Sterilization Law (1933), 75
Strasser, Gregor, 70
Streicher, Julius, 91, 98, 103, 106
Streim, 101
Streit, Christian, 98, 103, 106
'Structuralist' approach, 7, 10, 17, 62, 65, 67, 69, 84, 85, 108ff, 115, 121, 123
Stufenplan, 110, 123, 129
Stürmer, Michael, 198, 201
Sudeten Crisis, 93, 120, 176

Table Talk (Hitler), 126
Taylor, A.J.P., 109
Thalheimer, August, 24, 44
Thies, Jochen, 110, 127
Toland, John, 83
Totalitarianism, 3, 11–13, 15, ch. 2
  passim, 42, 58, 83, 135, 139, 193
Trades unions, 165
'Traditionalist' historians, 2
Treblinka, 83
Trevor-Roper, Hugh R., 109, 110
Trivialization, of Nazism, 15, 19, 88
Trotsky, Leon, 25
Trott, zu Solz, Adam, 152, 155
Trustees of Labour, 74
Turner, Henry Ashby Jr., 40 nl, 137–9

Unification, ch. 10 *passim*, 210
Universities in Germany, 7, 12–13
Urals, 98
USA, 7, 17, 54, 110, 125ff
USSR, 19, 32, 33, 39, 54, 85, 97, 99,
  102, 105, 112, 115, 118, 124, 125, 126,
  128, 152, 176, 207, 211

Vergangenheitsbewältigung, 1
Versailles settlement, 117, 127
'*Verstehen*' philosophy of history, 14
Vierteljahrshefte für Zeitgeschichte, 12
Vietnam, 190
Voges, Michael, 165
Volkmann, Hans-Erich, 48, 49
Volkmann, Heinrich, 138
Volksgemeinschaft, 140
Volksgenossen, 135

Wagener, Otto, 142
Wagner, Gerhard, 92
Wannsee Conference, 99

War, 3–4, 43, 53ff, 75ff, 123ff, 142
  of annihilation, 54, 100, 125
  economy, 41
  First World, 7, 37, 39, 84, 102, 112,
    113, 121, 127
  Second World, 78, 79, 132, 179
War against the Jews, The (Lucy
  Dawidowicz), 83
Wartenburg, Paul Graf Yorck von, 155
Warthegau, 95, 96
Weber, Max, 215
Wehler, Hans-Ulrich, 8–10, 38, 199
Wehrmacht, 52, 53, 100, 103, 106, 111,
  118, 143, 146, 171, 186 *See also*
  Army
Weimar Republic, 40, 50, 135, 140, 143,
  146, 154, 158, 179
Weinberg, Gerhard, 111
Weisenborn, Günther, 156
Weizsäcker, Ernst Freiherr von, 117
West Germany, 2, 5, 11–14, 16, 18, 77,
  81, 146, 149, 156
  see also Federal Republic of Germany
White-collar workers, 188
White Rose resistance group, 152, 155
Widerstandpraxis, 161
Wilhelm, Hans-Heinrich, 98, 100
Wilheimine era, 116
Winkler, Heinrich August, 36, 139, 143
Women, 77, 144, 189
Working class 143, 145, 147–8, 165
  resistance of, 157, 165, 166, 172

Youth, 146–7, 210

Zetkin, Clara, 24
Zitelmann, Rainer, 203, 204, 205, 206
Z-plan, 128